Advances in Soft Tissue and Bone Sarcoma

Advances in Soft Tissue and Bone Sarcoma

Guest Editor

Catrin Sian Rutland

MDPI

Basel • Beijing • Wuhan • Barcelona • Belgrade • Novi Sad • Cluj • Manchester

Guest Editor
Catrin Sian Rutland
School of Veterinary Medicine
and Science
University of Nottingham
Nottingham
UK

Editorial Office
MDPI AG
Grosspeteranlage 5
4052 Basel, Switzerland

This is a reprint of the Special Issue, published open access by the journal *Cancers* (ISSN 2072-6694), freely accessible at: https://www.mdpi.com/journal/cancers/special_issues/C8D9NZ20HW.

For citation purposes, cite each article independently as indicated on the article page online and as indicated below:

Lastname, A.A.; Lastname, B.B. Article Title. *Journal Name* **Year**, *Volume Number*, Page Range.

ISBN 978-3-7258-3373-3 (Hbk)
ISBN 978-3-7258-3374-0 (PDF)
https://doi.org/10.3390/books978-3-7258-3374-0

Cover image courtesy of Catrin Sian Rutland

Contents

About the Editor

Catrin Sian Rutland

Prof. Dr. Catrin Sian Rutland (BSc PGCHE MSc MMedSci PhD DSc SFHEA FAS) has over 25 years worth of experience researching comparative medicine and science. She is an internationally renowned Researcher with more than 140 published research papers, books and book chapters, and 200+ conference presentations to date. As Professor of molecular medicine, she specializes in techniques encompassing areas such as molecular biology, anatomy, histology, and imaging. She works on osteosarcomas, and breast and prostate cancers in human and comparative oncology. She is Sub-Dean for Postgraduate Research, a Postgraduate Supervisor of over 60 MSci, MRes, and PhD students, 51 vacation scholars, and 46 undergraduate research projects. Catrin has a passion for teaching the next generation of scientists and clinicians.

Preface

This Special Issue on bone and soft tissue sarcomas delves into the clinical and translational research being conducted on these cancers. Despite advances in clinical and basic research, soft tissue and bone sarcomas remain clinically challenging. This Special Issue highlights the latest discoveries in soft tissue and bone cancers from the laboratory through to the clinics, bench to bedside, and beyond. It brings together original research and reviews covering bone and soft tissue sarcomas. The topics range from detection and diagnosis methods using histopathology, imaging, and molecular advances through to treatment and prognosis, including epidemiological studies, outcome analysis, and surgical and adjunct treatments. Sarcomas from human and veterinary patients, as well as models, are investigated.

The topics covered will be of interest to scientists, clinicians and of course the public and patients. It will also provide valuable reading for medical and science undergraduates and postgraduates.

I would like to thank the authors, the teams at MDPI and of course the patients, hospital and veterinary practice staff, and everyone else involved in the studies.

Catrin Sian Rutland
Guest Editor

cancers

Editorial

Advances in Soft Tissue and Bone Sarcoma

Catrin S. Rutland [1,2]

[1] School of Veterinary Medicine and Science, Faculty of Medicine and Health Sciences, University of Nottingham, Nottingham NG7 2RD, UK; catrin.rutland@nottingham.ac.uk
[2] Faculty of Medicine and Health Science, Biodiscovery Institute, University of Nottingham, Nottingham NG7 2RD, UK

1. Introduction

This *Cancers* Special Issue on bone and soft tissue sarcomas highlights the latest discoveries in soft tissue and bone cancers from the laboratory through to the clinics, from bench to bedside, and beyond. Despite advances in clinical and basic research, soft tissue and bone sarcomas remain clinically challenging. This Special Issue brings together original research and reviews covering bone and soft tissue sarcomas on topics ranging from detection and diagnosis methods using histopathology, imaging, and molecular advances through to treatment and prognosis techniques, including treatment efficacy and survival outcome analysis.

Both primary bone cancers and soft tissue sarcomas are relatively rare, each accounting for less than 1% of malignancies [1]. Not only are sarcomas relatively rare, but they have differing locations in the body and different characteristics and subtypes; more than 100 subtypes of soft tissue sarcomas have been classified to date [2,3]. Sarcomas are therefore often difficult to diagnose, characterise and treat.

The research being conducted is as diverse as the sarcomas themselves. Patient samples and datasets, both in vivo and in vitro, models including organoids and organ chips [4–8], mathematical and bioinformatics models [9,10], and clinical trials along with cohort studies are being used alongside machine-aided learning, including in areas such as radiomics, biomarkers and next-generation sequencing-based methods [11–15]. Advances in imaging techniques such as surgery, magnetic resonance imaging (MRI) and positron emission tomography (PET), and molecular imaging technology such as PET tracers [8,16,17], are also improving diagnostic, prognostic, surgical and drug development tools and approaches. Advances are being achieved in drug discovery and personalised medicine, including in targeted therapies, immunotherapy, chimeric antigen receptor (CAR) T-cell therapy, tumour-infiltrating lymphocytes (TILs), vaccines and combination therapies, to name just a few [10,17,18], and developments and innovations in genetic testing, molecular profiling and epigenetic aspects [3,19] of sarcomas are needed. A deeper understanding of mechanisms of resistance and research into differing sub-types and the tumour microenvironment is also essential to move sarcoma research and clinical approaches forward. More novel approaches to diagnostics, prognostics and therapeutics are essential looking forward, as scientific discoveries are translated into the clinic.

This "Bone and Soft Tissue Sarcomas" *Cancers* Special Issue (https://www.mdpi.com/journal/cancers/special_issues/C8D9NZ20HW accessed on 15 August 2024) consists of 14 papers with article submissions accepted until 30th April 2024. It explores the advances in diagnosis, prognosis, mechanisms and treatment outcomes for bone and soft tissue sarcomas. The research covers bone sarcomas such as chordomas, which are rare malignant neoplasms, through to the more common osteosarcoma. This Special Issue also covers soft tissue sarcomas, such as one of the most common sub-types of soft tissue sarcoma—undifferentiated pleomorphic sarcoma—through to rare sarcomas within the peritoneal cavity.

Citation: Rutland, C.S. Advances in Soft Tissue and Bone Sarcoma. *Cancers* **2024**, *16*, 2875. https://doi.org/10.3390/cancers16162875

Received: 7 August 2024
Accepted: 13 August 2024
Published: 19 August 2024

2. An Overview of the Articles Published in This Special Issue

The key themes of this special issue are classification, diagnostic and prognostic advances and indicators (Contributions 1, 2, 3, 5, 9, 10, 12, 13 and 14), key areas of therapeutic development (6, 7, 8, 12 and 14) and treatment performance and outcomes (contributions 4, 7, 8 and 11). Contributions 1–6 represent a systematic review and reviews of the field, whereas Contributions 7–14 are original research articles focusing on in vitro and in vivo models or patient cohorts.

Contribution 1 in this Special Issue is 'Soft Tissue Sarcoma Mimicking Melanoma: A Systematic Review' by Cassalia and coauthors. Contribution 2, 'Predictors of Symptomatic Venous Thromboembolism in Patients with Soft Tissue Sarcoma in the Lower Extremity', is a review by Kamalapathy and coauthors. Kim et al., present 'Classification of Chondrosarcoma: From Characteristic to Challenging Imaging Findings' in Contribution 3, while Contribution 4 outlines the 'Current Landscape of Immunotherapy for Advanced Sarcoma' as reviewed by Albarrán and colleagues. Contribution 5, by Costci et al., is 'Gender Differences in Soft Tissue and Bone Sarcoma: A Narrative Review' and the final review, Contribution 6 by Lesovaya and coauthors, is 'Genetic, Epigenetic and Transcriptome Alterations in Liposarcoma for Target Therapy Selection'.

Contribution 7 is a research article on 'Treatment Pathways and Prognosis in Advanced Sarcoma with Peritoneal Sarcomatosis' by Klingler and colleagues. Sarcomas within the peritoneal cavity are not only rare but remain difficult to treat due to their differing subtypes and characteristics. This research, focusing on surgical procedures, presents 19 patients with peritoneal sarcomatosis, outlining their journeys from diagnosis and treatment through to outcomes, in order to share management practices with others.

Contribution 8, by Polera et al., also looks at potential treatment pathways in the paper 'The First-In-Class Anti-AXL×CD3ε Pronectin™-Based Bispecific T-Cell Engager Is Active in Preclinical Models of Human Soft Tissue and Bone Sarcomas'. This article explores AXL, a TAM family tyrosine kinase receptor, as a target for an innovative immunotherapeutic strategy. These in vitro and murine in vivo studies have indicated the antitumor efficacy of pAXL×CD3ε against sarcoma cells, which potentially represents a new-generation strategy for managing sarcomas.

Brookes and coauthors explored 'What Is the Significance of Indeterminate Pulmonary Nodules in High-Grade Soft Tissue Sarcomas? A Retrospective Cohort Study' in Contribution 9. Understanding the roles of indeterminate pulmonary nodules in high-grade soft tissue sarcoma, whether they may be benign or malignant, may impact clinical decision making. The important conclusions of this research, looking at 389 patients, indicate that in patients with grade 3 soft tissue sarcomas, significantly worse overall survival was observed, as was an increased risk of developing lung metastases. These significant differences were not observed in grade 2 patients presenting with indeterminate pulmonary nodules. Clinically, the authors indicated that the primary tumour must be considered alongside indeterminate pulmonary nodules when considering risk progression, and determined that monitoring via CT scans at 6 and 12 months would be advisable.

Contribution 10, by Iiuz et al., investigated 'Rapid Classification of Sarcomas Using Methylation Fingerprint: A Pilot Study'. This research article explored the potential for methylation and copy-number variation data in terms of rapid point-of-care sarcoma classification. The end goal was to reduce the time taken to classify sarcomas in order to commence appropriate treatments in a more time efficient manner, and potentially to expand the tools available for classification.

Contribution 11, 'High-Grade Pleomorphic Sarcomas Treated with Immune Checkpoint Blockade: The MD Anderson Cancer Center Experience', by Nasr and coauthors, investigated one of the most common soft tissue sarcomas. Their work included 26 undifferentiated pleomorphic sarcoma patients and 10 patients with other high-grade pleomorphic sarcomas. This retrospective study indicated that immune checkpoint blockade (ICB) treatment resulted in significantly improved progression-free survival. Toxicity was manageable, with no patient deaths. Notably, their data also indicated that compared to previous clinical trials, their data showed that a combination treatment seemed inferior to standalone ICB regarding progression-free survival.

Contribution 12, by de Brot and colleagues, investigated three potential diagnostic and prognostic indicators, highlighting prospective therapeutic targets in their article 'Immunohistochemical Investigation into Protein Expression Patterns of FOXO4, IRF8 and LEF1 in Canine Osteosarcoma'. Using immunohistochemistry alongside quantitative H-scoring and qualitative analysis, their research showed the expression of FOXO4, IRF8 and LEF1 in osteosarcoma. Their work highlighted IRF8 and LEF1 as particularly promising biomarker candidates and therapeutic targets, given their expression patterns, mechanisms and involvement in a number of molecular pathways.

Contribution 13 by Yoon et al. researched 65 patients in 'Quantitative Bone SPECT/CT of Central Cartilaginous Bone Tumors: Relationship between SUVmax and Radiodensity in Hounsfield Unit'. This contribution focused on accurately grading cartilaginous bone tumors using SPECT/CT. Their research revealed a negative correlation between SUVmax and radiodensity in HU measurements in central cartilaginous bone tumours, and patients with a higher SUVmax and lower HUSD were more likely to have a malignant cartilaginous bone tumour. This highlighted the diagnostic and prognostic uses of this technique in central cartilaginous bone tumours.

The final article published, Contribution 14, was 'Conventional spinal chordomas: investigation of SMARCB1/INI1 protein expression, genetic alterations in SMARCB1 gene and clinicopathological features in 89 patients' by Maioli and coauthors. This research determines the immunohistochemical expression of SMARCB1/INI1 and fluorescence in situ hybridisation (FISH) to understand the underlying genetic alterations in the SMARCB1 gene in 89 patients with conventional spinal chordomas. This supports the information relating to the potential of SMARCB1/INI1 as a target for molecular therapy.

3. Conclusions

Many of the studies published in this Special Issue made advances and recommendations relating to diagnosis, prognosis and therapeutics. Directions for future research included the need for research into differing subtypes of sarcomas and for more varied cohorts in general. Work completed in vitro and in vivo has highlighted the need for the development of promising tools and for more translational research to be conducted (such as Contributions 8 and 12). Furthermore, research such as that presented in Contribution 11 has demonstrated the need for cohort studies following clinical trials. Contribution 7 highlights the need for large soft tissue sarcoma databases, especially in relation to future research on subtypes and progression, to support evidence-based approaches for management and tailored treatment plans. The papers in this *Cancers* Special Issue also highlight the complexities faced in classifying and characterising sarcomas, which in turn complicate the discovery of molecular mechanisms and pathways, which makes finding biomarkers more difficult, and additionally makes it challenging to discover and develop effective treatment options.

Generalised 5-year survival rates for soft tissue sarcomas (2000–2018) ranged from 82% for localised to 59.6% for regionalised and 16.7% for distant sarcomas; for bone and joint, these survival rates ranged from 82.6% for localised to 67.4 for regional and 30.8% for distant [20]. With these survival rates in mind, there is still much research needed into the diagnosis, prognosis and treatment tools required to enhance healthcare options.

Acknowledgments: As the Guest Editor of this Special Issue, I wish to thank all of the authors who published their valuable research within this issue, and to also thank our reviewers. My gratitude and thanks also go to the *Cancers* editorial office and the publications team who have supported this Special Issue.

Conflicts of Interest: The author declares no conflicts of interest.

List of Contributions

1. Cassalia, F.; Cavallin, F.; Danese, A.; Del Fiore, P.; Di Prata, C.; Rastrelli, M.; Belloni Fortina, A.; Mocellin, S. Soft Tissue Sarcoma Mimicking Melanoma: A Systematic Review. *Cancers* **2023**, *15*, 3584.
2. Kamalapathy, P.N.; Kline, A.; Hollow, H.; Raskin, K.; Schwab, J.H.; Lozano-Calderón, S. Predictors of Symptomatic Venous Thromboembolism in Patients with Soft Tissue Sarcoma in the Lower Extremity. *Cancers* **2023**, *15*, 315.
3. Kim, J.-H.; Lee, S.K. Classification of Chondrosarcoma: From Characteristic to Challenging Imaging Findings. *Cancers* **2023**, *15*, 1703.
4. Albarrán, V.; Villamayor, M.L.; Pozas, J.; Chamorro, J.; Rosero, D.I.; San Román, M.; Guerrero, P.; Pérez de Aguado, P.; Calvo, J.C.; García de Quevedo, C.; et al. Current Landscape of Immunotherapy for Advanced Sarcoma. *Cancers* **2023**, *15*, 2287.
5. Cosci, I.; Del Fiore, P.; Mocellin, S.; Ferlin, A. Gender Differences in Soft Tissue and Bone Sarcoma: A Narrative Review. *Cancers* **2024**, *16*, 201.
6. Lesovaya, E.A.; Fetisov, T.I.; Bokhyan, B.Y.; Maksimova, V.P.; Kulikov, E.P.; Belitsky, G.A.; Kirsanov, K.I.; Yakubovskaya, M.G. Genetic, Epigenetic and Transcriptome Alterations in Liposarcoma for Target Therapy Selection. *Cancers* **2024**, *16*, 271.
7. Klingler, F.; Ashmawy, H.; Häberle, L.; Esposito, I.; Schimmöller, L.; Knoefel, W.T.; Krieg, A. Treatment Pathways and Prognosis in Advanced Sarcoma with Peritoneal Sarcomatosis. *Cancers* **2023**, *15*, 1340.
8. Polerà, N.; Mancuso, A.; Riillo, C.; Caracciolo, D.; Signorelli, S.; Grillone, K.; Ascrizzi, S.; Hokanson, C.A.; Conforti, F.; Staropoli, N.; et al. The First-In-Class Anti-AXL×CD3ε Pronectin™-Based Bispecific T-Cell Engager Is Active in Preclinical Models of Human Soft Tissue and Bone Sarcomas. *Cancers* **2023**, *15*, 1647.
9. Brookes, M.J.; Chan, C.D.; Crowley, T.P.; Ragbir, M.; Beckingsale, T.; Ghosh, K.M.; Rankin, K.S. What Is the Significance of Indeterminate Pulmonary Nodules in High-Grade Soft Tissue Sarcomas? A Retrospective Cohort Study. *Cancers* **2023**, *15*, 3531.
10. Iluz, A.; Maoz, M.; Lavi, N.; Charbit, H.; Or, O.; Olshinka, N.; Demma, J.A.; Adileh, M.; Wygoda, M.; Blumenfeld, P.; et al. Rapid Classification of Sarcomas Using Methylation Fingerprint: A Pilot Study. *Cancers* **2023**, *15*, 4168.
11. Nasr, L.F.; Zoghbi, M.; Lazcano, R.; Nakazawa, M.; Bishop, A.J.; Farooqi, A.; Mitra, D.; Guadagnolo, B.A.; Benjamin, R.; Patel, S.; et al. High-Grade Pleomorphic Sarcomas Treated with Immune Checkpoint Blockade: The MD Anderson Cancer Center Experience. *Cancers* **2024**, *16*, 1763. https://doi.org/10.3390/cancers16091763.
12. Brot, S.D.; Cobb, J.; Alibhai, A.A.; Jackson-Oxley, J.; Haque, M.; Patke, R.; Harris, A.E.; Woodcock, C.L.; Lothion-Roy, J.; Varun, D.; et al. Immunohistochemical Investigation into Protein Expression Patterns of FOXO4, IRF8 and LEF1 in Canine Osteosarcoma. *Cancers* **2024**, *16*, 1945. https://doi.org/10.3390/cancers16101945.
13. Yoon, H.; Lee, S.K.; Kim, J.-Y.; Joo, M.W. Quantitative Bone SPECT/CT of Central Cartilaginous Bone Tumors: Relationship between SUVmax and Radiodensity in Hounsfield Unit. *Cancers* **2024**, *16*, 1968. https://doi.org/10.3390/cancers16111968.
14. Maioli, M.; Cocchi, S.; Gambarotti, M.; Benini, S.; Magagnoli, G.; Gamberi, G.; Griffoni, C.; Gasbarrini, A.; Ghermandi, R.; Noli, L.E.; et al. Conventional Spinal Chordomas: Investigation of SMARCB1/INI1 Protein Expression, Genetic Alterations in *SMARCB1* Gene, and Clinicopathological Features in 89 Patients. *Cancers* **2024**, *16*, 2808. https://doi.org/10.3390/cancers16162808.

References

1. Siegel, R.L.; Miller, K.D.; Jemal, A. Cancer statistics, 2018. *CA Cancer J. Clin.* **2018**, *68*, 7–30. [CrossRef] [PubMed]
2. Ressing, M.; Wardelmann, E.; Hohenberger, P.; Jakob, J.; Kasper, B.; Emrich, K.; Eberle, A.; Blettner, M.; Zeissig, S.R. Strengthening health data on a rare and heterogeneous disease: Sarcoma incidence and histological subtypes in Germany. *BMC Public Health* **2018**, *18*, 235. [CrossRef]
3. Nacev, B.A.; Sanchez-Vega, F.; Smith, S.A.; Antonescu, C.R.; Rosenbaum, E.; Shi, H.; Tang, C.; Socci, N.D.; Rana, S.; Gularte-Mérida, R.; et al. Clinical sequencing of soft tissue and bone sarcomas delineates diverse genomic landscapes and potential therapeutic targets. *Nat. Commun.* **2022**, *13*, 3405. [CrossRef] [PubMed]

4. Simpson, S.; Dunning, M.D.; de Brot, S.; Grau-Roma, L.; Mongan, N.P.; Rutland, C.S. Comparative review of human and canine osteosarcoma: Morphology, epidemiology, prognosis, treatment and genetics. *Acta Vet. Scand.* **2017**, *59*, 71. [CrossRef] [PubMed]
5. Dodd, R.D.; Mito, J.K.; Kirsch, D.G. Animal models of soft-tissue sarcoma. *Dis. Models Mech.* **2010**, *3*, 557–566. [CrossRef] [PubMed]
6. Costa, A.; Gozzellino, L.; Nannini, M.; Astolfi, A.; Pantaleo, M.A.; Pasquinelli, G. Preclinical Models of Visceral Sarcomas. *Biomolecules* **2023**, *13*, 1624. [CrossRef] [PubMed]
7. Nolan, J.; Pearce, O.M.T.; Screen, H.R.C.; Knight, M.M.; Verbruggen, S.W. Organ-on-a-Chip and Microfluidic Platforms for Oncology in the UK. *Cancers* **2023**, *15*, 635. [CrossRef] [PubMed]
8. Tanaka, M.; Nakamura, T. Modeling fusion gene-associated sarcoma: Advantages for understanding sarcoma biology and pathology. *Pathol. Int.* **2021**, *71*, 643–654. [CrossRef] [PubMed]
9. Fu, Z.; Sun, G.; Li, J.; Yu, H. Identification of hub genes related to metastasis and prognosis of osteosarcoma and establishment of a prognostic model with bioinformatic methods. *Medicine* **2024**, *103*, e38470. [CrossRef] [PubMed]
10. Teng, J.S.; Wang, Y. An immune-related eleven-RNA signature-drived risk score model for prognosis of osteosarcoma metastasis. *Sci. Rep.* **2024**, *14*, 13401. [CrossRef] [PubMed]
11. Patton, A.; Dermawan, J.K. Current updates in sarcoma biomarker discovery: Emphasis on next-generation sequencing-based methods. *Pathology* **2024**, *56*, 274–282. [CrossRef] [PubMed]
12. Zhang, N.; Han, Y.; Cao, H.; Wang, Q. Inflammasome-related gene signatures as prognostic biomarkers in osteosarcoma. *J. Cell. Mol. Med.* **2024**, *28*, e18286. [CrossRef] [PubMed]
13. Borghi, A.; Gronchi, A. Sarculator: How to improve further prognostication of all sarcomas. *Curr. Opin. Oncol.* **2024**, *36*, 253–262. [CrossRef] [PubMed]
14. Zhu, N.; Meng, X.; Wang, Z.; Hu, Y.; Zhao, T.; Fan, H.; Niu, F.; Han, J. Radiomics in Diagnosis, Grading, and Treatment Response Assessment of Soft Tissue Sarcomas: A Systematic Review and Meta-analysis. *Acad. Radiol.* **2024**, *in press*. [CrossRef] [PubMed]
15. Alkhatib, O.; Miles, T.; Jones, R.P.; Mair, R.; Palmer, R.; Winter, H.; McDermott, F.D. Current and future genomic applications for surgeons. *Ann. R. Coll. Surg. Engl.* **2024**, *106*, 321–328. [CrossRef] [PubMed]
16. Shomal Zadeh, F.; Pooyan, A.; Alipour, E.; Hosseini, N.; Thurlow, P.C.; Del Grande, F.; Shafiei, M.; Chalian, M. Dynamic contrast-enhanced magnetic resonance imaging (DCE-MRI) in differentiation of soft tissue sarcoma from benign lesions: A systematic review of literature. *Skelet. Radiol.* **2024**, *53*, 1343–1357. [CrossRef] [PubMed]
17. Guja, K.E.; Ganjoo, K.N.; Iagaru, A. Molecular Imaging in Soft-tissue Sarcoma: Evolving Role of FDG PET. *Semin. Nucl. Med.* **2024**, *54*, 332–339. [CrossRef] [PubMed]
18. Pilavaki, P.; Gahanbani Ardakani, A.; Gikas, P.; Constantinidou, A. Osteosarcoma: Current Concepts and Evolutions in Management Principles. *J. Clin. Med.* **2023**, *12*, 2785. [CrossRef] [PubMed]
19. Brownlie, J.; Kulkarni, S.; Algethami, M.; Jeyapalan, J.N.; Mongan, N.P.; Rakha, E.A.; Madhusudan, S. Targeting DNA damage repair precision medicine strategies in cancer. *Curr. Opin. Pharmacol.* **2023**, *70*, 102381. [CrossRef] [PubMed]
20. Surveillance Research Program. SEER*Explorer: An interactive website for SEER Cancer Statistics. Available online: https://seer.cancer.gov/statistics-network/explorer/ (accessed on 11 June 2024).

cancers

MDPI

Systematic Review

Soft Tissue Sarcoma Mimicking Melanoma: A Systematic Review

Fortunato Cassalia [1,†], Francesco Cavallin [2,†], Andrea Danese [3], Paolo Del Fiore [4,*], Claudia Di Prata [5], Marco Rastrelli [4,5], Anna Belloni Fortina [6] and Simone Mocellin [4,5]

1 Unit of Dermatology, Department of Medicine, University of Padua, 35121 Padua, Italy; fortunato1287@gmail.com
2 Independent Researcher, 36020 Solagna, Italy; cescocava@libero.it
3 Unit of Dermatology, Department of Integrated Medical and General Activity, University of Verona, 37100 Verona, Italy; adanese4@gmail.com
4 Soft-Tissue, Peritoneum and Melanoma Surgical Oncology Unit, Veneto Institute of Oncology IOV-IRCCS, 35128 Padua, Italy; marco.rastrelli@unipd.it (M.R.); simone.mocellin@unipd.it (S.M.)
5 Department of Surgical, Oncological and Gastroenterological Sciences (DISCOG), University of Padua, 35128 Padua, Italy; claudia.diprata@gmail.com
6 Pediatric Dermatology Unit, Department of Medicine DIMED, University of Padua, 35121 Padua, Italy; anna.bellonifortina@unipd.it
* Correspondence: paolo.delfiore@iov.veneto.it; Tel.: +39-49-8212714
† These authors contributed equally to this work.

Simple Summary: Sarcoma may show similarities to malignant melanoma, making it difficult to differentiate between these two neoplasms. This systematic review summarizes evidence on cases of sarcoma that were initially diagnosed as melanoma to help clinicians in the diagnostic process. A comprehensive search of key databases identified 23 case reports and 4 case series with a total of 34 patients. Heterogeneous clinical presentation and frequent immunohistochemistry positivity contributed to the initial misdiagnosis. The second assessment was performed due to unusual presentation or uncertain diagnosis, and the final diagnosis was clear cell sarcoma (50%) or other soft tissue sarcomas (50%). EWSR1 translocation was investigated in 50% of cases, among which 94% were found to be positive. This systematic review suggests that a second diagnosis should be considered in cases of atypical lesions, and ESWR1 translocation should be investigated.

Abstract: Background: Sarcoma may show similarities to malignant melanoma in terms of morphologic and immunohistochemical aspects, making it difficult to differentiate between these two neoplasms during the diagnostic process. This systematic review aims to summarize available evidence on cases of sarcoma that were initially diagnosed as melanoma. Methods: A comprehensive search of the MEDLINE/Pubmed, EMBASE, and SCOPUS databases was conducted through March 2023. We included case series and case reports of sarcoma patients that were initially diagnosed as malignant melanoma. PRISMA guidelines were followed. Results: Twenty-three case reports and four case series with a total of 34 patients were included. The clinical presentation was heterogeneous, and the most involved anatomical regions were lower limbs (24%), head/neck (24%), and upper limbs (21%). IHC positivity was reported for S100 (69%), HMB45 (63%), MelanA (31%), and MiTF (3%). The main reasons for a second assessment were unusual presentation (48%) and uncertain diagnosis (28%). EWSR1 translocation was investigated in 17/34 patients (50%) and found to be positive in 16/17 (94%). The final diagnosis was clear cell sarcoma (50%) or other soft tissue sarcomas (50%). Conclusions: Melanoma and some histotypes of sarcoma share many similarities. In cases of atypical lesions, a second diagnosis should be considered, and ESWR1 translocation should be investigated.

Keywords: melanoma; sarcoma; mimicking

Citation: Cassalia, F.; Cavallin, F.; Danese, A.; Del Fiore, P.; Di Prata, C.; Rastrelli, M.; Belloni Fortina, A.; Mocellin, S. Soft Tissue Sarcoma Mimicking Melanoma: A Systematic Review. *Cancers* **2023**, *15*, 3584. https://doi.org/10.3390/cancers15143584

Academic Editor: Catrin Sian Rutland

Received: 20 June 2023
Revised: 8 July 2023
Accepted: 10 July 2023
Published: 12 July 2023

1. Introduction

The 2020 World Health Organization (WHO) Classification of Soft Tissue Tumors indicates sarcomas as rare tumors that are further subclassified into approximately 70 subtypes, each characterized by a distinct morphology that often translates into a specific clinical behavior, as well as into specific therapeutic approaches [1,2]. They can occur anywhere in the body, affecting the extremities in 50% of cases, the trunk and retroperitoneum in 40% of cases, and the head and neck in 10% of cases [3].

Some of these neoplasms can be confused with cutaneous melanoma, since they share similar clinical, histological, and immunohistochemical features [4]. In addition, they are rare tumors that are often overlooked by clinicians who do not pose the diagnostic suspicion to the pathologist [2,5,6]. Therefore, healthcare specialists may encounter some difficulties in differentiating between sarcoma and melanoma during the diagnostic process. For example, clear cell sarcoma (CCS) clinically presents as a deep and small (<5 cm) soft tissue lesions, often juxtaposed with tendons, fascia, or aponeurosis, which may mistakenly suggest some forms of melanoma, such as acral melanoma, nodular melanoma, or amelanotic melanoma [7,8]. Moreover, CCS shows a phenotype identical to that of conventional melanoma characterized by strong expression of S100 protein in 100% of cases and variable expression of HMB-45, Melan A, and MiTF. As a matter of fact, CCS can be genetically differentiated from melanoma due to some peculiarities, including (i) the typical reciprocal translocation t(12;22) (q13;q12) that gives rise to the EWSR1-ATF1 oncogene and (ii) the absence of BRAF/NRAS mutations that can often characterize melanoma [4]. Other examples include malignant peripheral nerve sheath tumor (MPNST) and Kaposi's sarcoma. MPNST usually arises from peripheral nerves and may be associated with patients with neurofibromatosis type 1 (NF1). The clinical presentation involves the development of a painful and/or rapidly expanding mass with associated neurological deficits. The biological behavior of MPNSTs has been described as unpredictable, and the differential diagnosis includes several tumors, particularly spindle cell/desmoplastic melanoma, which serves as the main differential because of its higher incidence, remarkably similar morphology, and overlapping immunochemical markers [9,10]. Kaposi's sarcoma is typical of immunocompromised patients and may also present as a single papular skin lesion, clinically mimicking melanoma, which can be distinguished by histologic and immunohistochemical appearance [11].

The clinical presentation of such a neoplasm may mislead the physician, with potential implications for therapeutic strategy and prognosis. For example, surgical widening of the margins varies according to the neoplasm under treatment (melanoma or sarcoma); chemoradiotherapy may be offered for sarcoma, but melanoma patients may benefit from a different first-line approach, such as immunotherapy or target therapy or both, whereas sentinel lymph node biopsy is routinely performed in the diagnostic workup of melanoma but is still under debate for soft tissue sarcoma [12–14].

Awareness of the similarities between sarcoma and melanoma and the ability to recognize the two entities play a crucial role in patient care. However, a clinician may have little to no direct experience in this matter and may retrieve only limited information from a case report or a small case series.

The aim of this systematic review was to summarize available evidence on cases of sarcoma that were initially diagnosed as malignant melanoma to help clinicians in the diagnostic process and to improve patient care.

2. Materials and Methods

2.1. Study Design

This is a systematic review of case series and case reports describing cases of sarcoma that were initially diagnosed as malignant melanoma. The review was conducted according to the Preferred Reporting Items for Systematic Reviews and Meta-Analyses (PRISMA) guidelines [15]. The review protocol was registered in PROSPERO (CRD42023403882).

2.2. Search Strategy

We systematically searched MEDLINE/PubMed, EMBASE, and SCOPUS to detect eligible studies. The search strategy was conducted without language restrictions through March 2023. In PubMed, the following search strategy was used: sarcoma mimicking melanoma OR sarcoma resembling melanoma OR melanoma-like. The search strategy was tailored to conform to the other electronic sources. The lists from each source were joined, and the duplicates were removed. Two investigators (F.C. and A.D.) separately evaluated titles and abstracts of the records and removed those that fell outside the scope of the review. The full texts of all potentially eligible records were examined to dismiss those not fulfilling the inclusion criteria. Finally, the reference lists of included records were hand-searched to detect further studies of interest. Any disagreement was solved by consensus with a third investigator (P.D.F.). Studies not including human subjects were excluded. No language restrictions were applied.

2.3. Data Collection

Two investigators (F.C. and A.D.) independently extracted relevant data from the included articles. For each article, study features, patient characteristics, tumor information, and outcome measures were collected. A third investigator (A.B.F.) checked the extracted data. Any inconsistency was solved by consensus.

2.4. Assessment of the Quality of Included Studies

The quality of the included studies was assessed according to eight criteria: (i) clear criteria for inclusion of the patient(s); (ii) valid methods for identification of the initial condition; (iii) valid methods for identification of the final condition; (iv) in a case series, consecutive inclusion of patients; (v) clear reporting of demographics; (vi) clear reporting of clinical information; (vii) reporting of the time of the second assessment; and (viii) reporting of the reason for the second assessment. The criteria were adapted from the Joanna Briggs Institute (JBI) critical appraisal tool [16] to fit the context under evaluation (case series and case reports describing cases of sarcoma that were initially diagnosed as malignant melanoma). Two investigators (F.C. and M.A.) independently appraised the risk of bias of the included studies, and any inconsistency was solved by consensus with all authors.

2.5. Data Synthesis

The selection procedure was presented in a flow chart. Pertinent data were extracted from included studies and summarized in tables. The inclusion of case reports and very small case series precluded the feasibility of a meaningful meta-analysis; hence, a narrative synthesis of included studies was conducted.

3. Results

3.1. Search Results

The comprehensive search of key databases yielded 365 non-duplicate records. We excluded 321 records according to the title or the abstract, and we identified 43 potentially eligible records for the full-text review. During this phase, 24 records satisfied the inclusion criteria, while 19 records were excluded due to different design (n = 4), different topic (n = 12), or different participants (n = 1) or because we could not find the full text (n = 2) (Supplementary Table S1). Three additional eligible records were identified via hand search. Finally, 27 records [9,11,17–41] were included in the narrative synthesis (Figure 1).

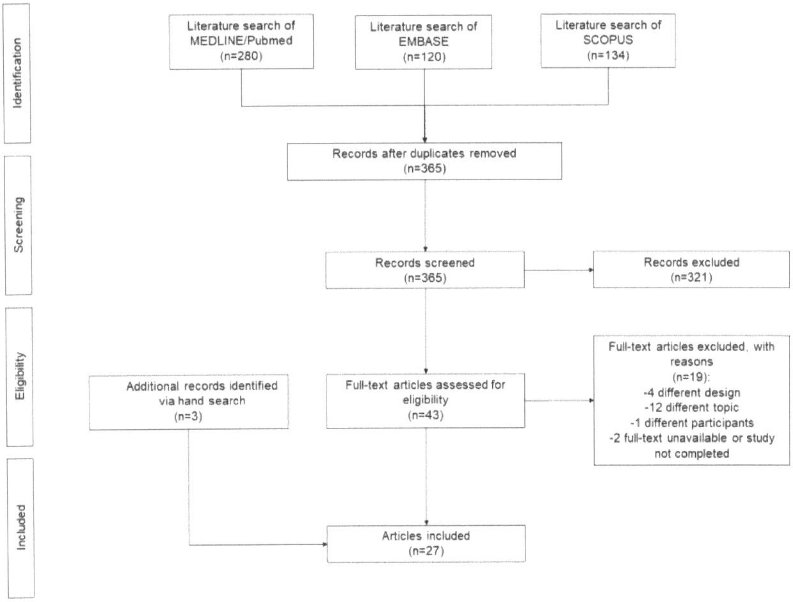

Figure 1. Flow chart of the selection process.

3.2. Narrative Synthesis of the Findings

The synthesis included 23 case reports (85%) and 4 case series (15%). A total of 3 studies (11%) were published in 1989–2000, 7 (26%) in 2001–2011, and 17 (63%) in 2012–2022. Study and patient characteristics are summarized in Tables 1 and 2.

Overall, the studies reported on a total of 34 patients (aged 12–86 years), including 25 males (74%) and 9 females (26%). The initial diagnosis was melanoma (24 patients, 71%) or suspected melanoma (10 patients, 29%). The clinical presentation was heterogeneous (Table 1), and the involved anatomical regions included lower limbs (8/33 patients, 24%), head/neck (8/33 patients, 24%), upper limbs (7/33 patients, 21%), visceral area (4/33 patients, 12%), trunk (3/33 patients, 9%), and genital area (3/33 patients, 9%) (the information was not available for one patient).

IHC positivity was reported for S100 (22/32 patients, 69%), HMB45 (20/32 patients, 63%), MelanA (10/32 patients, 31%), and MiTF (1/32 patients, 3%).

The reasons for a second assessment and/or diagnostic re-evaluation included unusual presentation (12/25, 48%), uncertain diagnosis (7/25, 28%), expert opinion (2/25, 8%), no response to treatment (1/25, 4%), search for EWSR1 translocation (1/25, 4%), review after surgery (1/25, 4%), and review after death (1/25, 4%), while the information was not reported in nine patients.

EWSR1 translocation was investigated in 17/34 patients (50%) and found to be positive in 16 CCS patients and 1 MPNST patient.

The final diagnosis was clear cell sarcoma in 17 patients (50%) and soft tissue sarcoma in 17 patients (50%). The latter included sarcoma of perivascular epithelioid cells (n = 4), malignant peripheral nerve sheath tumor (n = 4), Kaposi's sarcoma (n = 1), chondroid syringoma (n = 1), cutaneous angiosarcoma (n = 1), cutaneous epithelioid angiosarcoma (n = 1), epithelioid malignant schwannoma (n = 1), malignant giant cell tumor of soft tissue (n = 1), malignant schwannoma (n = 1), myeloid sarcoma (n = 1), and pleomorphic sarcoma (n = 1).

An overview of the main findings is displayed in Figure 2.

Table 1. Characteristics of included studies: patient characteristics and clinical presentation.

First Author	Year	Type of Study	N pts	Sex	Age (Years)	Initial Diagnosis	Anatomical Region	Site	Clinical Presentation	IHC Positivity
Potter AJ [17]	2022	Case report	1	M	30	Melanoma	Lower limbs	Toe	Ulcerated, nodular cutaneous lesion on the distal third toe, which had been present for several years	S100, HMB45, SOX10
Tahiri EL [18]	2022	Case report	1	M	31	Melanoma	Lower limbs	Heel	Heel mass nodule	S100, HMB45, Melan A
Biglow LR [9]	2021	Case report	1	F	47	Melanoma	Upper limbs	Finger	Subcutaneous nodule at the finger without any obvious nevus or skin color changes	S100, SOX10, Vimentin, BCL2
Zhang X [19]	2021	Case report	1	M	68	Melanoma	Visceral	Pleura	Dyspnea and cough following a dental abscess that was treated with root canal procedure; imaging studies revealed a large right pleural effusion, raising the concern of an empyema	SOX10, S100, HMB45, Melan-A
Nawrocki S [20]	2020	Case report	1	M	25	Melanoma	Lower limbs	Left inguinal region	Raised blood blister that changed colors	S100, HMB45, Melan A
Obiorah IE [21]	2019	Case series	2	F	37	Melanoma	Head and neck	Left neck	Complaint of left jaw pain and swelling	HMB45, S100, CD31, CD34, CD68
				M	33	Melanoma	Trunk	Back	Mid-back pain radiating to the flanks, as well as leg weakness and numbness, with gait abnormalities	HMB45, S100, Vimentin
Donzel M [22]	2019	Case report	1	M	27	Melanoma	Head and neck	Palate	Palatal ulcerations; ill-defined and erythematous, with a friable center, superficial erosions, and irregular, raised edges	HMB45, SOX10, Melan A
Obiorah IE [23]	2018	Case report	1	F	43	Melanoma	Head and neck	Right neck	Small nodule on the right side of her neck	S100, HMB45, Vimentin
Curry JL [24]	2018	Case report	1	M	68	Melanoma	Upper limbs	Left upper harm	Primary tumor not told; recurrence: new; slightly tender; 1.0 cm purpuric cutaneous nodule within the lymphatic drainage field of his previous primary melanoma of his left upper arm	CD11, CD43, CD68
Leon-Castillo A [25]	2017	Case report	1	M	65	Melanoma	Head and neck	Occipital scalp	Large multinodular, cutaneous occipital scalp lesion with erythematous background	CD31, ERG, D2-40, factor VIII–related antigen, Tyrosinase, HMB45, Melan A
Zivanovic M [26]	2017	Case report	1	M	20	Melanoma	Lower limbs	Foot	N/A	S100, Melan A, HMB45
Jackson CR [27]	2016	Case report	1	M	56	Melanoma	Trunk	Chest	Flesh-colored chest lesion for 7 years	S100, SOX10, CD34
Castriconi M [28]	2015	Case report	1	M	56	Melanoma	Upper limbs	Right axilla	Giant mass located on the right axilla	N/A
Sayah M [29]	2015	Case report	1	F	54	Suspect of melanoma	Visceral	Cecum	Severe iron deficiency anemia and hematochezia	S100, Cytokeratins, HMB45

Table 1. *Cont.*

First Author	Year	Type of Study	N pts	Sex	Age (Years)	Initial Diagnosis	Anatomical Region	Site	Clinical Presentation	IHC Positivity
Liu C [30]	2014	Case series	2	M	29	Suspect of melanoma	Upper limbs	Left thumb	Solid gray–white tumor	HMB45, Melan A, CD56, S100, Vimentin, NSE
				M	76	Suspect of melanoma	Visceral	Jejunum	Complaining of bowel obstruction, macroscopic examination: tumor (2.5 cm × 2.2 cm × 1.5 cm) with a whitish–gray surface	S100, Vimentin, NSE
Sidiropoulos M [31]	2013	Case report	1	M	13	Suspect of melanoma	Head and neck	Lower lip	Symptomatic papule on the lower lip that was suggestive of a mucocele	S100, CD99, sinaptofisina, HMB45, MiTF
Falconieri G [32]	2012	Case series	3	M	12	Melanoma	Lower limbs	Left foot	Lesion in the dorsal aspect of foot	S100, Melan A
				M	60	Melanoma	Lower limbs	Upper thigh	Slowly growing pigmented nodular lesion	S100
				F	29	Melanoma	Lower limbs	Right foot	Lesion in the sole of the foot	S100, Melan A, HMB45
Rodríguez MM [33]	2009	Case report	1	M	53	Melanoma	Upper limbs	Right harm	Painful erythematous, dome-shaped, nodular lesion 1.3 cm in diameter, firm to palpation and movable, with a serohemorrhagic crust on its surface	S100, HMB45
Tanas MR [34]	2009	Case report	1	M	67	Melanoma	Trunk	Abdomen	Abdominal mass	S100, HMB45, MiFT, Melan A
Zoufaly A [1[illegible]]	2007	Case report	1	M	69	Melanoma	N/A	N/A	N/A	N/A
Brightman LA [35]	2006	Case report	1	M	86	Melanoma	Head and neck	Scalp	Large irregular dark gray–blue plaque with an adjacent speckled tan nodule	S100, CD31, CD34
Matsuda Y [36]	2005	Case report	1	M	75	Suspect of melanoma	Lower limbs	Left thigh	Oval-shaped mass; elastic, soft, and adherent to the left thigh on palpation	S-100, NSE, GFAP, MBP, Chromogranin A and synaptophysin
Demir Y [37]	2003	Case report	1	M	80	Suspect of melanoma	Head and neck	Scalp	Painless ulceration on his scalp	S100
Bonetti F [38]	2001	Case series	4	F	28	Suspect of melanoma	Visceral	Ileum	Abdominal pain	HMB45, MART 1
				F	19	Suspect of melanoma	Genital	Uterus	Abdominal pain	HMB45
				F	40	Suspect of melanoma	Genital	Uterus	Surgery because of uterine leiomyomas; during the operation, a 2.5 cm × 12 cm × 1.5 cm pelvic nodule was accidentally found and thought to represent endometriosis	HMB45
				F	41	Suspect of melanoma	Genital	Myometrium	Presumed fibroids in uterus	HMB45, MART 1

Table 1. *Cont.*

First Author	Year	Type of Study	N pts	Sex	Age (Years)	Initial Diagnosis	Anatomical Region	Site	Clinical Presentation	IHC Positivity
Ferreiro JA [39]	1995	Case report	1	M	75	Melanoma	Head and neck	Face	Non-painful mass of the face	keratin, S100, Vimentin
Honma K [40]	1989	Case report	1	M	65	Melanoma	Upper limbs	Left axillary fossa	N/A	Leu7, NSE
Gould E [41]	1989	Case report	1	M	78	Melanoma	Upper limbs	Left arm	Black nodule above the elbow	a1 anti-chymotrypsin (AACT), a1 antitrypsin (AAC)

Table 2. Characteristics of included studies: second assessment and final diagnosis.

First Author	Year	Why Second Assessment and/or Diagnostic Re-Evaluation	EWSR1 Translocation	Final Diagnosis
Potter AJ [17]	2022	Unresponsive to treatment	Positive	Clear cell sarcoma
Tahiri EL [18]	2022	Unusual presentation	Positive	Clear cell sarcoma
Biglow LR [9]	2021	Unusual presentation	Negative	Malignant peripheral nerve sheath tumor
Zhang X [19]	2021	Uncertain diagnosis	Positive	Clear cell sarcoma
Nawrocki S [20]	2020	Uncertain diagnosis	Positive	Clear cell sarcoma
Obiorah IE [21]	2019	Review after death	Positive	Clear cell sarcoma
		Uncertain diagnosis	Positive	Clear cell sarcoma
Donzel M [22]	2019	Uncertain diagnosis	Positive	Clear cell sarcoma
Obiorah IE [23]	2018	Uncertain diagnosis	N/A	Clear cell sarcoma
Curry JL [24]	2018	Review after surgery	N/A	Myeloid sarcoma
Leon-Castillo A [25]	2017	N/A	N/A	Cutaneous angiosarcoma
Zivanovic M [26]	2017	N/A	Positive	Clear cell sarcoma
Jackson CR [27]	2016	Expert opinion	N/A	MPNST—malignant peripheral nerve sheath tumors
Castriconi M [28]	2015	N/A	N/A	Pleomorphic sarcoma
Sayah M [29]	2015	Unusual presentation	Positive	Clear cell sarcoma

Table 2. *Cont.*

First Author	Year	Why Second Assessment and/or Diagnostic Re-Evaluation	EWSR1 Translocation	Final Diagnosis
Liu C [30]	2014	Unusual presentation	Positive	Clear cell sarcoma
		Unusual presentation	Positive	Clear cell sarcoma
Sidiropoulos M [31]	2012	Research for EWSR1 translocation	Positive	Clear cell sarcoma
Falconieri G [32]	2012	Unusual presentation	Positive	Clear cell sarcoma
		Unusual presentation	Positive	Clear cell sarcoma
		Unusual presentation	Positive	Clear cell sarcoma
Rodríguez MM [33]	2009	Uncertain diagnosis	Positive	Clear cell sarcoma
Tanas MR [34]	2009	N/A	N/A	Malignant peripheral nerve sheath tumor
Zoufaly A [11]	2007	N/A	N/A	Kaposi's sarcoma
Brightman LA [35]	2006	Uncertain diagnosis	N/A	Cutaneous epithelioid angiosarcoma
Matsuda Y [36]	2005	N/A	N/A	Malignant peripheral nerve sheath tumor
Demir Y [37]	2003	Expert opinion	N/A	Malignant schwannoma
Bonetti F [38]	2001	Unusual presentation	N/A	Sarcoma of perivascular epithelioid cells
		Unusual presentation	N/A	Sarcoma of perivascular epithelioid cells
		Unusual presentation	N/A	Sarcoma of perivascular epithelioid cells
		Unusual presentation	N/A	Sarcoma of perivascular epithelioid cells
Ferreiro JA [39]	1995	N/A	N/A	Chondroid syringoma
Honma K [40]	1989	N/A	N/A	Epithlioid malignant schwannoma
Gould E [41]	1989	N/A	N/A	Malignant giant cell tumor of soft tissue

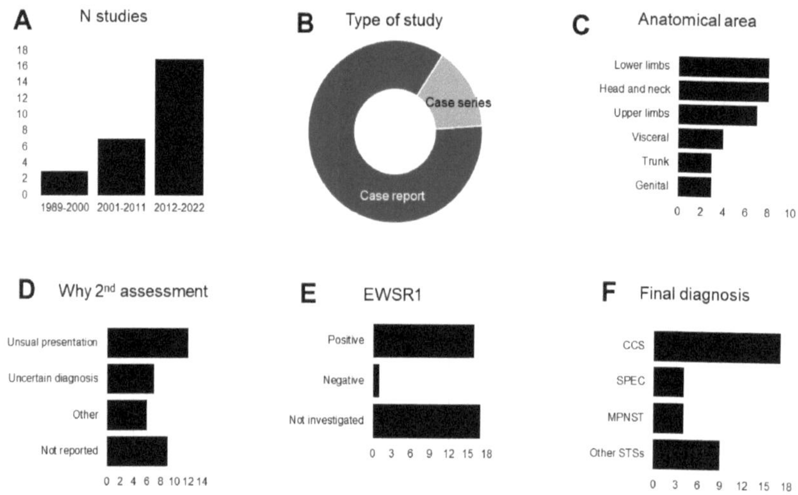

Figure 2. Overview of main findings. CCS: clear cell sarcoma; MPNST: malignant peripheral nerve sheath tumor; SPEC: sarcoma of perivascular epithelioid cells; STS: soft tissue sarcoma.

3.3. *Critical Appraisal of the Quality of Included Studies*

Table 3 summarizes the quality assessment of the included studies. All studies (27/27, 100%) reported clear criteria for inclusion of the patient(s). Valid methods for identification of the initial (melanoma) and final (sarcoma) conditions were described by 16/27 (59%) and 18/27 (67%) studies, respectively. All case series (4/4, 100%) included consecutive patients. Clear reporting of demographics and clinical information were found in 27/27 (100%) and 24/27 (89%) studies, respectively. Only one study (4%) specified the timing of the second assessment, and 17/27 studies (63%) clearly reported the reason for the second assessment.

Table 3. Summary of the quality assessment of the included studies.

First Author	Year	Clear Criteria for Inclusion	Valid Methods for the Identification of the Initial Condition	Valid Methods For Identification of the Final Condition	In a Case Series, Consecutive Inclusion of Participants	Clear Reporting of Demographics	Clear Reporting of Clinical Information	Reporting of Time of Second Assessment	Reporting of Reason for Second Assessment
Potter AJ [17]	2022	Yes	Yes	Yes	N/A	Yes	Yes	No	Yes
Tahiri EL [18]	2022	Yes	Yes	yes	N/A	Yes	Yes	No	Yes
Biglow LR [9]	2021	Yes	Yes	Unclear	N/A	Yes	Yes	No	Yes
Zhang X [19]	2021	Yes	Yes	Yes	N/A	Yes	Yes	Unclear	Yes
Nawrocki S [20]	2020	Yes	Yes	Yes	N/A	Yes	Yes	No	Yes
Obiorah IE [21]	2019	Yes	Yes	Yes	Yes	Yes	Yes	Yes	Yes
Donzel M [22]	2019	Yes	Yes	Yes	N/A	Yes	Yes	No	Yes
Obiorah IE [23]	2018	Yes	Yes	Yes	N/A	Yes	Yes	No	Yes
Curry JL [24]	2018	Yes	Yes	Yes	N/A	Yes	Yes	No	Yes
Leon-Castillo A [25]	2017	Yes	Yes	Yes	N/A	Yes	Yes	No	No
Zivanovic M [26]	2017	Yes	Yes	Yes	N/A	Yes	No	No	No
Jackson CR [27]	2016	Yes	Yes	Unclear	N/A	Yes	Yes	No	Yes
Castriconi M [28]	2015	Yes	Yes	Unclear	N/A	Yes	Yes	No	No
Sayah M [29]	2015	Yes	Unclear	Yes	N/A	Yes	Yes	No	Yes
Liu C [30]	2014	Yes	Unclear	Yes	Yes	Yes	Yes	No	Yes
Sidiropoulos M [31]	2012	Yes	Unclear	Yes	N/A	Yes	Yes	No	Yes
Falconieri G [32]	2012	Yes	Yes	Yes	Yes	Yes	Yes	No	Yes
Rodríguez MM [33]	2009	Yes	Unclear	Yes	N/A	Yes	Yes	No	Unclear
Zoufaly A [11]	2007	Yes	Yes	No	N/A	Yes	No	No	No
Tanas MR [34]	2009	Yes	Unclear	Unclear	N/A	Yes	Yes	No	No
Brightman LA [35]	2006	Yes	Yes	Yes	N/A	Yes	Yes	No	Yes
Matsuda Y [36]	2005	Yes	Unclear	Unclear	N/A	Yes	Yes	No	No
Demir Y [37]	2003	Yes	Unclear	Yes	N/A	Yes	Yes	No	Yes
Bonetti F [38]	2001	Yes	Unclear	Yes	Yes	Yes	Yes	No	Yes

Table 3. *Cont.*

First Author	Year	Clear Criteria for Inclusion	Valid Methods for the Identification of the Initial Condition	Valid Methods For Identification of the Final Condition	In a Case Series, Consecutive Inclusion of Participants	Clear Reporting of Demographics	Clear Reporting of Clinical Information	Reporting of Time of Second Assessment	Reporting of Reason for Second Assessment
Ferreiro JA [39]	1995	Yes	Unclear	Unclear	N/A	Yes	Yes	No	No
Honma K [40]	1989	Yes	Unclear	Unclear	N/A	Yes	No	No	No
Gould E [41]	1989	Yes	Unclear	unclear	N/A	Yes	Yes	No	No

4. Discussion

This systematic review evaluated the available evidence on cases of sarcoma that were initially diagnosed as malignant melanoma. Our search yielded only case reports and small case series [9,11,17–41], which, individually, can provide sparce information to healthcare providers; however, summarizing data from such sources may allow for a better understanding of the topic. Most of the studies were published in the last decade, which may suggest a rising interest in differentiating sarcoma from melanoma during the diagnostic process.

Overall, the clinical presentation of such cases was heterogeneous, and some sarcomas were initially misdiagnosed because several aspects, such as clinical factors, localization of the lesion, and histologic appearance, suggested a malignant melanoma [9,17–28,32–35,39–41]. In some cases, the clinician considered other diagnoses but finally opted for melanoma [29–31,36–38]. Furthermore, immunohistochemistry is not helpful for differentiating sarcoma from melanoma, which was suggested by the positivity of some markers, such as S100, HBMG-45, and MelanA [9,17–23,26,27,29–39]. Hence, the rarity of cases of sarcoma mimicking melanoma likely played an important role in opting for melanoma as the reasonable initial diagnosis of choice. The reader should be aware that we assume such rarity given the few cases in the literature, but we do not have robust information about the real magnitude of cases of sarcoma mimicking melanoma.

In fact, the main reasons for the second assessment leading to a diagnosis of sarcoma were unusual presentation and uncertainty about the diagnosis, which suggested further investigations to the clinicians [9,18–23,29,30,32,33,35,38]. In a few cases, the second assessment was performed because the patient did not respond to the treatment [17] or during a retrospective review of cases [21,24].

This implies that the correct identification of a sarcoma mimicking melanoma relies on the healthcare provider being aware of the possibility of such a case and being able to identify when unusual features merit further investigation.

This also means that the prevalence of such cases is unknown because the literature does not include episodes when the healthcare provider did not feel the need for further investigations, and no systematic investigations have been conducted in large series of melanoma patients.

We believe that misdiagnosing a sarcoma as a melanoma may have potential implications for patient care because of the use of different therapeutic approaches, including sentinel node biopsy, first-line therapy, and surgical therapy [12–14]. Unfortunately, available information is insufficient to assess the prognostic effect of such misdiagnosis.

Interestingly, half of the sarcomas found at the second assessment were CCS [17–23,26,29–33]. We believe that the common features shared by CCS and melanoma [7,8] and the lower incidence of CCS were likely to be responsible for the initial misdiagnosis. When investigated, EWSR1 translocation was found to be positive in almost all cases [17–22,26,29–33]; hence, clinicians may benefit from the investigation of EWSR1 translocation in the initial diagnostic process.

This systematic review has some limitations that should be considered by the reader. First, the research topic was prone to be described in case reports and very small case series, limiting the available information and the potential for further analyses. Second, the lack of epidemiological studies prevented any considerations of the prevalence of cases of sarcoma that were initially diagnosed as melanoma. Third, information about the timing of the second assessment would provide interesting information but was largely missing in the literature. Fourth, the role of EWSR1 translocation in the identification of CCS could not be investigated because of selected reporting (the included studies reported some CCS cases with positive EWSR1 and one MPNST cases with negative EWSR1).

Within its limitations, our systematic review underlines an underreported problem in the diagnosis of melanoma and sarcoma, informs physicians about features that can make differential diagnosis difficult, and highlights the importance of searching for EWSR1 translocation in the diagnostic process. Due to the rarity of sarcoma, healthcare providers possess a heterogeneous level of experience and expertise in managing such diseases. Therefore, it is crucial for physicians to ensure that pathologists are appropriately guided to relevant diagnostic procedures, especially when excising suspicious melanoma lesions in centers without specialized knowledge of sarcoma. The histologic and immunohistochemical similarities between melanoma and sarcoma can occasionally present a challenge for less experienced pathologists, increasing the risk of misdiagnosis. Therefore, effective collaboration among physicians, surgeons, and pathologists is essential to accurately guide the diagnostic process and assist pathologists in reaching a definitive histologic diagnosis. Physicians should consider sarcoma—particularly CCS, as mentioned—as a plausible differential diagnosis when encountering lesions that lack the typical clinical features of melanoma, especially those located deep within or near tendons and/or aponeurosis, particularly in young patients. It is noteworthy that the definitive diagnosis of CCS often relies on identifying EWSR1 translocation. Therefore, physicians should provide explicit guidance to pathologists, enabling them, when necessary, to actively search for EWSR1 translocation to definitively confirm the diagnosis. Alternatively, in the absence of clear guidance from the physician, a less experienced pathologist facing difficulties in reaching a definitive diagnosis for a suspected melanoma lesion should seek a second opinion from more experienced colleagues. This proactive approach may facilitate the timely implementation of appropriate therapeutic interventions, ultimately leading to improved patient outcomes and, potentially, prognosis.

5. Conclusions

Atypical skin lesions may be misdiagnosed as melanomas if they share many similarities. Physicians should be aware of such a possibility in the diagnostic process, as it may have potential implications for the treatment strategy. In the case of atypical skin lesions, it may be useful to investigate the presence of EWSR1 translocation, since CSS are the most common histology to be found in case of re-evaluation. Referral to tertiary expert centers may be recommended. Further investigations are required to better understand the epidemiology of misleading diagnosis and to raise awareness of the issue.

Supplementary Materials: The following supporting information can be downloaded at: https://www.mdpi.com/article/10.3390/cancers15143584/s1, Table S1: List of excluded records after reading the full-text.

Author Contributions: Conceptualization, F.C. (Fortunato Cassalia) and P.D.F.; methodology, F.C. (Francesco Cavallin); formal analysis, F.C. (Francesco Cavallin) and F.C. (Fortunato Cassalia); investigation, F.C. (Fortunato Cassalia) and A.D.; writing—original draft preparation, F.C. (Fortunato Cassalia), F.C. (Francesco Cavallin), A.D. and P.D.F.; writing—review and editing, C.D.P., M.R., S.M. and A.B.F.; supervision, M.R., S.M. and A.B.F. All authors have read and agreed to the published version of the manuscript.

Funding: This research received "Current Research" funds from the Italian Ministry of Health to cover publication costs.

Data Availability Statement: The data presented in this study are available within the article.

Acknowledgments: The authors wish to thank "Piccoli Punti ONLUS" and "Fondazione Lucia Valentini Terrani" for their long-lasting support.

Conflicts of Interest: The authors declare no conflict of interest.

References

1. Gatta, G.; van der Zwan, J.M.; Casali, P.G.; Siesling, S.; Tos, A.P.D.; Kunkler, I.; Otter, R.; Licitra, L.; Mallone, S.; Tavilla, A.; et al. Rare cancers are not so rare: The rare cancer burden in Europe. *Eur. J. Cancer* **2011**, *47*, 2493–2511. [CrossRef] [PubMed]
2. Sbaraglia, M.; Bellan, E.; Tos, A.P.D. The 2020 WHO Classification of Soft Tissue Tumours: News and perspectives. *Pathologica* **2021**, *113*, 70–84. [CrossRef]
3. American Cancer Society Key Statistics for Soft Tissue Sarcomas. Available online: https://www.cancer.org/cancer/soft-tissue-sarcoma/about/key-statistics.html (accessed on 10 April 2023).

4. Hisaoka, M.; Ishida, T.; Kuo, T.T.; Matsuyama, A.; Imamura, T.; Nishida, K.; Kuroda, H.; Inayama, Y.; Oshiro, H.; Kobayashi, H.; et al. Clear cell sarcoma of soft tissue: A clinicopathologic, immunohistochemical, and molecular analysis of 33 cases. *Am. J. Surg. Pathol.* **2008**, *32*, 452–460. [CrossRef] [PubMed]
5. Panagopoulos, I.; Mertens, F.; Isaksson, M.; Mandahl, N. Absence of mutations of the BRAF gene in malignant melanoma of soft parts (clear cell sarcoma of tendons and aponeuroses). *Cancer Genet. Cytogenet.* **2005**, *156*, 74–76. [CrossRef] [PubMed]
6. Patel, R.M.; Downs-Kelly, E.; Weiss, S.W.; Folpe, A.L.; Tubbs, R.R.; Tuthill, R.J.; Goldblum, J.R.; Skacel, M. Dual-color, break-apart fluorescence in situ hybridization for EWS gene rearrangement distinguishes clear cell sarcoma of soft tissue from malignant melanoma. *Mod. Pathol.* **2005**, *18*, 1585–1590. [CrossRef] [PubMed]
7. Dim, D.C.; Cooley, L.D.; Miranda, R.N. Clear cell sarcoma of tendons and aponeuroses: A review. *Arch. Pathol. Lab Med.* **2007**, *131*, 152–156. [CrossRef] [PubMed]
8. Pavlidis, N.A.; Fisher, C.; Wiltshaw, E. Clear-cell sarcoma of tendons and aponeuroses: A clinicopathologic study. Presentation of six additional cases with review of the literature. *Cancer* **1984**, *54*, 1412–1417. [CrossRef] [PubMed]
9. Biglow, L.R.; Cuda, J.; Dotson, J. A Rare Case of Epithelioid Malignant Peripheral Nerve Sheath Tumor Mimicking Malignant Melanoma. *Cureus* **2021**, *13*, e13424. [CrossRef]
10. Gaspard, M.; Lamant, L.; Tournier, E.; Valentin, T.; Rochaix, P.; Terrier, P.; Ranchere-Vince, D.; Coindre, J.M.; Filleron, T.; Le Guellec, S. Evaluation of eight melanocytic and neural crest-associated markers in well-characterized series of 124 malignant peripheral nerve sheath tumours (MPNST): Useful to distinguish MPNST from melanoma? *Histopathology* **2018**, *73*, 969–982. [CrossRef] [PubMed]
11. Zoufaly, A.; Schmiedel, S.; Lohse, A.; van Lunzen, J. Intestinal Kaposi's sarcoma may mimic gastrointestinal stromal tumor in HIV infection. *World J. Gastroenterol.* **2007**, *13*, 4514–4516. [CrossRef]
12. Shafiq, M.B.; Rafi, I.; Shoaib, A.; Ali, S.; Iqbal, F.; Latif, T.; Mushtaq, U. The Outcome of Extremity Soft Tissue Sarcomas in Terms of Resection Margins: A Study from a Cancer Dedicated Center. *Cureus* **2022**, *14*, e26086. [CrossRef]
13. Kollender, R.; Merimsky, O.; Sternheim, A.; Gortzak, Y.; Dadia, S.; Doron, A.; Novikov, I.; Kollender, Y.; Soyfer, V. Radiation Therapy Before Repeat Wide Resection for Unplanned Surgery of Soft Tissue Sarcoma ("Oops" Operation) Results in Improved Disease-Free Survival. *Adv. Radiat. Oncol.* **2022**, *7*, 101007. [CrossRef] [PubMed]
14. Keung, E.Z.; Krause, K.J.; Maxwell, J.; Morris, C.D.; Crago, A.M.; Houdek, M.T.; Kane, J.; Lewis, V.; Callegaro, D.; Miller, B.; et al. Sentinel Lymph Node Biopsy for Extremity and Truncal Soft Tissue Sarcomas: A Systematic Review of the Literature. *Ann. Surg. Oncol.* **2023**, *30*, 958–967. [CrossRef] [PubMed]
15. Page, M.J.; McKenzie, J.E.; Bossuyt, P.M.; Boutron, I.; Hoffmann, T.C.; Mulrow, C.D.; Shamseer, L.; Tetzlaff, J.M.; Akl, E.A.; Brennan, S.E.; et al. The PRISMA 2020 statement: An updated guideline for reporting systematic reviews. *BMJ* **2021**, *372*, n71. [CrossRef] [PubMed]
16. Munn, Z.; Barker, T.H.; Moola, S.; Tufanaru, C.; Stern, C.; McArthur, A.; Stephenson, M.; Aromataris, E. Methodological quality of case series studies: An introduction to the JBI critical appraisal tool. *JBI Evid. Synth.* **2020**, *18*, 2127–2133. [CrossRef] [PubMed]
17. Potter, A.J.; Dimitriou, F.; Karim, R.Z.; Mahar, A.; Chan, C.; Long, G.V.; Scolyer, R.A. Cutaneous clear cell sarcoma with an epidermal component mimicking melanoma. *Pathology* **2022**, *54*, 369–371. [CrossRef]
18. Elousrouti, L.T.; Hammas, N.; Elmernissi, F.Z.; Elfatemi, H.; Chbani, L. Clear-Cell Sarcoma with an Unusual Presentation Mimicking Metastatic Melanoma. *Cureus* **2022**, *14*, e32010. [CrossRef]
19. Zhang, X.; Zhang, P.J.; Sussman, R.; Litzky, L.A.; Kucharczuk, J.C.; Deshpande, C. Atypical clear cell sarcoma of the pleura presenting as large pleural effusion with 22q12 abnormality: A challenging case with twists and turns. *Hum. Pathol. Case Rep.* **2021**, *24*, 200489. [CrossRef]
20. Nawrocki, S.; Fitzhugh, V.A.; Groisberg, R.; Aviv, H.A.; Maghari, A. A rare case of primary dermal clear cell sarcoma with focal epidermotropism: An entity difficult to distinguish from melanoma. *J. Cutan. Pathol.* **2020**, *47*, 621–624. [CrossRef] [PubMed]
21. Obiorah, I.E.; Ozdemirli, M. Clear cell sarcoma in unusual sites mimicking metastatic melanoma. *World J. Clin. Oncol.* **2019**, *10*, 213–221. [CrossRef] [PubMed]
22. Donzel, M.; Zidane-Marinnes, M.; Paindavoine, S.; Breheret, R.; de la Fouchardière, A. Clear cell sarcoma of the soft palate mimicking unclassified melanoma. *Pathology* **2019**, *51*, 331–334. [CrossRef]
23. Obiorah, I.E.; Brenholz, P.; Özdemirli, M. Primary Clear Cell Sarcoma of the Dermis Mimicking Malignant Melanoma. *Balkan Med. J.* **2018**, *35*, 203–207. [CrossRef] [PubMed]
24. Curry, J.L.; Tetzlaff, M.T.; Wang, S.A.; Landon, G.; Alouch, N.; Patel, S.P.; Nagarajan, P.; Gupta, S.; Aung, P.P.; Devine, C.E.; et al. Case Report of Myeloid Sarcoma Masquerading as In-Transit Metastasis at a Previous Melanoma Site: Avoiding a Diagnostic Pitfall. *Am. J. Dermatopathol.* **2018**, *40*, 831–835. [CrossRef] [PubMed]
25. Leon-Castillo, A.; Chrisinger, J.S.A.; Panse, G.; Samdani, R.T.; Ingram, D.R.; Ravi, V.; Prieto, V.G.; Wang, W.-L.; Lazar, A.J. Index report of cutaneous angiosarcomas with strong positivity for tyrosinase mimicking melanoma with further evaluation of melanocytic markers in a large angiosarcoma series. *J. Cutan. Pathol.* **2017**, *44*, 692–697. [CrossRef] [PubMed]
26. Zivanovic, M.; Luzar, B.; Bacchi, C.E.; Calonje, J.E. Cutaneous clear cell sarcoma with intraepidermal component mimicking spitzoid melanoma. *Virchows Arch.* **2017**, *471* (Suppl. S1), S1–S352.
27. Jackson, C.R.; Minca, E.C.; Kapil, J.P.; Smith, S.C.; Billings, S.D. Superficial malignant peripheral nerve sheath tumor with overlying intradermal melanocytic nevus mimicking spindle cell melanoma. *J. Cutan. Pathol.* **2016**, *43*, 1220–1225. [CrossRef]

28. Castriconi, M.; Antropoli, M.; Grillo, M.; Andreano, M.; Santoro, M.; Villamania, E. Unusual bleeding of a giant cell fibroblastoma: A soft tissue sarcoma of the skin mimicking metastatic melanoma. *Ann. Ital. Chir.* **2015**, *86*, S2239253X15023701.
29. Sayah, M.; Hammer, S. Colonic Clear Cell Sarcoma Associated with Neurofibromatosis Type 2. *Am. J. Clin. Pathol.* **2015**, *144* (Suppl. S2), A409. [CrossRef]
30. Liu, C.; Ren, Y.; Li, X.; Cao, Y.; Chen, Y.; Cui, X.; Li, L.; Li, F. Absence of 19 known hotspot oncogenic mutations in soft tissue clear cell sarcoma: Two cases report with review of the literature. *Int. J. Clin. Exp. Pathol.* **2014**, *7*, 5242–5249. [PubMed]
31. Sidiropoulos, M.; Busam, K.; Guitart, J.; Laskin, W.B.; Wagner, A.M.; Gerami, P. Superficial paramucosal clear cell sarcoma of the soft parts resembling melanoma in a 13-year-old boy. *J. Cutan. Pathol.* **2013**, *40*, 265–268. [CrossRef] [PubMed]
32. Falconieri, G.; Bacchi, C.E.; Luzar, B. Cutaneous clear cell sarcoma: Report of three cases of a potentially underestimated mimicker of spindle cell melanoma. *Am. J. Dermatopathol.* **2012**, *34*, 619–625. [CrossRef] [PubMed]
33. Rodriguez-Martin, M.; Saez-Rodriguez, M.; Esquivel, B.; Gonzalez, R.S.; Cabrera, A.N.; Herrera, A.M. Clear cell sarcoma: A case mimicking primary cutaneous malignant melanoma. *Indian J. Dermatol.* **2009**, *54*, 168–172. [CrossRef] [PubMed]
34. Tanas, M.R.; Rubin, B.P. Malignant neuroectodermal tumor with melanocytic and rhabdomyoblastic differentiation. *Rare Tumors* **2009**, *1*, e26. [CrossRef] [PubMed]
35. Brightman, L.A.; Demierre, M.-F.; Byers, H.R. Macrophage-rich epithelioid angiosarcoma mimicking malignant melanoma. *J. Cutan. Pathol.* **2006**, *33*, 38–42. [CrossRef] [PubMed]
36. Matsuda, Y.; Saoo, K.; Hosokawa, K.; Yamakawa, K.; Yokohira, M.; Zeng, Y.; Takeuchi, H.; Iwai, J.; Shirai, T.; Obika, K.; et al. Epithelioid malignant peripheral nerve sheath tumor. Report of a case with inflammatory infiltration. *Pathol. Res. Pract.* **2005**, *201*, 355–360. [CrossRef]
37. Demir, Y.; Tokyol, C. Superficial malignant schwannoma of the scalp. *Dermatol. Surg.* **2003**, *29*, 879–881. [PubMed]
38. Bonetti, F.; Martignoni, G.; Colato, C.; Manfrin, E.; Gambacorta, M.; Faleri, M.; Bacchi, C.; Sin, V.-C.; Wong, N.-L.; Coady, M.; et al. Abdominopelvic sarcoma of perivascular epithelioid cells. Report of four cases in young women, one with tuberous sclerosis. *Mod. Pathol.* **2001**, *14*, 563–568. [CrossRef] [PubMed]
39. Ferreiro, J.A.; Nascimento, A.G. Hyaline-cell rich chondroid syringoma. A tumor mimicking malignancy. *Am. J. Surg. Pathol.* **1995**, *19*, 912–917. [CrossRef] [PubMed]
40. Honma, K.; Watanabe, H.; Ohnishi, Y.; Tachikawa, S.; Tachikawa, K. Epithelioid malignant schwannoma. A case report. *Acta Pathol. Jpn.* **1989**, *39*, 195–202.
41. Gould, E.; Albores-Saavedra, J.; Rothe, M.; Mnaymneh, W.; Menendez-Aponte, S. Malignant giant cell tumor of soft parts presenting as a skin tumor. *Am. J. Dermatopathol.* **1989**, *11*, 197–201. [CrossRef]

cancers

Review

Predictors of Symptomatic Venous Thromboembolism in Patients with Soft Tissue Sarcoma in the Lower Extremity

Pramod N. Kamalapathy, Adam Kline, Hannah Hollow, Kevin Raskin, Joseph H. Schwab and Santiago Lozano-Calderón *

Department of Orthopaedic Surgery, Massachusetts General Hospital, Boston, MA 02114, USA
* Correspondence: slozanocalderon@mgh.harvard.edu

Simple Summary: 28 patients, or 4.36%, were diagnosed with venous thromboembolism after soft tissue sarcoma surgery. The most significant risk factors for this complication were pre-operative (PTT) partial thromboplastin time, post-operative PTT, post-op chemotherapy, metastasis at diagnosis, additional surgery for metastasis or local recurrence, and tumor size larger than 10 cm. Risk of wound complications and infection increased in those who received prophylaxis medications.

Abstract: Orthopedic surgery and soft-tissue sarcoma (STS) both independently increase the risk of developing symptomatic venous thromboembolic events (SVTE), but there are no established risk factors or guidelines for how to prophylactically treat patients with STS undergoing surgery. The objectives of this study were to (1) identify the prevalence of SVTE in patients undergoing STS surgery, (2) identify risk factors for SVTE, and (3) determine the risk of wound complications associated with VTE prophylaxis. This retrospective study was conducted in a tertiary level, academic hospital. A total of 642 patients were treated for soft-tissue sarcoma in the lower extremity with follow up for at least 90 days for the development of SVTE such as deep venous thrombosis and pulmonary embolism. Multivariate logistic regression was used to identify predictors for these events by controlling for patient characteristics, surgical characteristics, and treatment variables, with significance held at $p < 0.05$. Twenty eight patients (4.36%) were diagnosed with SVTE. Multivariate analysis found six significant predictors ordered based on standardized coefficients: pre-operative (PTT) partial thromboplastin time ($p < 0.001$), post-operative PTT ($p = 0.010$), post-op chemotherapy ($p = 0.013$), metastasis at diagnosis ($p = 0.025$), additional surgery for metastasis or local recurrence ($p = 0.004$), and tumor size larger than 10 cm ($p < 0.001$). The risk of wound complications ($p = 0.04$) and infection ($p = 0.017$) increased significantly in patients who received chemical prophylaxis. Our study identifies risk factors for patients at increased risk of developing VTE. Further prospective research is necessary to identify which protocols would be beneficial in preventing SVTE in high-risk patients with a low profile of wound complications.

Keywords: orthopedic surgery; sarcoma; symptomatic venous thromboembolism (SVTE); DVT prophylaxis

Citation: Kamalapathy, P.N.; Kline, A.; Hollow, H.; Raskin, K.; Schwab, J.H.; Lozano-Calderón, S. Predictors of Symptomatic Venous Thromboembolism in Patients with Soft Tissue Sarcoma in the Lower Extremity. *Cancers* **2023**, *15*, 315. https://doi.org/10.3390/cancers15010315

Academic Editor: Catrin Sian Rutland

Received: 7 November 2022
Revised: 22 December 2022
Accepted: 28 December 2022
Published: 3 January 2023

1. Introduction

Affecting nearly 350,000–600,000 Americans annually, symptomatic venous thromboembolism (SVTE) represents one of the most common preventable causes of hospital deaths [1–4]. Venous thromboembolism (VTE) is a blood clot arising in a deep vein, including both deep vein thrombosis (DVT) and pulmonary embolism (PE). Cancer and orthopedic surgery have been independently identified as risk factors for SVTE [5–9]. The literature reports that the incidence of SVTE in this patient population is between 1.4 and 21% with the use of mechanical and/or chemical prophylaxis [8,10–13]. Through the expression of thrombin and the release of microparticles that influence the solidity of the blood, cancer cells increase coagulation [14,15] and subsequent platelet formation, which is then thought to facilitate the metastatic process [16].

Although all types of surgery increase the risk of clotting and therefore SVTE, orthopedic surgery involves many prothrombotic processes such as coagulation activation from tissue and bone injury; venous injuries; and long periods of immobilization, which further increase the risk and occurrence of SVTE [17]. Thus, this population with STS are at increased risk of undergoing a postoperative thromboembolic event compared to patients after a soft-tissue orthopedic procedure with non-oncologic characteristics [8].

There are currently a few existing studies that evaluate the unique risk factors of an STS patient undergoing orthopedic surgery [18–20]. However, these studies are limited by their older case series, small sample sizes, and reliance on national databases. The American Academy of Orthopaedic Surgeons (AAOS) guidelines indicate there is a lack of evidence on the predictors of SVTE in orthopedic oncology patients undergoing surgery, resulting in unclear protocols for post-surgical prophylaxis treatment [11]. Prophylactic anticoagulation medication may help to reduce SVTE and the resulting morbidity and mortality; however, it can lead to surgical bed bleeding, hematoma, and wound complications [21,22]. Identifying SVTE predictors may be extremely beneficial for clarifying and understanding a proper treatment regimen for patients at risk, while helping surgeons to minimize the risks related to anticoagulation. The objectives of this study were to (1) identify the prevalence of SVTE in patients undergoing soft tissue sarcoma surgery, (2) identify risk factors for SVTE in this population, and (3) determine the risk of wound complications associated with VTE prophylaxis.

2. Materials and Methods

This retrospective study was conducted at a single large referral institution in order to assess the predictors of soft-tissue high-grade sarcoma of the lower extremity. A total of 642 patients were found using a research patient data registry search that included patients older than 18 years of age who have been surgically treated at our institution for soft tissue sarcomas from January 1992 to December 2017. All tumors were microscopically confirmed to be high-grade soft-tissue sarcomas in the lower extremity, excluding the pelvis, buttocks, and genitals. Exclusion criteria included patients with primary bone sarcomas extending and/or metastasizing to soft tissue or those without follow-up for at least 90 days after surgery. The average age at diagnosis of the population was 53.2 years old, and 56% of patients were male. Body mass index was similar, at 28.1 in no-SVTE patients and 28.5 in SVTE patients (Table 1).

Table 1. Patient characteristics.

Variable	No VTE (614)	VTE (28)	Odds Ratio (95% CI)	p-Value
Age at diagnosis	53.2 (40.8–66)	53.4 (44–60.8)	0.99 (0.97–1.02)	0.984
BMI [a]	28.1 (23–31.4)	28.5 (26.1–30.7)	1.00 (0.93–1.06)	0.918
		Sex		
Male	344 (94.8%)	19 (5.2%)	Reference	
Female	270 (96.8%)	9 (3.2%)	2.65 (1.02–8.21)	0.07
		Smoking status [a]		
Never smoked	273 (95.8%)	12 (4.2%)	Reference	
Current smoker	65 (94.2%)	4 (5.8%)	2.22 (0.83–6.32)	0.114
Quit	195 (94.2%)	12 (5.8%)	2.31 (0.58–7.98)	0.196

[a]: Age at diagnosis is available for 606 patients; smoking status is available for 564 people.

2.1. Outcomes of Interest

The outcome event was defined as a radiographically (CT scan or ultrasound) confirmed clinically symptomatic DVT or PE within 90 days of the index surgery. Within the scope of this research, patients were followed up at least at two weeks, six weeks, or three months after surgery. Imaging was only obtained in patients with clinical presentation of

SVTE. Routine imaging was not used, as previous literature has now recognized an increase in the prevalence of thromboembolic events if asymptomatic patients are included [19]. The treatment of asymptomatic patients is also controversial [10].

In our practice, DVT prophylaxis is given to patients for at least four weeks. If the patient remains immobile, DVT prophylaxis extends to six weeks or as needed. The protocol period of immobilization after surgery is two days bed-rest after resection if preoperative radiation therapy is given and the condition of the soft tissues is suboptimal; or five days of bed rest with limited dangling if preoperative radiation therapy is given and a rotational flap ± skin graft is used for closure (there is no restriction if radiation was not used preoperatively); or seven days of bed rest and limited dangling, independent of the use of radiation therapy, with a free flap was used for closure.

Information on variables such as patient characteristics, details about the outcome event, and tumor characteristics; treatment variables; preoperative and postoperative clinical variables; and any complications was collected. Tumor site, gross histology, tumor size, and depth were confirmed by the pathology reports and operation notes. Our institution determines grade based on a 1–3 scale, and stage was determined using the American Joint Committee on Cancer staging system (AJCC), which utilizes tumor size, metastases, and grade [23,24]. If a patient presented with an STS tumor in the upper extremity that metastasized to the lower extremity, metastasis was recorded as presenting at diagnosis. Wound complications were defined as a broad category that included infections and conditions such as wound dehiscence, delayed wound healing, local thrombosis, and soft-tissue reconstruction failures. Additional procedures were confirmed by additional operative notes for irrigation and debridement, new soft-tissue reconstructions with grafts or flaps, revisions of these, and other complications. Pre-operative and post-operative variables within a week before and after the surgery were also used. Follow-up varied between patients; while some patients exhibited symptoms of concern and were assessed for DVT quickly, others were only assessed for DVT during regular scheduled follow-ups based on the guidelines stated above.

2.2. Statistical Analysis

Patients with SVTE events were compared to those without them to identify any potential predictors of SVTE. All variables were assessed with a multivariate regression model by controlling for patient characteristics, surgical characteristics, and treatment variables. Odds ratios and 95% confidence intervals were reported for all outcomes. Due to the large nature of this retrospective study, there were missing data for a small percentage of patients. Each analysis included only those patients with the variable of interest available. Statistical significance was held at $p < 0.05$. STATA 15 by StataCorp (College Station, TX, USA: StataCorp LLC) was used for all statistical analyses.

3. Results

A total of 28 patients (4.36%) out of 642 were diagnosed with SVTE within 90 days of surgery (27 DVT and 1 PE). The average age at diagnosis of the population was 53.2 years old. Fifty six percent of patients were male. There were no differences in the age of patients, body mass index, or smoking status between no-SVTE patients and SVTE patients.

A multivariate logistic model found that tumor size larger than 10 cm (OR 2.41, 95% CI 1.07–5.21), post-surgical chemotherapy (OR 2.98, 95% CI 1.35–6.42), pre-op PTT (OR 0.77, 95% CI 0.68–0.89), post-op PTT (OR 0.91, 95% CI 0.75–0.98), metastasis at diagnosis (OR 3.18, 95% CI 1.11–8.59), and additional surgery for metastasis or local recurrence (OR 2.89, 95% CI 1.24–6.97) were significant predictors of SVTE (Tables 2–4).

Table 2. Tumor characteristics.

Variable	No VTE (614)	VTE (28)	Odds Ratio (95% CI)	p-Value
		Histology		
Undifferentiated/general	91	4	Reference	
Leiomyosarcoma	150	2	0.30 (0.04–1.57)	0.170
Fibrosarcoma	169	5	0.67 (0.17–2.76)	0.556
Angiosarcoma	47	0	-	0.992
Liposarcoma	50	5	2.23 (0.56–9.36)	0.247
Malignant peripheral nerve Sheath tumor	26	0	-	0.996
Rhabdomyosarcoma	10	2	4.13 (0.53–2.39)	0.124
Synovial sarcoma	71	10	3.15 (1.01–1.18)	0.060
		Site		
Thigh	378 (95.2%)	19 (4.8%)	Reference	
Leg	198 (97.1%)	6 (2.9%)	0.94 (0.32–2.42)	0.901
Foot	35 (9.2%)	3 (90.8%)	1.92 (0.28–7.78)	0.416
		Grade [a]		
1/3	53 (94.6%)	3 (5.4%)	Reference	
2/3	202 (96.2%)	8 (3.8%)	0.50 (0.13–2.48)	0.348
3/3	182 (95.3%)	9 (4.7%)	0.51 (0.14–2.68)	0.408
1–2/3	26 (96.3%)	1 (3.7%)	-	0.988
2–3/3	128 (94.8%)	7 (5.2%)	0.70 (0.16–3.60)	0.637
		Stage [a]		
I	54 (94.7%)	3 (5.3%)	Reference	
II	206 (97.2%)	6 (2.8%)		0.247
IIIA	156 (94.5%)	9 (5.5%)		0.993
IIIB	100 (97.1%)	3 (2.9%)		0.
IV	76 (91.6%)	7 (8.4%)		0.989
Dimension larger than 10 cm [a]	**279 (94.9%)**	**15 (5.1%)**	**2.41 (1.07–5.21)**	**<0.001**
Vascular Invasion [a]	55 (94.8%)	3 (5.2%)	1.53 (0.35–4.84)	0.515
Metastasis at Diagnosis [a]	**67 (89.3%)**	**8 (10.7%)**	**3.18 (1.11–8.59)**	**0.025**

[a]: Grade and stage are available for 619 people, dimension for 623, vascular invasion for 604, and metastasis at diagnosis for 500 people. Significant values are bolded.

Table 3. Surgery and other treatment.

Variable	No VTE (614)	VTE (28)	Odds Ratio (95% CI)	p-Value (Multivariate)
		Surgery		
Operative time	2.0 (1.5–2.6)	2.7 (2.1–3.3)	1.11 (0.72–2.87)	0.387
Vascular injury	10 (1.4%)	0 (0%)	-	-
Positive margin [a]	113 (94.2%)	7 (5.8%)	1.64 (0.57–4.38)	0.483
Blood loss [a]	374.1 (50–350)	777.7 (125–778)	1.22 (0.74–3.20)	0.519
Reconstruction	396 (97.3%)	11 (2.7%)	1.34 (0.84–2.45)	0.221
Graft [a]	112 (96.6%)	4 (3.4%)	1.00 (0.28–2.79)	0.982
Tourniquet use	121 (97.6%)	3 (2.4%)	1.2 (0.43–4.10)	0.456
		Chemotherapy		
Pre-op [a]	173 (94%)	11 (6%)	1.50 (0.59–3.68)	0.278
Post-op [a]	**137 (91.3%)**	**13 (8.7%)**	**2.98 (1.35–6.42)**	**0.013**
		Radiation		
Pre-op [a]	373 (95.6%)	17 (4.4%)	1.33 (0.54–3.56)	0.838
Post-op [a]	138 (95.2%)	7 (4.8%)	1.13 (0.44–2.59)	0.733
		VTE prophylaxis		
None	156 (98.7%)	2 (1.3%)	Reference	Reference
ASA	51 (98.1%)	1 (1.9%)	8.37 (0.36–9.97)	0.099
LMWH	244 (93.8%)	16 (6.2%)	4.54 (0.99–2.92)	0.057
Warfarin	149 (94.3%)	9 (5.7%)	3.49 (0.73–2.48)	0.142
Multiple	14 (100%)	0	-	0.994

[a]: Margin information is available for 629 patients, blood loss for 619, graft for 620, pre-op chemo for 603, pre-op radiation for 631, and post-op radiation information for 625. Significant values are bolded.

Table 4. Pre-operative and post-operative blood values and complications.

Variable	No VTE (614)	VTE (28)	Odds Ratio (95% CI)	*p*-Value (Multivariate)
		Pre-Op		
Partial thromboplastin [a]	30.39 (24.9–30.25)	27.11 (24.5–29.2)	**0.77 (0.68–0.89)**	**<0.001**
PT(INR) [a]	1.08 (1–1.1)	1.119 (1–1.1)	1.41 (0.20–5.24)	0.648
WBC [a]	7.40 (5.56–8.46)	7.74 (5.15–8.35)	0.978 (0.83–1.08)	0.754
PLT [a]	284.60 (206–332)	298.4 (229–298)	1.00 (0.98–1.10)	0.734
HGB [a]	13.11 (11.3–14.6)	13.47 (10.95–14.8)	1.03 (0.91–1.14)	0.375
		Post Op		
Partial thromboplastin [a]	38.62 (25.9–39.2)	33.58 (25.7–33.6)	**0.91 (0.75–0.98)**	**0.010**
PT(INR) [a]	1.17 (1.01–1.20)	1.13 (1.1–1.2)	0.36 (0.01–4.24)	0.228
WBC [a]	9.21 (7–10.7)	9.94 (7.1–13.78)	1.07 (0.79–1.14)	1.304
		Complications		
Infection	116 (92.8%)	9 (7.2%)	1.21 (0.41–3.11)	0.216
Wound Complication	120 (92.3%)	10 (7.7%)	2.25 (1.07–5.21)	0.124
Additional Surgery for metastasis or local recurrence	**141 (92.2%)**	**12 (7.8%)**	**2.89 (1.24–6.97)**	**0.004**

[a]: Pre-op values: PTT values is available for 529 patients, PT/INR (383 patients), PLT (573 people), WBC and HGB (576 people), and glucose (456 people). Post-op information: PTT is available for 506 patients, PT/INR (540 people), WBC (509 people). Significant values are bolded.

Of the 642 people, 484 received at least one VTE prophylactic agent. The most common agent was low-molecular-weight heparin (LMWH), which was given to 244 people (Table 3). Patients had no differences in VTE rates based on their chemoprophylaxis ($p > 0.05$). Moreover, patients that received chemoprophylaxis were associated with increased risk of wound complications (OR 1.20, CI 1.01–1.43, $p = 0.04$) and infection (OR 1.24, CI 1.04–1.48, $p = 0.017$). However, no specific chemical prophylaxis was found to be associated with increased wound complication risk (Table 5).

Table 5. DVT prophylaxis and wound complication risk.

Variable	*p*-Value (Multivariate)	Odds Ratio	CI Interval
None	Reference		
Aspirin	0.098	2.49	0.1–42.0
Warfarin	0.089	4.25	0.38–9.46
LMWH	0.066	7.68	0.22–15.2
Multiple treatments	0.078	4.60	0.4–10.3

4. Discussion

Orthopedic surgery and cancer are both independently associated with an increased risk of developing SVTE [25]. Currently, there are no guidelines that take into account the unique risk factors of this population for prescribing DVT prophylaxis. This study was designed to identify potential predictors of SVTE in STS patients and complications associated with prophylaxis treatments. This study identified the prevalence rate of SVTE following soft tissue sarcoma surgery to be 4.36%. Six significant predictors—post-op PTT, pre-op PTT, post-op chemotherapy, metastasis at diagnosis, additional surgery for metastasis or local recurrence, and tumor size larger than 10 cm—were found to associated with an increased risk of developing SVTE after surgery while adjusting for patient characteristics, tumor characteristics, treatments, and laboratory values.

The prevalence of SVTE in our cohort was slightly lower than that reported in the literature. The percentage of reported SVTE incidence rates in orthopedic surgery varies considerably, ranging from 0.6 to 21% [8,10–13]. One possible reason for this considerable range could be attributed to the lack of standard protocol used to diagnose SVTE in the published literature. This variation can also be attributed to the variation in the diagnosis of subclinical VTE, which often goes undiagnosed. Our specific cohort did not include imaging studies in asymptomatic patients, reflecting the current clinical practice in which only symptomatic patients are tested. Patients with clinical concern of DVT or PE underwent further imaging, while those with ailments such as unilateral swelling underwent lower-extremity DVT ultrasound. Yet, this is one of the largest cohorts of patients with soft-tissue sarcoma followed to date.

The increase in the risk of SVTE due to metastasis is in line with the idea that the diffuse nature of the tumor leads to hypercoagulability, increasing the risk of thrombosis [26,27]. Metastatic cancers are usually known to be larger and to release more procoagulant factors, requiring a multitude of treatments and leading to shorter survival times. One such treatment, postoperative chemotherapy, is also a significant predictor of SVTE and major bleeding complications [5,28]. Postoperative chemotherapy puts stress on the body and exacerbates any irregular clotting abnormalities [28]. Similarly, additional surgery due to local recurrence or metastasis, which itself is known to be positively predictive of developing DVTs, is significantly correlated with SVTEs, as surgery increases the risk of immobilization and other prothrombic factors. Radiation therapy was likewise expected to be a predictor of SVTE, but it was not found to be significantly associated [26,29]. Radiation therapy has been shown to cause endothelial prothrombotic response, influencing the thrombomodulin complex and various cytokines [10,30] in oncologic patients, but its effects on microcirculation is one possible reason why this discrepancy exists [31].

Our study found that preoperative and postoperative PTT were significantly associated with an increased risk of DVT. Activated PTT is a measure of intrinsic coagulation pathways, meaning PTT levels can be used to measure the rate of coagulation. A low level of PTT indicates a procoagulation tendency because of a greater number of clotting proteins; therefore, it follows that low PTT levels are predictive factors for DVT [32].

Our study did not find any specific chemoprophylaxis associated with significantly decreased risk of SVTE. Currently, AAOS does not have a standard recommendation of DVT prophylaxis for patients undergoing soft-tissue sarcoma patients. Regarding elective hip and knee arthroplasty, AAOS only has evidence sufficient to screen patients with a previous history of SVTE as a risk factor [33]. The American College of Clinical Pharamacology (ACCP) recommends some form of chemical prophylaxis for all patients undergoing major orthopedic surgery; however, there is no standard guideline for which prophylaxis can be employed [34].

Heparin was used as the standard for DVT prophylaxis in the early 1920s until the introduction of warfarin in 1948 [35]. In most cases, heparin was followed by warfarin as a treatment regimen. Eventually, LMWH was issued to alleviate the need to consistently monitor the patient [21]. Many studies identifying risk factors have low statistical power due to the rare occurrence of SVTE and the lack of large data collection in this specific group of patients [36]. Levine et al. demonstrated that LMWH is an equally effective alternative to the unfractionated heparin delivered in the hospital [22]. Singh et al. found that the incidence of DVT in patients undergoing orthopedic oncology lower-limb surgery was low even without prophylaxis, but noted that further investigation with larger sample sizes was necessary [36].

All patients at our institution receive mechanical prophylaxis, either compression stockings or intermittent pneumatic compression devices. In total, 484 of the patients at our institution received at least one prophylactic treatment (Table 3). Chemical prophylaxis is positively associated with wound complications and infection, but no specific prophylactic agent was found to lead to significantly increased risk [36]. This suggests that patients might be over anticoagulated and placed at risk of hematoma formation, with subsequent wound complications and infections. This compounds the necessity of analyzing risk factors for developing SVTE in order to prescribe a patient the proper treatment and minimize their overall complications.

This study had a number of limitations. First, due to its retrospective design, there was incomplete information available for different patients. The large cohort size reduced the effect of the loss of data, as we excluded patients with missing data in each of the independent calculations to reduce concerns. Second, assessing for SVTE was not routine unless the patient was symptomatic and treated at our institution, explaining the lower rate of SVTE recorded in this study compared to in the literature. A prospective study would be helpful to further evaluate the conclusions of this study. However, our design reflects the current standard clinical practice or screening. Third, the study was limited to our institution. This could potentially lead to a more homogenous patient population and lack of generalizability. Our large referral cancer center treats a wide variety of patients, making this limitation less of a concern. Fourth, the present paper attempts to provide a comprehensive analysis of the factors that could contribute to venous thromboembolism postoperatively; however, there are a number of factors, such as prothrombotic agents or previous thromboembolism, that could not be analyzed due to the limited sample size or power of the study. Despite these challenges, the conclusions reached in this study provide clinicians valuable information about orthopedic oncology patients with soft-tissue sarcomas of the lower extremities and how to assess their risk for SVTE.

5. Conclusions

Six variables were found to be significant predictors of SVTE in orthopedic oncology patients undergoing surgery: tumor size greater than 10 cm, metastasis of tumor at diagnosis, postoperative chemotherapy, preoperative and postoperative PTT, and additional surgeries. Surgeons and healthcare professionals could minimize the risk of developing SVTE for STS patients by actively following patients with increased risk factors and reducing complications associated with their surgery and recovery. Thrombophylaxis is a gray area in cancer patients, with further prospective studies being required in order to determine which protocols in high-risk patients would be beneficial in preventing SVTE with a low profile of complications in terms of wound healing, postoperative hematoma, and infections.

Author Contributions: Conceptualization, S.L.-C.; methodology, S.L.-C. and P.N.K.; software, S.L.-C.; validation, P.N.K., formal analysis, P.N.K.; investigation, P.N.K.; resources, S.L.-C.; data curation, P.N.K., A.K. and H.H.; writing—original draft preparation, S.L.-C. and P.N.K.; writing—review and editing, P.N.K., A.K., H.H., K.R., J.H.S. and S.L.-C.; visualization, P.N.K.; supervision, S.L.-C.; project administration, P.N.K.; funding acquisition, S.L.-C. All authors have read and agreed to the published version of the manuscript.

Funding: This research received no external funding.

Data Availability Statement: Data is unavailable to be made public due to privacy or ethical restrictions, but can be considered based on request to corresponding author.

Conflicts of Interest: The authors declare no conflict of interest.

References

1. Maynard, G.A. Medical Admission Order Sets to Improve Deep Vein Thrombosis Prevention: A Model for Others or a Prescription for Mediocrity? *J. Hosp. Med.* **2009**, *4*, 77–80. [CrossRef] [PubMed]
2. O'Connor, C.; Adhikari, N.K.J.; DeCaire, K.; Friedrich, J.O. Medical Admission Order Sets to Improve Deep Vein Thrombosis Prophylaxis Rates and Other Outcomes. *J. Hosp. Med.* **2009**, *4*, 81–89. [CrossRef] [PubMed]
3. Office of the Surgeon General (US); National Heart, Lung, and Blood Institute (US). *The Surgeon General's Call to Action to Prevent Deep Vein Thrombosis and Pulmonary Embolism*; Publications and Reports of the Surgeon General; Office of the Surgeon General (US): Rockville, MD, USA, 2008.
4. Monreal, M.; Kakkar, A.K.; Caprini, J.A.; Barba, R.; Uresandi, F.; Valle, R.; Suarez, C.; Otero, R.; RIETE Investigators. The Outcome after Treatment of Venous Thromboembolism Is Different in Surgical and Acutely Ill Medical Patients. Findings from the RIETE Registry. *J. Thromb. Haemost. JTH* **2004**, *2*, 1892–1898. [CrossRef] [PubMed]
5. Falanga, A.; Zacharski, L. Deep Vein Thrombosis in Cancer: The Scale of the Problem and Approaches to Management. *Ann. Oncol. Off. J. Eur. Soc. Med. Oncol.* **2005**, *16*, 696–701. [CrossRef] [PubMed]

6. Falanga, A.; Marchetti, M.; Russo, L. The Mechanisms of Cancer-Associated Thrombosis. *Thromb. Res.* **2015**, *135* (Suppl. S1), S8–S11. [CrossRef] [PubMed]
7. Imberti, D.; Bianchi, C.; Zambon, A.; Parodi, A.; Merlino, L.; Gallerani, M.; Corrao, G. Venous Thromboembolism after Major Orthopaedic Surgery: A Population-Based Cohort Study. *Intern. Emerg. Med.* **2012**, *7*, 243–249. [CrossRef]
8. Kaiser, C.L.; Freehan, M.K.; Driscoll, D.A.; Schwab, J.H.; Bernstein, K.D.A.; Lozano-Calderon, S.A. Predictors of Venous Thromboembolism in Patients with Primary Sarcoma of Bone. *Surg. Oncol.* **2017**, *26*, 506–510. [CrossRef]
9. Kim, S.M.; Park, J.M.; Shin, S.H.; Seo, S.W. Risk Factors for Post-Operative Venous Thromboembolism in Patients with a Malignancy of the Lower Limb. *Bone Jt. J.* **2013**, *95*, 558–562. [CrossRef]
10. Mitchell, S.Y.; Lingard, E.A.; Kesteven, P.; McCaskie, A.W.; Gerrand, C.H. Venous Thromboembolism in Patients with Primary Bone or Soft-Tissue Sarcomas. *J. Bone Jt. Surgery. Am. Vol.* **2007**, *89*, 2433–2439. [CrossRef]
11. Morii, T.; Mochizuki, K.; Tajima, T.; Aoyagi, T.; Satomi, K. Venous Thromboembolism in the Management of Patients with Musculoskeletal Tumor. *J. Orthop. Sci.* **2010**, *15*, 810–815. [CrossRef]
12. Ramo, B.A.; Griffin, A.M.; Gill, C.S.; McDonald, D.J.; Wunder, J.S.; Ferguson, P.; Bell, R.S.; Phillips, S.E.; Schwartz, H.S.; Holt, G.E. Incidence of Symptomatic Venous Thromboembolism in Oncologic Patients Undergoing Lower-Extremity Endoprosthetic Arthroplasty. *J. Bone Jt. Surg. Am. Vol.* **2011**, *93*, 847–854. [CrossRef]
13. Tuy, B.; Bhate, C.; Beebe, K.; Patterson, F.; Benevenia, J. IVC Filters May Prevent Fatal Pulmonary Embolism in Musculoskeletal Tumor Surgery. *Clin. Orthop. Relat. Res.* **2009**, *467*, 239–245. [CrossRef]
14. Gil-Bernabé, A.M.; Lucotti, S.; Muschel, R.J. Coagulation and Metastasis: What Does the Experimental Literature Tell Us? *Br. J. Haematol.* **2013**, *162*, 433–441. [CrossRef]
15. Kesieme, E.; Kesieme, C.; Jebbin, N.; Irekpita, E.; Dongo, A. Deep Vein Thrombosis: A Clinical Review. Available online: https://www.dovepress.com/deep-vein-thrombosis-a-clinical-review-peer-reviewed-article-JBM (accessed on 8 October 2020).
16. Cavanaugh, P.G.; Sloane, B.F.; Honn, K.V. Role of the Coagulation System in Tumor-Cell-Induced Platelet Aggregation and Metastasis. *Haemostasis* **1988**, *18*, 37–46. [CrossRef]
17. Cionac Florescu, S.; Anastase, D.-M.; Munteanu, A.-M.; Stoica, I.C.; Antonescu, D. Venous Thromboembolism Following Major Orthopedic Surgery. *Maedica (Buchar)* **2013**, *8*, 189–194.
18. Krzyzaniak, H.; You, D.Z.; Mosca, G.; Monument, M.J.; Schneider, P.S. Venous Thromboembolism Rates in Patients with Bone and Soft Tissue Sarcoma of the Extremities Following Surgical Resection: A Systematic Review. *J. Surg. Oncol.* **2021**, *124*, 390–399. [CrossRef]
19. Alcindor, T.; Al-Fakeeh, A.; Goulding, K.; Solymoss, S.; Ste-Marie, N.; Turcotte, R. Venous Thromboembolism in Patients with Sarcoma: A Retrospective Study. *Oncologist* **2019**, *24*, e111–e114. [CrossRef]
20. Shantakumar, S.; Connelly-Frost, A.; Kobayashi, M.G.; Allis, R.; Li, L. Older Soft Tissue Sarcoma Patients Experience Increased Rates of Venous Thromboembolic Events: A Retrospective Cohort Study of SEER-Medicare Data. *Clin. Sarcoma Res.* **2015**, *5*, 18. [CrossRef]
21. Oduah, E.I.; Linhardt, R.J.; Sharfstein, S.T. Heparin: Past, Present, and Future. *Pharmaceuticals* **2016**, *9*, 38. [CrossRef]
22. Levine, M.; Gent, M.; Hirsh, J.; Leclerc, J.; Anderson, D.; Weitz, J.; Ginsberg, J.; Turpie, A.G.; Demers, C.; Kovacs, M. A Comparison of Low-Molecular-Weight Heparin Administered Primarily at Home with Unfractionated Heparin Administered in the Hospital for Proximal Deep-Vein Thrombosis. *N. Engl. J. Med.* **1996**, *334*, 677–681. [CrossRef]
23. AJCC Cancer Staging Manual | Mahul B. Amin | Springer. Available online: https://www.springer.com/gp/book/9783319406176 (accessed on 8 October 2020).
24. Soft Tissue Sarcoma Stages. Available online: https://www.cancer.org/cancer/soft-tissue-sarcoma/detection-diagnosis-staging/staging.html (accessed on 21 December 2022).
25. Falck-Ytter, Y.; Francis, C.W.; Johanson, N.A.; Curley, C.; Dahl, O.E.; Schulman, S.; Ortel, T.L.; Pauker, S.G.; Colwell, C.W. Prevention of VTE in Orthopedic Surgery Patients: Antithrombotic Therapy and Prevention of Thrombosis, 9th Ed: American College of Chest Physicians Evidence-Based Clinical Practice Guidelines. *Chest* **2012**, *141*, e278S–e325S. [CrossRef] [PubMed]
26. A Cherkashin, M.; A Berezina, N. Venous Thromboembolism Incidence in Radiation Oncology: Retrospective Trial. *J. Clin. Oncol.* **2017**, *35*, e18289. [CrossRef]
27. Robinson, K.S.; Anderson, D.R.; Gross, M.; Petrie, D.; Leighton, R.; Stanish, W.; Alexander, D.; Mitchell, M.; Flemming, B.; Gent, M. Ultrasonographic Screening before Hospital Discharge for Deep Venous Thrombosis after Arthroplasty: The Post-Arthroplasty Screening Study. A Randomized, Controlled Trial. *Ann. Intern. Med.* **1997**, *127*, 439–445. [CrossRef] [PubMed]
28. Lyman, G.H.; Eckert, L.; Wang, Y.; Wang, H.; Cohen, A. Venous Thromboembolism Risk in Patients with Cancer Receiving Chemotherapy: A Real-World Analysis. *Ononcologist* **2013**, *18*, 1321–1329. [CrossRef] [PubMed]
29. Guy, J.-B.; Bertoletti, L.; Magné, N.; Rancoule, C.; Mahé, I.; Font, C.; Sanz, O.; Martín-Antorán, J.M.; Pace, F.; Vela, J.R.; et al. Venous Thromboembolism in Radiation Therapy Cancer Patients: Findings from the RIETE Registry. *Crit. Rev. Oncol./Hematol.* **2017**, *113*, 83–89. [CrossRef]
30. Effect of Ionizing Radiation on Cellular Procoagulability and Co-Ordinated Gene Alterations | Haematologica. Available online: https://www.haematologica.org/article/view/4538 (accessed on 8 October 2020).
31. Halle, M.; Ekström, M.; Farnebo, F.; Tornvall, P. Endothelial Activation with Prothrombotic Response in Irradiated Microvascular Recipient Veins. *J. Plast. Reconstr. Aesthetic Surg.* **2010**, *63*, 1910–1916. [CrossRef]

32. Zakai, N.A.; Ohira, T.; White, R.; Folsom, A.R.; Cushman, M. Activated Partial Thromboplastin Time and Risk of Future Venous Thromboembolism. *Am. J. Med.* **2008**, *121*, 231–238. [CrossRef]
33. Mont, M.A.; Jacobs, J.J.; Boggio, L.N.; Bozic, K.J.; Della Valle, C.J.; Goodman, S.B.; Lewis, C.G.; Yates, A.J.; Watters, W.C.; Turkelson, C.M.; et al. Preventing Venous Thromboembolic Disease in Patients Undergoing Elective Hip and Knee Arthroplasty. *J. Am. Acad. Orthop. Surg.* **2011**, *19*, 768–776. [CrossRef]
34. Flack, C.K.; Monn, M.F.; Patel, N.B.; Gardner, T.A.; Powell, C.R. National Trends in the Performance of Robot-Assisted Sacral Colpopexy. *J. Endourol.* **2015**, *29*, 777–783. [CrossRef]
35. Galanaud, J.-P.; Laroche, J.-P.; Righini, M. The History and Historical Treatments of Deep Vein Thrombosis. *J. Thromb. Haemost. JTH* **2013**, *11*, 402–411. [CrossRef]
36. Singh, V.A.; Yong, L.M.; Vijayananthan, A. Is DVT Prophylaxis Necessary after Oncology Lower Limb Surgery? A Pilot Study. *SpringerPlus* **2016**, *5*, 943. [CrossRef]

cancers

MDPI

Review

Classification of Chondrosarcoma: From Characteristic to Challenging Imaging Findings

Jun-Ho Kim [1] and Seul Ki Lee [2,*]

[1] Department of Orthopaedic Surgery, Center for Joint Diseases, Kyung Hee University Hospital at Gangdong, Seoul 05278, Republic of Korea
[2] Department of Radiology, St. Vincent's Hospital, College of Medicine, The Catholic University of Korea, Seoul 06591, Republic of Korea
* Correspondence: beneffy@catholic.ac.kr

Simple Summary: Chondrosarcomas are a very heterogeneous group of cartilage-forming tumors that comprise approximately one-third of all malignant bone tumors. The World Health Organization classifies chondrosarcomas as benign, intermediate, or malignant cartilaginous tumors. Clinical management is guided by characteristic imaging findings and histopathological grade. However, the differentiation between enchondromas and low-grade chondrosarcomas and between low-grade and high-grade chondrosarcomas is challenging for radiologists and pathologists. Many potentially helpful advanced imaging modalities exist for diagnosing chondroid tumors and multidisciplinary discussions of all modalities should be combined when making treatment decisions.

Abstract: Chondrosarcomas can be classified into various forms according to the presence or absence of a precursor lesion, location, and histological subtype. The new 2020 World Health Organization (WHO) Classification of Tumors of Soft Tissue and Bone classifies chondrogenic bone tumors as benign, intermediate (locally aggressive), or malignant, and separates atypical cartilaginous tumors (ACTs) and chondrosarcoma grade 1 (CS1) as intermediate and malignant tumors. respectively. Furthermore, the classification categorizes chondrosarcomas (including ACT) into eight subtypes: central conventional (grade 1 vs. 2–3), secondary peripheral (grade 1 vs. 2–3), periosteal, dedifferentiated, mesenchymal, and clear cell chondrosarcoma. Most chondrosarcomas are the low-grade, primary central conventional type. The rarer subtypes include clear cell, mesenchymal, and dedifferentiated chondrosarcomas. Comprehensive analysis of the characteristic imaging findings can help differentiate various forms of chondrosarcomas. However, distinguishing low-grade chondrosarcomas from enchondromas or high-grade chondrosarcomas is radiologically and histopathologically challenging, even for experienced radiologists and pathologists.

Keywords: chondrosarcoma; classification; 2020 World Health Organization classification of tumors of soft tissue and bone; atypical cartilaginous tumor; high-grade chondrosarcoma; plain radiograph; computed tomography; magnetic resonance imaging

Citation: Kim, J.-H.; Lee, S.K. Classification of Chondrosarcoma: From Characteristic to Challenging Imaging Findings. *Cancers* **2023**, *15*, 1703. https://doi.org/10.3390/cancers15061703

Academic Editor: Catrin Sian Rutland

Received: 24 January 2023
Revised: 8 March 2023
Accepted: 8 March 2023
Published: 10 March 2023

1. Introduction

Chondrosarcomas are malignant tumors that produce a chondroid (cartilaginous) matrix [1,2]. They can be classified as either primary or secondary. Primary chondrosarcomas, which arise de novo, are the third most common primary malignant tumors of the bone after myelomas and osteosarcomas and account for 20–27% of all primary malignant bone tumors [1]. Conversely, secondary chondrosarcomas are associated with pre-existing cartilaginous lesions, such as enchondroma or osteochondroma [3,4]. Chondrosarcomas can also be classified based on the osseous location in which they arise; namely, central (within the intramedullary cavity), peripheral (within the cartilage cap of a pre-existing osteochondroma), or periosteal (juxtacortical; on the surface of the bone) [5]. Further classification of chondrosarcomas is based on histological subtypes, including conventional (grades 1–3), clear cell, mesenchymal, and dedifferentiated [6]. Finally, the 2020 World Health Organization (WHO) classification categorizes chondrosarcomas into eight subtypes: central conventional (grade 1 vs. 2–3), secondary peripheral (grade 1 vs. 2–3), periosteal, dedifferentiated, mesenchymal, and clear cell [7]. The characteristic imaging features of numerous categories of chondrosarcomas may aid in accurate diagnosis and classification. Radiography can support the diagnosis of chondroid tumors as enchondromas with characteristic findings including typical chondroid matrix mineralization [8]. Computed tomography (CT) and magnetic resonance imaging (MRI) can reveal imaging features of malignancy to distinguish between chondrosarcomas and enchondromas [8]. This review article summarizes the various classifications of chondrosarcomas and provides the characteristic to challenging imaging findings to differentiate among the various forms of chondrosarcoma.

2. 2020 WHO Classification of Chondrosarcomas

The 2020 WHO classification categorizes chondrogenic bone tumors as benign, intermediate (locally aggressive), or malignant (Figure 1) [7]. The term "atypical cartilaginous tumor (ACT)", which was first introduced in the 2013 WHO classification, refers to low-grade chondrosarcomas located in the appendicular skeleton (long and short tubular bones) that are considered the intermediate group (chondrosarcoma grade 0.5) [9]. Other chondrosarcomas are assigned to the malignant group. It is important to note that "chondrosarcoma grade 1 (CS1)" is histologically the same as ACT but is assigned to the malignant group; CS1 should be applied separately to tumors of the axial skeleton (including the pelvic bones and the skull base). Chondrosarcomas located in the axial skeleton have a worse outcome and require more aggressive treatment compared to those in the appendicular skeleton [7,10–12].

2013 WHO classification of chondrogenic bone tumors

Benign	Intermediate	Malignant
Subungual exostosis	Chondroblastoma	Chondrosarcoma, grade 2
Bizarre parosteal chondromatous proliferation	Chondromyxoid fibroma	Chondrosarcoma, grade 3
Periosteal chondroma	Atypical cartilaginous tumor/Chondrosarcoma, grade 1	Clear cell chondrosarcoma
Enchondroma		Mesenchymal chondrosarcoma
Osteochondroma		Dedifferentiated chondrosarcoma
Osteochondromyxoma		
Synovial Chondromatosis		

2020 WHO classification of chondrogenic bone tumors

Benign	Intermediate	Malignant
Subungual exostosis	Synovial Chondromatosis	Chondrosarcoma, grade 1
Bizarre parosteal chondromatous proliferation	Atypical cartilaginous tumor	Chondrosarcoma, grade 2
Periosteal chondroma		Chondrosarcoma, grade 3
Enchondroma		Clear cell chondrosarcoma
Osteochondroma		Mesenchymal chondrosarcoma
Chondroblastoma		Dedifferentiated chondrosarcoma
Chondromyxoid fibroma		
Osteochondromyxoma		

Figure 1. Comparison of the 2013 and 2020 World Health Organization (WHO) classifications of chondrogenic bone tumor. Diseases highlighted are those that are subject to change from 2013 to 2020 WHO classification.

Finally, the 2020 WHO classification categorizes chondrosarcomas (including ACT) into eight subtypes (Table 1): central conventional (grade 1 vs. 2–3), secondary peripheral (grade 1 vs. 2–3), periosteal, dedifferentiated, mesenchymal, and clear cell chondrosarcoma [7]. We discuss four stages used to determine the classification of chondrosarcomas, as well as the characteristic to challenging features of various chondrosarcomas.

Table 1. 2020 WHO classification of chondrosarcomas [7,13].

	Entity	Remarks
Conventional chondrosarcomas	Central atypical cartilaginous tumor (ACT)/chondrosarcoma grade 1 (CS1)	De novo or secondary (possible precursor: enchondroma)
	Secondary peripheral ACT/CS1	Precursor: osteochondroma
	Central chondrosarcoma grades 2 and 3 (CS2,3)	De novo or secondary (possible precursor: enchondroma)
	Secondary peripheral CS2,3	Precursor: osteochondroma
	Periosteal chondrosarcoma	
Rare subtypes	Dedifferentiated chondrosarcoma	Precursor: conventional chondrosarcoma
	Mesenchymal chondrosarcoma	
	Clear cell chondrosarcoma	

2.1. First Stage: Histological Grading

The biological behavior of chondrosarcomas is graded as 1 to 3 based on nuclear size, staining pattern (hyperchromasia), mitotic activity, and cellularity degree [14]. CS1 refers to low-grade tumors containing chondrocytes with small dense nuclei, although some slightly enlarged nuclei (>8 μm) and a few multinucleated cells (most commonly binucleated) may be present [1]. The stroma is predominantly chondroid with sparse or absent myxoid areas [1]. Chondrosarcoma grade 2 (CS2) tumors are intermediate-grade tumors containing less chondroid matrix and an increased cellular portion compared to CS1 tumors [1]. Chondrocyte nuclei are enlarged, either vesicular or hyperchromatic, and are also binucleated and multinucleated [1]. The stroma is frequently myxoid [1]. Chondrosarcoma grade 3 (CS3) tumors are high-grade tumors exhibiting greater cellularity than CS1 and CS2 tumors and nuclear pleomorphism with sparse or absent chondroid matrix [1]. The nuclei are typically vesicular, often spindle-shaped, and may be 5–10-fold larger than normal [1]. The non-mineralized tissue in chondrosarcomas has high water content, varying histologically from mature hyaline cartilage to a more myxoid stroma [1]. The edges of chondrosarcomas are characterized by chondroid tissue invading the trabecular bone [15]. Once this morphological feature has been identified, the degree of cellularity is used to determine the chondrosarcoma grade [1]. Invasion of the endosteal surface marks the beginning of extraosseous extension as the first step toward high-grade chondrosarcoma [10].

Most chondrosarcomas are conventional, with 60% classified as CS1 or CS2 [16]. Conventional chondrosarcomas tend to occur in older people, and more than 50% of patients are >50 years of age [13]. These chondrosarcomas are referred to as central chondrosarcomas. The 5-year survival rate is 88% for patients with CS1 and 57% for patients with CS2 and CS3 with local recurrence and metastasis rates of 20% and 14%, respectively [17]. The most common skeletal location for conventional chondrosarcomas is the long tubular bone, accounting for approximately 45% of cases [1,18,19]. The femur is the single most commonly affected long bone, representing approximately 20–35% of cases, while the upper extremity is involved in 10–20% of cases, most frequently the proximal humerus [1,18,19]. Long tubular bone lesions most commonly involve the metaphysis (49% of cases) [15]. Conventional chondrosarcomas can also occur in flat bones such as the pelvic bones; however, higher-grade tumors more frequently occur in the axial skeleton than in the appendicular skeleton. For instance, the prevalence of CS2-3 in the iliac bone is 70% vs. 45% in the femur [20].

Radiographs of conventional chondrosarcomas typically reveal a mixed lytic and sclerotic appearance [1,5]. The sclerotic areas represent chondroid matrix mineralization, which is seen in 60–78% of lesions [1]. Well-differentiated tumors tend to have a characteristic ring-and-arc pattern (Figure 2), whereas higher-grade chondrosarcomas often contain relatively less matrix mineralization and have a more amorphous or stippled appearance [1,5,19,21–23]. It is vital to differentiate benign from malignant cartilage tumors;

increased biological activity presents as deep and extensive endosteal scalloping as an attempt of tumor cell extension to a second compartment [1].

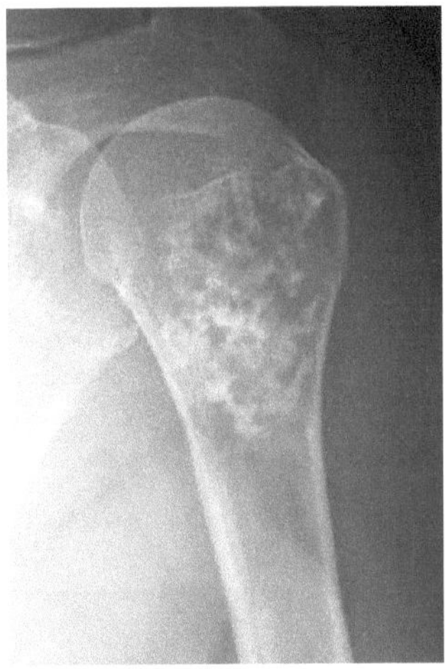

Figure 2. Atypical cartilaginous tumor of the humerus in a 59-year-old woman. Anteroposterior shoulder radiograph shows a mixed lytic and sclerotic lesion in the humerus. The sclerotic component represents typical chondroid ring-and-arc calcification.

Sensitive radiographic features differentiating enchondromas from chondrosarcomas include deep endosteal scalloping ≥2/3 of the normal cortical thickness [5,15] (Figure 3). Extensive longitudinal endosteal scalloping over ≥2/3 of the lesion length is also strong evidence of chondrosarcoma (although a somewhat less reliable criterion) [1,24] (Figure 4).

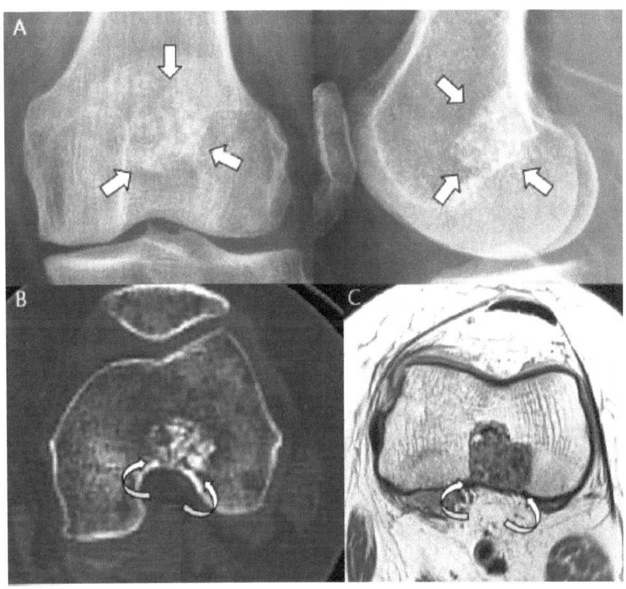

Figure 3. Atypical cartilaginous tumor of the distal femur in a 50-year-old woman. (**A**) Anteroposterior and lateral radiographs reveal a mixed lytic and sclerotic lesion in the distal femur (arrows) with typical ring-and-arc calcifications. (**B**) Computed tomography and (**C**) axial T2-weighted image demonstrate a lobulated chondroid tumor with deep endosteal scalloping (curved arrows) despite the small tumor size (1.7 cm).

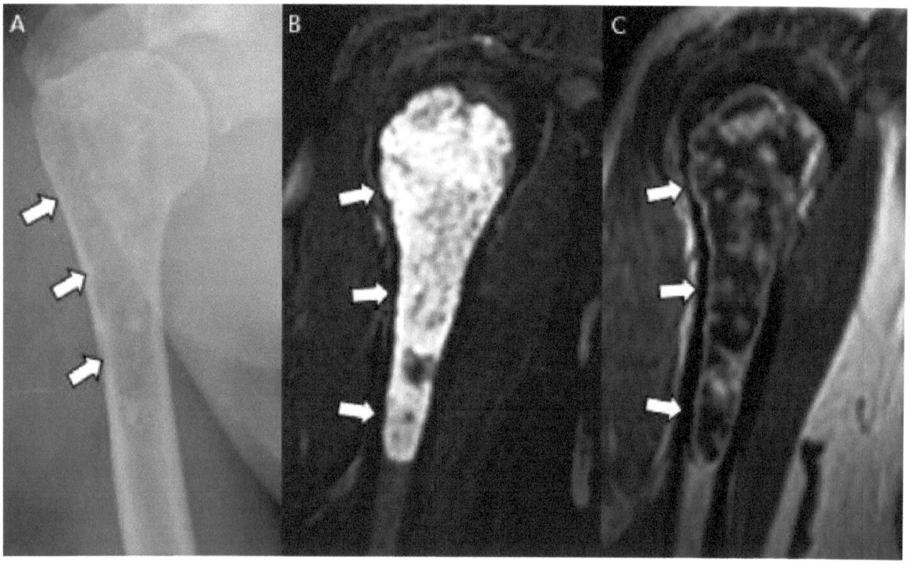

Figure 4. A typical cartilaginous tumor of the humerus in a 43-year-old woman. (**A**) Radiographs reveal a mixed lytic and sclerotic lesion in the humerus (arrows) with typical ring-and-arc calcifications. (**B**) Coronal T2-weighted image with fat suppression and (**C**) T1-weighted enhanced image demonstrate a lobulated chondroid tumor with longitudinal endosteal scalloping (arrows) along the 9 cm length of the tumor.

Chondrosarcomas frequently grow slowly, and the cortex responds to maintain the tumor in the medullary cavity. This attempt leads to the maintenance of a chondrosarcoma margin presenting as cortical remodeling, cortical thickening, and periosteal reaction [1,5] (Figure 5).

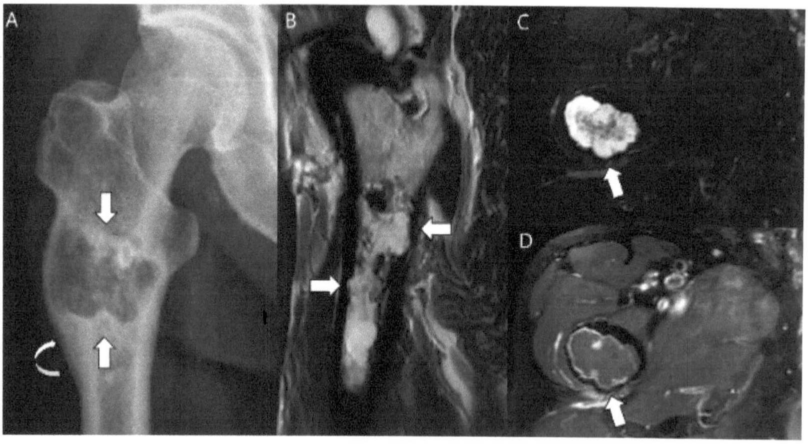

Figure 5. Chondrosarcoma grade 2 of the proximal femur in a 71-year-old man. (**A**) Anteroposterior radiograph reveals a lytic lesion in the proximal femur (arrows) resulting in cortical thickening and periosteal reaction (curved arrow). (**B**) Sagittal T2-weighted image shows a markedly high-signal lesion with deep endosteal scalloping (arrows). (**C**) Axial T2-weighted image with fat suppression and (**D**) axial T1-weighted enhanced image demonstrates a lobulated chondroid tumor with focal bone expansion (arrows).

Cortical destruction and soft tissue masses are further findings that can indicate an aggressive process with a perfect specificity of 100% [8] (Figure 6). A more aggressive moth-eaten and permeative bone appearance with more ill-defined margins may be seen in higher-grade chondrosarcomas and is frequently associated with mesenchymal and dedifferentiated subtypes [1,18]. CT allows the optimal detection of matrix mineralization, particularly when it is subtle or in a complex anatomic area, in addition to the accurate evaluation of the length and depth of endosteal scalloping [1,15,24]. Cortical response or cortical destruction with extraosseous soft tissue extension can also be well visualized by CT [15,24,25]. The enhanced CT findings for chondrosarcoma include a mild peripheral rim and septal enhancement [1,24]. MRI is the best method for evaluating the extent of marrow replacement and soft tissue extension [1,24]. Conventional chondrosarcomas have water-rich hyaline cartilage, which presents as a bright signal surrounded by low-signal septa on T2-weighted images (T2WI) [15]. Areas of matrix mineralization have a low signal in all MR pulse sequences [15]. This feature often creates marked heterogeneity in T2WI [1]. On T1-weighted images (T1WI), marrow-replacing lesions show a low-to-intermediate signal with possible entrapped areas of pre-existing fat marrow, presenting with high signal intensity on T1WI [1]. Soft tissue extension is well demonstrated on MRI and the characteristics of soft tissue extension are identical to those of the intraosseous component [1,24]. The contrast enhancement pattern is typically mild in degree and peripheral and septal in pattern [1]. Higher-grade lesions appear, with larger soft tissue masses showing more prominent diffuse or nodular contrast enhancement [1]. Higher-grade conventional chondrosarcomas occur more frequently in the axial skeleton. The prevalence of CS2 and CS3 in the iliac bone is 70%, with a predilection for the area around the previous region of the triradiate cartilage (Figure 7). For comparison, the prevalence of CS2 and CS3 in the entire femur is 45% [20].

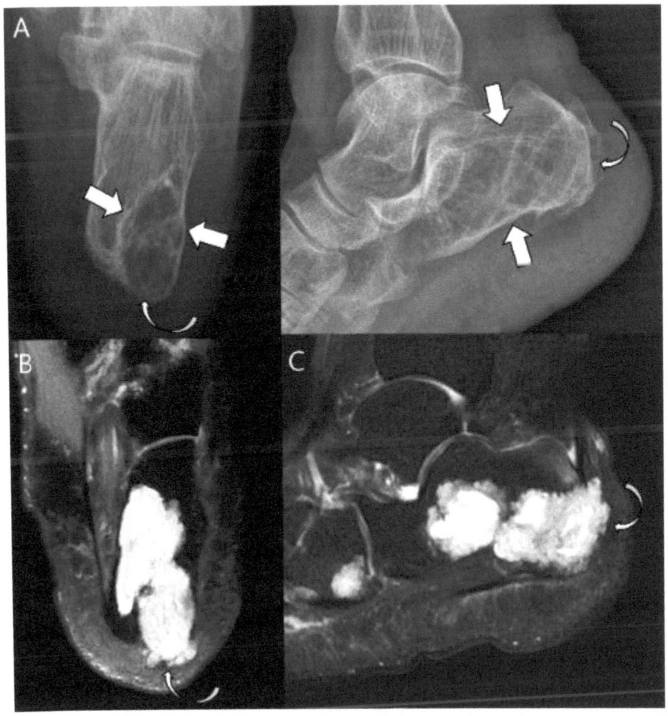

Figure 6. Atypical cartilaginous tumor of the calcaneus in a 74-year-old woman. (**A**) Plain radiographs reveal a lytic lesion in the calcaneus (arrows) with a partially destructed cortex (curved arrow). (**B**) Axial and (**C**) sagittal T2-weighted images with fat suppression show a lesion with marked high-signal intensity with focal extraosseous soft tissue extension (curved arrows).

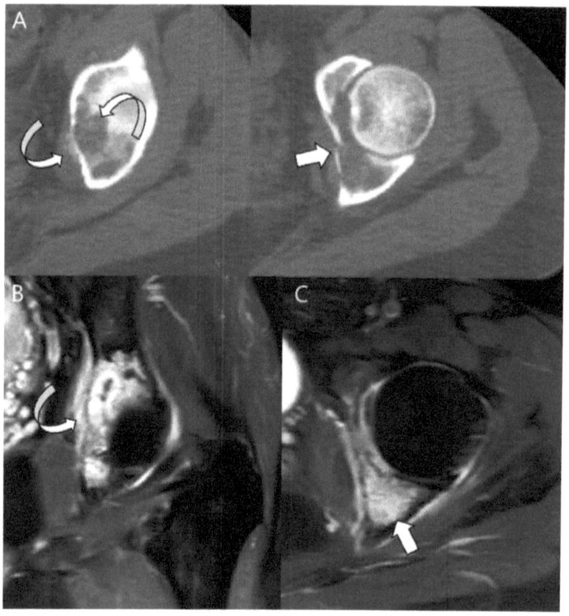

Figure 7. High-grade conventional chondrosarcoma of the acetabulum in a 49-year-old woman. (**A**) Axial CT scans reveal cortical breakage (thin arrow) with extraosseous extension containing matrix mineralization (curved arrow) in the left acetabulum. (**B**) Axial and (**C**) axial T1-weighted enhanced images with fat suppression show diffusely enhancing intraosseous (arrow) and extraosseous tumor components (curved arrow).

2.2. Second Stage: Primary vs. Secondary

Chondrosarcomas arising de novo are called primary chondrosarcomas (>90%), of which ≥80% are conventional (see Section 2.1) [16]. Conversely, chondrosarcomas superimposed on pre-existing benign cartilaginous neoplasms such as enchondromas or osteochondromas, those complicating enchondromatosis (Ollier's disease, Maffucci syndrome), and hereditary multiple exostoses (HME) are referred to as secondary chondrosarcomas (<10%) [1,5,17]. Their reported incidence rates are 0.4% to 2.2% in patients with solitary osteochondroma or enchondroma [17] and increase to 27.3% in patients with HME [3,26,27], 30–50% in patients with Ollier's disease, and up to 100% in patients with Maffucci syndrome [28–30]. Enchondromas are considered precursor lesions for 'secondary central chondrosarcomas', while osteochondromas are considered precursor lesions for 'peripheral chondrosarcomas'. The terms 'central' and 'peripheral' relate to the location of the tumor in the affected bone [13,17]. Underlying genetic differences exist between primary and secondary chondrosarcomas and induce clinical variations in presentation and behavior [31]. Patients with secondary chondrosarcomas are generally younger than those with primary chondrosarcomas, with a mean age of 34 years. The tumors are also generally low-grade [17,31,32] Changes in clinical symptoms in patients with known precursor lesions herald the development of chondrosarcomas [5,17]. The most common site of involvement is the pelvis, followed by the proximal femur. The scapula and proximal humerus are also relatively common sites [31].

Secondary peripheral chondrosarcomas occur in the cartilage cap, and the diagnosis of malignant transformation depends on the measurement of cartilage cap thickness [33–35]. The radiographic features of malignant transformation include (1) growth of a previously unchanged osteochondroma in a skeletally mature patent; (2) irregular or indistinct lesion surface; (3) focal areas of osteolysis within the osseous component of the lesion; (4) erosion or destruction of the adjacent bone; and (5) a significant soft tissue mass containing scattered or irregular calcifications [36]. The thickness of the cartilage cap can be assessed critically by CT and MRI [33,35]. Bernard et al. recently concluded that a cartilage cap thickness > 2 cm strongly suggested malignant transformation of osteochondroma in skeletally mature patients [37] (Figure 8). The MRI appearance of chondrosarcoma arising from the cartilage cap is as expected for well-differentiated hyaline chondral tissue, with low signal on T1WI and markedly high signal on T2WI, showing peripheral and septal enhancement with a lobular growth pattern. Matrix mineralization appears as punctate or curvilinear low-signal foci [34,35]. Some authors have stressed the qualitative evaluation of the cartilage cap rather than the absolute measurement of cartilage cap thickness. Irregularity of the surface of the cartilage cap may reflect an increase in the invasive nature of the tumor [31].

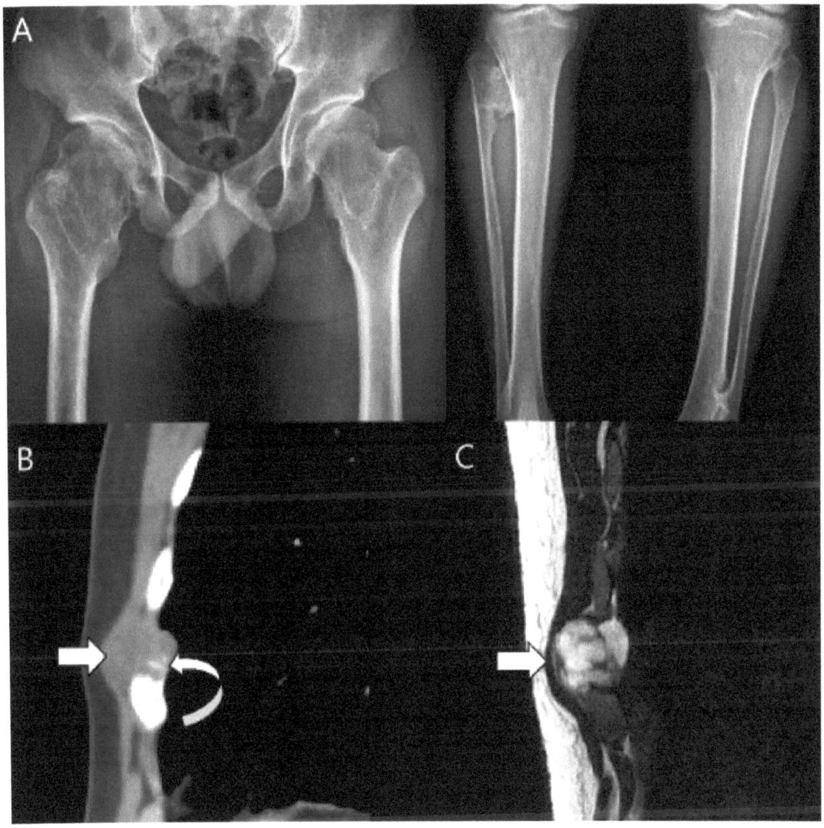

Figure 8. Secondary peripheral chondrosarcoma of the rib in a 34-year-old man. (**A**) Hip and tibial plain radiographs reveal underlying multiple exostoses. (**B**) Sagittal CT scan shows a lobulated mass with soft-tissue density (arrow) arising from the rib containing matrix mineralization (curved arrow), suggestive of a cartilage cap of sessile osteochondroma. (**C**) Sagittal T2-weighted image shows a mass of 2.3 cm in thickness with high signal intensity (arrow).

Secondary central chondrosarcomas present extended endosteal scalloping, cortical remodeling, cortical destruction, and periosteal reaction on plain radiographs, especially when compared to previous images of the underlying enchondroma [38,39]. On CT, the characteristic features of malignancy are lytic areas, endosteal scalloping on $\geq 2/3$ of the cortex, or extension to soft tissue [38]. If one of the following criteria is present on MRI, malignant transformation of the underlying enchondroma can be assumed: cortical destruction, spontaneous pathologic fracture, periosteal reaction, peritumoral edema, and soft tissue mass [38] (Figure 9). However, the conversion of a solitary enchondroma to a chondrosarcoma remains controversial, mainly due to the need for radiologic evidence for an enchondroma showing its eventual transformation into chondrosarcoma over several decades of follow-up [18]. Recently, Brien et al. [18] reported the criteria for secondary central chondrosarcoma within a single lesion site at any time, even if no serial follow-up radiologic films are available. They reported that the features of conventional chondrosarcomas (endosteal scalloping, expansion of the affected bone, cortical thickening, and amorphous calcification) in association with the features of typical benign enchondromas (well-defined ring-and-arc calcifications) justify the diagnosis of secondary central chondrosarcoma even without prior demonstration of underlying silent enchondroma [18] (Figure 10). Most central chondrosarcomas are thought to be primary and constitute approximately 75% of all chondrosarcomas. However, remnants of pre-existing enchondromas were found in 40% of central chondrosarcomas, suggesting that most central chondrosarcomas could be secondary to a pre-existing enchondroma [18].

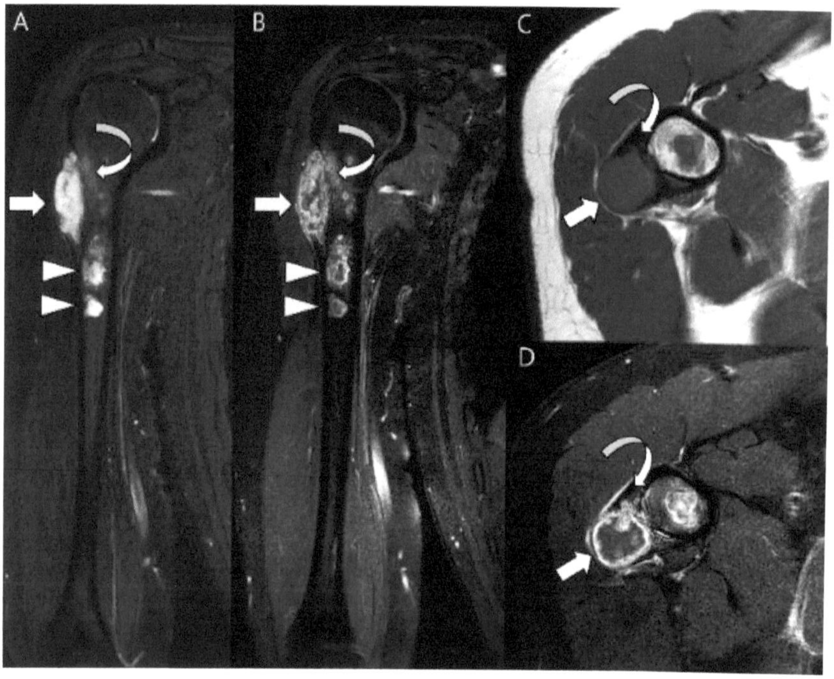

Figure 9. Secondary central chondrosarcoma of the humerus in a 19-year-old man. (**A**) Coronal T2-weighted image and (**B**) coronal T1-weighted enhanced images with fat suppression show multiple intramedullary chondroid tumors (arrowheads) with peripheral and septal enhancement in the humerus, suggesting enchondromatosis. The major lesion shows bone expansion at the metaphysis (arrow) with peritumoral edema and enhancement (curved arrow). (**C**) Axial T1-weighted and (**D**) enhanced images show a peripherally enhancing major lesion (arrow) with cortical remodeling (curved arrow).

2.3. Third Stage: Central vs. Peripheral vs. Periosteal

Chondrosarcomas are also categorized as central, peripheral, or periosteal (juxtacortical), depending on the osseous location [1]. Central chondrosarcomas are intramedullary in origin (see Section 2.1), while peripheral chondrosarcomas arise within the cartilage caps of osteochondromas (see Section 2.2). Periosteal (juxtacortical) chondrosarcomas rarely (<2%) arise on the bone surface [5,17]. On gross pathologic examination, periosteal chondrosarcoma is covered by a fibrous pseudocapsule that is continuous with the periosteum [1]. Extrinsic erosion of the cortex is often present [1]. The histological appearance is identical to that of conventional central chondrosarcoma [1]. Periosteal chondrosarcomas most frequently affect adults in the 3rd to 4th decades of life and have a mild male predilection [1]. Of 59 cases reported in the literature, 29 (49%) were located in the femur, 14 (24%) in the humerus, and eight (14%) in the tibia, with more rarely reported sites including the ilium, fibula, and ribs [40–42]. Most cases involved a low-grade tumor with local recurrence rates of 13–28% and an overall disease-free 5-year survival of 83% [41,42].

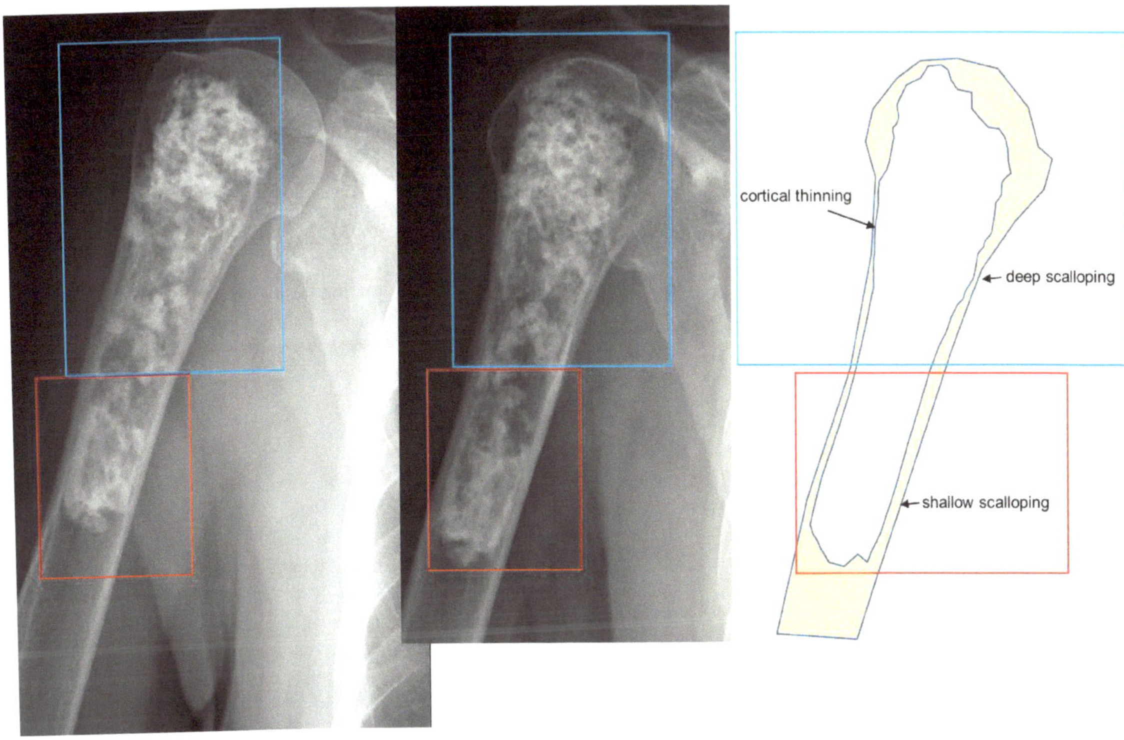

Figure 10. Secondary central chondrosarcoma of the humerus in an 81-year-old woman. The residual enchondroma in the red box (narrow scalloping) is combined with the additional features of chondrosarcoma in the blue box (cortical thinning and deep scalloping).

Radiographs show a round to oval lobulated soft tissue mass on the surface of the bone, lifting the periosteum over the tumor as a fibrous pseudocapsule [1,5]. The underlying cortex is almost invariable, presenting as either thickened or thinned, while complete cortical destruction is rare [5]. A Codman triangle may be seen where the periosteum is lifted [1]. Typical chondroid matrix mineralization is usually present and metaplastic ossification is often seen to a variable extent [1]. The medullary canal is typically not involved, although extension has been observed on MRI [1,40,41] (Figure 11). Periosteal chondroma and periosteal osteosarcoma are the most difficult tumors to differentiate from periosteal chondrosarcoma [43,44]. Tumor size is the only differentiating feature between periosteal chondroma (median size 2.5 cm) and periosteal chondrosarcoma (median size 4 cm) [40]. Periosteal osteosarcomas and chondrosarcomas both contain cartilage, but chondrosarcomas show no osteoid formation on histological examination [41,43].

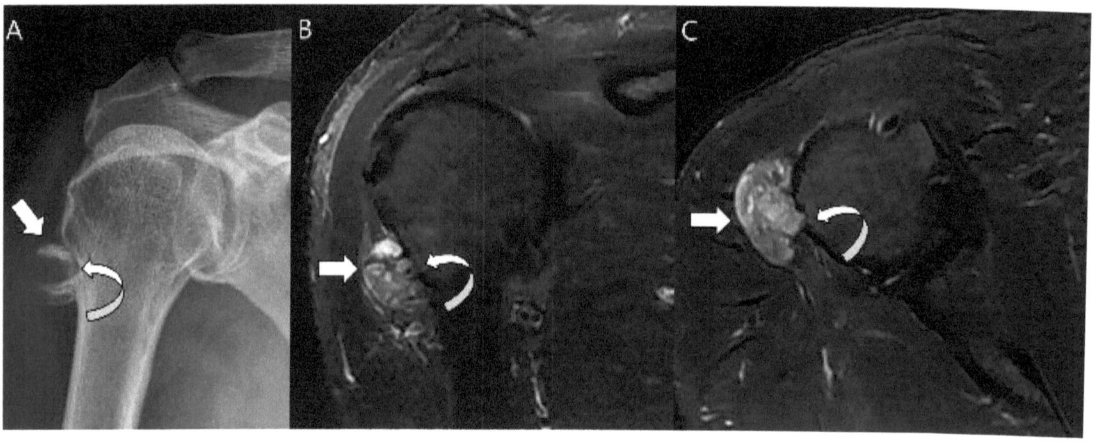

Figure 11. Periosteal chondrosarcoma grade 1 of the humerus in a 66-year-old man. (**A**) Radiograph shows a juxtacortical mass with Codman's triangles (arrow) in the humerus. Note the associated cortical thinning (curved arrow). (**B**) Coronal and (**C**) axial T2-weighted images with fat suppression show a juxtacortical mass with high signal intensity and lobular margins (arrows). The mass has caused cortical erosion (curved arrow) but no evident marrow invasion.

2.4. Fourth Stage: Conventional vs. Subtypes

Various histological subtypes of chondrosarcomas have been described, including conventional, mesenchymal, clear cell, and dedifferentiated [1]. Most chondrosarcomas are pathologically classified as conventional (80–85%; see Section 2.1). Several subtypes exist that differ in location, appearance, treatment, and prognosis [17]. These include clear cell (1–2%), mesenchymal (3–10%), and dedifferentiated (5–10%) chondrosarcomas [16].

Clear cell chondrosarcomas are low-grade variants characterized by an epiphyseal location in long bones [45]. On histological analysis, these lesions have numerous cells with abundant clear vacuolated cytoplasm [1,5]. Patients are most commonly affected in the 3rd to 5th decades of life [1]. Long bones are affected in 85–90% of cases with the proximal femur (68%) and proximal humerus (23%) the most commonly involved long bones [45]. Radiographs reveal a predominantly lytic epiphyseal lesion with distinct sclerotic margins that simulate a benign lesion [5,45] (Figure 12). Matrix mineralization is not as frequently apparent in clear cell chondrosarcomas (approximately 30% of cases) as in conventional chondrosarcomas [46–48]. In approximately 30% of cases, mild bone expansion may be apparent, but soft tissue extension is rare (<10% of cases) [1,5]. Because of their epiphyseal location, clear cell chondrosarcomas can be difficult to distinguish from chondroblastomas [1]. Clinically, clear cell chondrosarcomas usually present one or two decades later than chondroblastomas [18]. On MRI, clear cell chondrosarcomas are heterogeneous due to areas of hemorrhage or cystic changes [45]. Peritumoral edema is unusual and always mild as opposed to that in chondroblastoma [45].

Mesenchymal chondrosarcomas are a rare high-grade variant that has a strong tendency to metastasize. They can originate from either bone or soft tissue [1]. The characteristic histological feature of this tumor type is a bimorphic pattern characterized by differentiated cartilage admixed with solid highly cellular areas composed of undifferentiated small round cells [1]. In the undifferentiated areas, small, round cells typically simulate Ewing's sarcoma and have a hemangiopericytomatous vascular pattern [49,50]. The prognosis of mesenchymal chondrosarcomas is poor, and they present in a younger age group than conventional chondrosarcomas (mean age ~25 years) [5]. In contrast to conventional chondrosarcomas, mesenchymal chondrosarcomas most commonly involve the axial skeleton; for example, the craniofacial region [1]. Radiographs usually show aggressive bone destruction with a moth-eaten to permeative bone pattern and an ill-defined periosteal reaction [51,52]. The tumor is often very large with extensive extraosseous components [1]. CT typically shows chondroid mineralization, and the lesion may appear heavily calcified, but more commonly shows "finely stippled" calcification [53]. Mesenchymal chondrosarcomas have a different pattern of contrast enhancement than conventional chondrosarcomas on MRI; often, diffuse and typical chondroid septal and peripheral enhancement is lacking [1]. Some areas show low-signal, serpentine, high-flow vessels, a feature not seen in other chondrosarcomas [1]. The diagnosis of mesenchymal chondrosarcoma is suggested by an aggressive osseous lesion with subtle chondroid matrix mineralization and an intermediate signal on T2WI (lower than that of conventional chondrosarcoma), with more dramatic enhancement than expected with conventional chondrosarcoma [1].

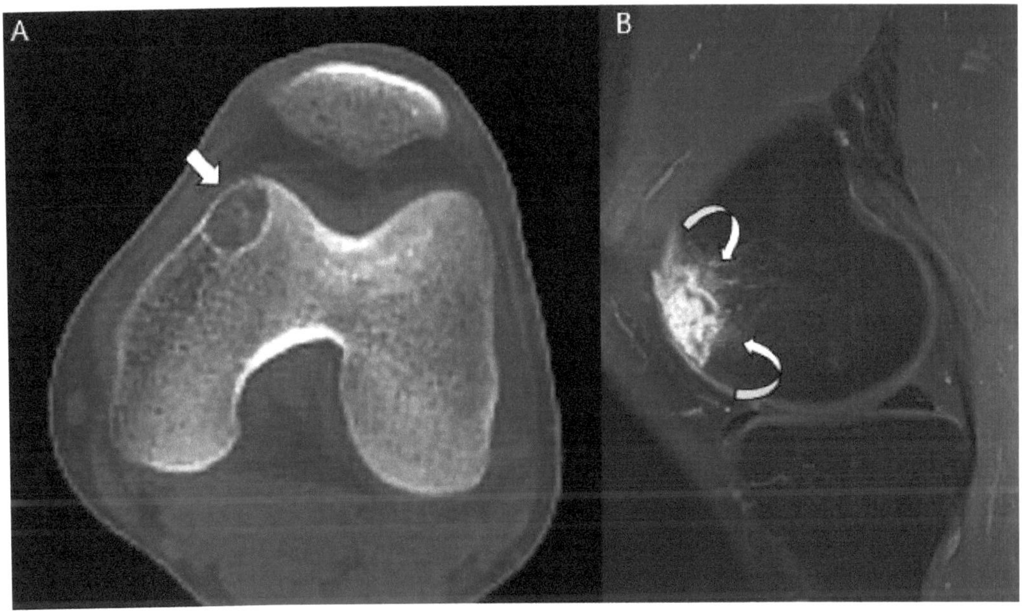

Figure 12. Clear cell chondrosarcoma of the distal femur in a 31-year-old man. (**A**) Axial CT scan shows an osteolytic lesion with a thin sclerotic margin at the distal femur (arrow). (**B**) Sagittal T1-weighted enhanced image with fat suppression shows a heterogeneously enhancing lesion with mild peritumoral enhancement at the distal femoral epiphysis (curved arrows).

Dedifferentiated chondrosarcoma is characterized by a conventional low-grade chondrosarcoma with an abrupt transition to foci that have dedifferentiated into a higher-grade, more aggressive component [1]. The non-cartilaginous portion is most frequently conventional osteosarcoma (70%) and less commonly malignant fibrous histiocytoma or fibrosarcoma [1,5]. Dedifferentiation can occur in 10–20% of conventional chondrosarcomas [1]. Patients with dedifferentiated chondrosarcomas are older than those with conventional lesions, usually 50–70 years of age (mean age: approximately 60 years) [54–56]. Dedifferentiated chondrosarcomas have a poor prognosis. A multicenter review of 337 patients reported that 21% had metastases at the time of diagnosis and the survival of these patients was 10% at 2 years [18,57]. The sites of involvement parallel those of conventional intramedullary chondrosarcoma, with common locations including the femur (35% of cases), pelvis (29%), humerus (16%), scapula (6%), rib (6%), and tibia (5%) [54–56]. The radiographic features of dedifferentiated chondrosarcomas are tumor bimorphism including aggressive bone destruction with extraosseous soft tissue extension, associated with an underlying cartilaginous lesion [17]. The imaging findings vary depending on the areas of high-grade transformation [1,58]. Tumors can be classified into three types based on radiographic findings: type 1, radiographic features the same as those of a central chondrosarcoma, with the addition of a suspected region with dedifferentiation; type 2, the tumor resembles the underlying benign enchondroma, but with destructive changes and/or a large soft tissue mass; and type 3, high-grade destructive lesions of the bone without signs of a cartilaginous component [56]. CT and MRI may reveal two distinct areas with differing intrinsic characteristics [1] (Figure 13). This bimorphic pattern is valuable in targeting the high-grade region during image-guided needle biopsy [59].

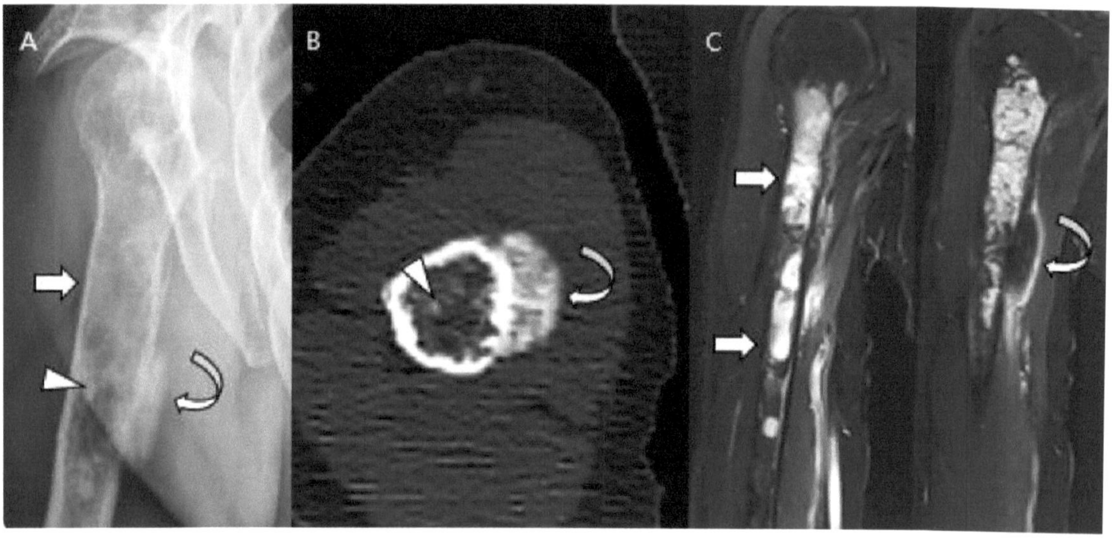

Figure 13. Dedifferentiated chondrosarcoma of the humerus in a 54-year-old man. (**A**) Plain radiograph shows an extensive mixed lytic and sclerotic lesion in the humerus with endosteal scalloping (arrow). Note the chondral-type mineralization in the intramedullary cavity (arrowhead) and the densely osteoid-type mineralization at the juxtacortical area (curved arrow). (**B**) Axial CT scan also reveals the intramedullary chondral-type (arrowhead) and the juxtacortical dense osteoid-type (curved arrow) mineralization. (**C**) Coronal T2-weighted images with fat suppression show high signal intramedullary lesion (arrows) with osteoblastic extraosseous extension (curved arrow), suggesting a dedifferentiated component of osteosarcoma.

Myxoid chondrosarcomas are now generally accepted as prominent myxoid changes of high-grade conventional chondrosarcomas [17]. However, extraskeletal myxoid chondrosarcoma (EMC) is a disease entity distinct from chondrosarcoma of the bone; these soft tissue sarcomas most commonly arise in the lower extremities [60,61] (Figure 14). The term "chondrosarcoma" used to describe EMC is a misnomer because well-formed hyaline cartilage is found only in a minority of EMCs, and S100 expression (which is present in all or most chondrosarcomas) is often very focal or absent [62,63]. The 2020 WHO classification categorizes EMC as "tumors of uncertain differentiation" [64]. Myxoid chondrosarcomas of the bone are also not designated as unique entities; rather, these tumors should be regarded as myxoid variants of conventional chondrosarcomas [7].

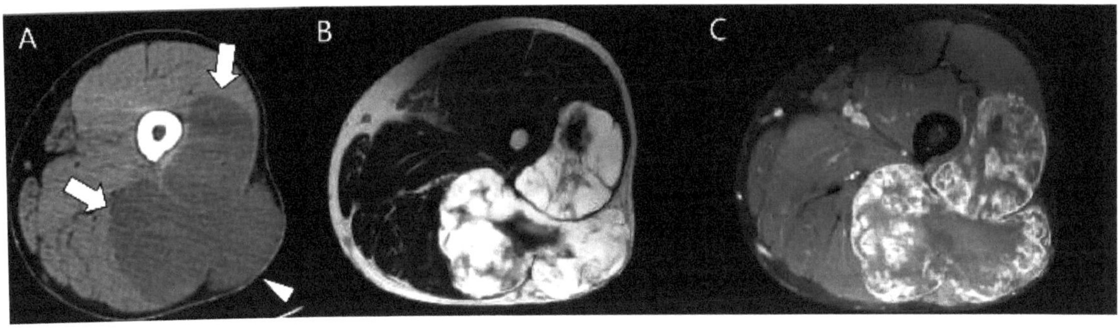

Figure 14. Extraskeletal myxoid chondrosarcoma of the right thigh in a 46-year-old man. (**A**) Axial CT scan reveals a lobulated, low-density soft tissue mass (arrows) without chondral-type mineralization between the vastus lateralis and biceps femoris muscles extending to the subcutaneous fat layer (arrowhead). (**B**,**C**) Axial T2-weighted and T1-weighted enhanced images show a soft tissue mass with high signal intensity and peripheral rim and septal enhancement.

3. Diagnostic Dilemma of Chondrosarcoma Classification

3.1. Distinction between Enchondroma and ACT

The differentiation between enchondromas and ACTs is crucial, as ACTs require curettage and watchful imaging follow-up, whereas most enchondromas require neither treatment nor follow-up [65]. Many imaging findings allow the differentiation between enchondromas and ACT, including cortical destruction, extraosseous soft tissue mass extension, periosteal reaction, size $\geq$ 5 cm, and endosteal scalloping (>2/3 of the cortical thickness) [66–68]. However, differentiating ACTs from enchondromas is challenging due to the lack of a gold standard for the diagnosis of ACT on histopathology [69,70]. While the presence of permeation and entrapment of pre-existing trabecular bone on histopathology are diagnostic for ACT, they may also result in a diagnostic conundrum, especially in cartilaginous lesions showing borderline imaging features in young patients, such as endosteal scalloping of approximately 50% of the cortex, lesion length of approximately 5 cm, or a change in the mineralization pattern with a lack of permeation [8]. In the absence of specific diagnostic criteria for histopathology, the differentiation between these two disease entities is often established by a consensus between radiologic, pathologic, and clinical findings [24].

The differentiation between enchondromas and ACT has been researched extensively because there remains low reliability in the clinical, radiological, and pathological distinctions between these two disease entities [70]. Choi et al. [66] identified some MRI features helpful for differentiating ACT from enchondroma, including the presence of a predominantly intermediate signal matrix on T1WI, multilobulated enhancement pattern on enhanced T1WI, cortical destruction, soft tissue mass, epiphyseal or flat bone involvement, and peritumoral edema (Figure 15), which favored a diagnosis of ACT. De Coninck el al. [71] evaluated the role of dynamic contrast-enhanced MRI (DCE-MRI) for the differentiation of enchondromas from chondrosarcomas and found that enhancement within the tumor, which was two times greater than that to muscle, combined with a 76° slope of the uptake curve, showed 100% sensitivity and 63% specificity for the detection of chondrosarcomas. However, the role of DCE-MRI in the differentiation of enchondroma from ACT remains ambiguous due to the lack of clear diagnostic histopathological criteria and the inclusion of low-grade and high-grade chondrosarcomas in previous studies [8]. In addition, diffusion-weighted imaging (DWI) is of no value in differentiating between enchondroma and ACT [72]. Studies quantifying tumor heterogeneity, including those applying MRI texture analysis, have shown improved diagnostic accuracy for the differentiation of benign and malignant cartilaginous tumors [68,73]. Assessing heterogeneity with imaging could provide important information on tumor characterization and might be a non-invasive biomarker for discrimination between tumor grades [68]. Pan et al. [74] developed three clinical radiomics nomograms to predict the malignancy risk of cartilaginous tumors based on radiomic signatures and clinical risk factors. All three nomograms demonstrated high performance for the differentiation of chondrosarcoma from enchondroma based on T1WI, fat-suppressed T2WI, and T1WI + T2WI fat-suppressed sequences with better accuracy than those of morphologic MRI analysis by musculoskeletal radiologists.

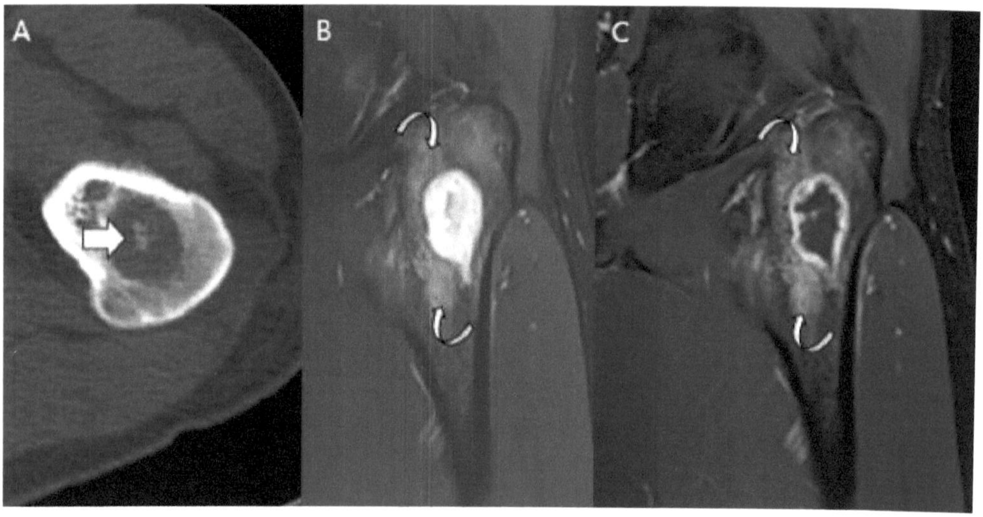

Figure 15. Atypical cartilaginous tumor of the proximal femur in a 22-year-old man. (**A**) Axial CT scan reveals a low-density intramedullary mass with chondral-type mineralization (arrow) in the proximal femur. (**B**,**C**) Coronal T2-weighted fat-suppressed and T1-weighted enhanced images show an intramedullary mass of 3 cm in size with high signal intensity and the peripheral rim and septal enhancement. Note the peritumoral edema with enhancement (curved arrows).

3.2. Biopsy or Follow-Up? Questions for Incidental Cartilage Lesions in the Long Bones

The increased use of MRI, which is now available in most healthcare systems, has resulted in the increased incidental identification of cartilage lesions in the long bones. Most of these lesions do not undergo biopsy and there is, typically, no histological confirmation of the diagnosis [75]. This may result in overtreatment of an enchondroma radiographically diagnosed as ACT or undertreatment if ACT is radiographically diagnosed as an enchondroma and the patient is erroneously discharged without follow-up [76]. However, a universal consensus on the management of these lesions is lacking; some centers recommend curettage, while others suggest surveillance with imaging [77,78]. Many authors have proposed radiographic follow-up protocols instead of biopsy for lesions without signs of local aggressiveness (cortical destruction and soft tissue extension), resulting in lower morbidity and costs [75,76,79,80]. The most recent studies on cartilaginous tumors have shifted toward active surveillance of ACTs to avoid unnecessary surgeries [80–82].

One study suggested distinguishing "active" lesions from "quiescent" lesions and recommended biopsy for the former (endosteal scalloping >2/3 of the cortex and >2/3 the length of the tumor, cortical thickening, and bone expansion) and radiological follow-up for the latter (in the absence of active findings) [77]. Kumar et al. [75] divided patients into "active" and "latent" groups based on the total growth of the cartilage lesion and advocated for biopsy in the active group with total growth > 6 mm, with surveillance with MRI every 3 years in the latent group. However, consensus evidence is lacking in the literature regarding follow-up frequency or duration, and no recommendations have been suggested for optimal imaging protocols. Deckers et al. [76] recommended annual MRI at least 2 years after diagnosis; if the findings remain stable, the frequency of MRI could be reduced to every 2 or 3 years. Herget et al. [38] recommended annual clinical and annual or biannual MRI for asymptomatic lesions > 5–6 cm and annual clinical and biannual imaging studies (radiographs or MRI if any doubts) for asymptomatic lesions < 5–6 cm. Patients with cartilage lesions ≤ 4 cm long with no endosteal scalloping can be discharged, with instructions to contact the hospital in case of new or increased pain [79]. In contrast, surgery is advised for tumors showing any aggressive features during follow-up, with curettage the preferred treatment for ACT [83]. Needle biopsies should not be recommended because they do not clearly differentiate enchondromas from ACT [83]. Several management protocols have been proposed [65,75,80,84]. We introduced the Birmingham Atypical Cartilaginous Tumor Imaging Protocol (Figure 16), which can be applied to cartilage lesions in the proximal humerus and around the knee [79]. As this protocol is only a guideline and has not been clinically validated, we cannot accept responsibility for any issues that may arise from its use [79].

Figure 16. Birmingham Atypical Cartilaginous Tumor Imaging Protocol applied to cartilage lesions in the proximal humerus and around the knee [79]. (**A**) Cartilage lesion < 4 cm, focal endosteal scalloping ≤10% or 36° of lesion circumference on the axial image with the greatest involvement; generalized endosteal scalloping ≥10% or 36° of lesion circumference on the axial image with the greatest involvement; MRI change = increase in longitudinal length of lesion ≥ 1 cm and/or development of aggressive features including increasing endosteal scalloping. (**B**) Cartilage lesion > 4 cm. (**C**) Cartilage lesion of any size with aggressive features (bone expansion and/or cortical thickening, periostitis, cortical destruction, and soft tissue mass).

3.3. Distinction between ACT/CS1 and High-Grade Chondrosarcoma

With the increasing incidence of ACT, the need for clear radiologic criteria to differentiate ACT from high-grade chondrosarcoma has become more important due to the different treatment options and prognoses [85]. High-grade chondrosarcoma requires wide resection with free surgical margins, whereas ACTs located in the long bones can be treated with intralesional curettage or regular follow-up [76]. However, the grading of chondrosarcoma based on imaging findings has shown low reliability; many diagnostic biopsies are unreliable owing to the heterogeneous composition of chondroid tumors (Figure 17) [69,86].

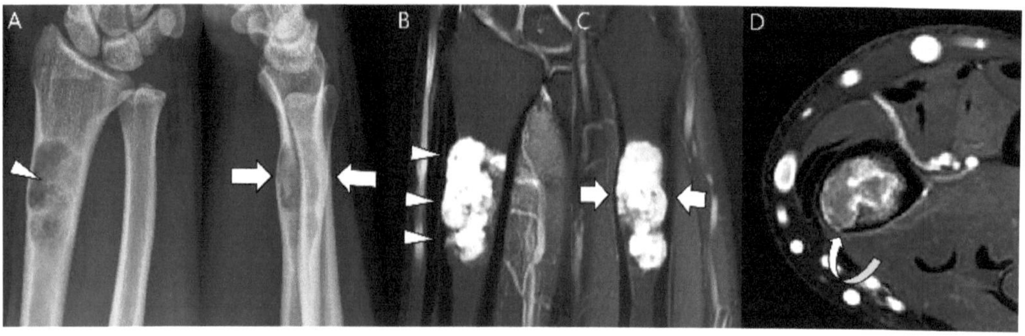

Figure 17. A 35-year-old man presenting with wrist pain. (**A**) Plain radiographs reveal a lobulated lytic lesion with chondroid matrix mineralization (arrowhead) and bone expansion (arrows) in the distal radius. (**B**,**C**) Coronal and sagittal T2-weighted fat-suppressed images show an intramedullary high signal mass with deep and extensive endosteal scalloping (arrowheads) and bone expansion (arrows). (**D**) Axial T1-weighted enhanced image shows peripheral rim and septal enhancement. Note the volar cortical thinning or defect (curved arrow). This lesion was noted as an atypical cartilaginous tumor at the initial incisional biopsy but was revealed as chondrosarcoma grade 2 at extended curettage.

High-grade chondrosarcoma may more often present with the following radiographic characteristics: moth-eaten or permeative bone destruction, less extensive matrix mineralization, loss of entrapped fatty marrow, cortical destruction, and a more aggressive periosteal reaction compared to ACT [1,85]. In addition, the histologic grades of lesions arising in the bones are poorer than those in the appendicular skeleton [87]. MRI is the modality of choice for identifying not only these radiographic features, but also the features of high-grade lesions, such as abundant (>50%) myxoid matrix, cortical destruction, soft-tissue extension, peritumoral edema, and periostitis (Figure 18) [88,89]. Jain et al. [87] reported that bone expansion did not differentiate between ACT/CS1 and high-grade chondrosarcoma unless the cortex was intact. Hemorrhagic necrosis and intra-articular extension are features of high-grade chondrosarcoma [87]. A biphasic pattern with a high-grade non-chondral sarcoma located adjacent to a typical chondral tumor is a characteristic feature of dedifferentiated chondrosarcoma [90] (Figure 13). Conversely, entrapped fat within the tumor and a characteristic lobular tumor morphology are highly indicative of ACT (Figure 19) [85,91].

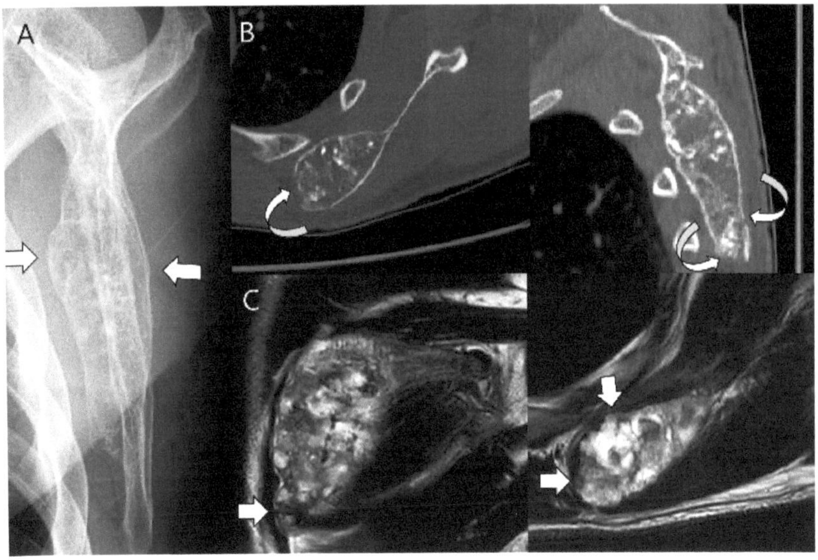

Figure 18. Chondrosarcoma grade 2 of the scapula in a 58-year-old man. (**A**) Plain radiograph shows a lobulated intramedullary mass with chondral-type mineralization and bone expansion (arrows) in the scapular body. (**B**) Axial and sagittal CT scans show a large intramedullary mass with cortical destruction (curved arrows). (**C**) Coronal and axial T2-weighted images show focal extraosseous soft tissue masses (arrows).

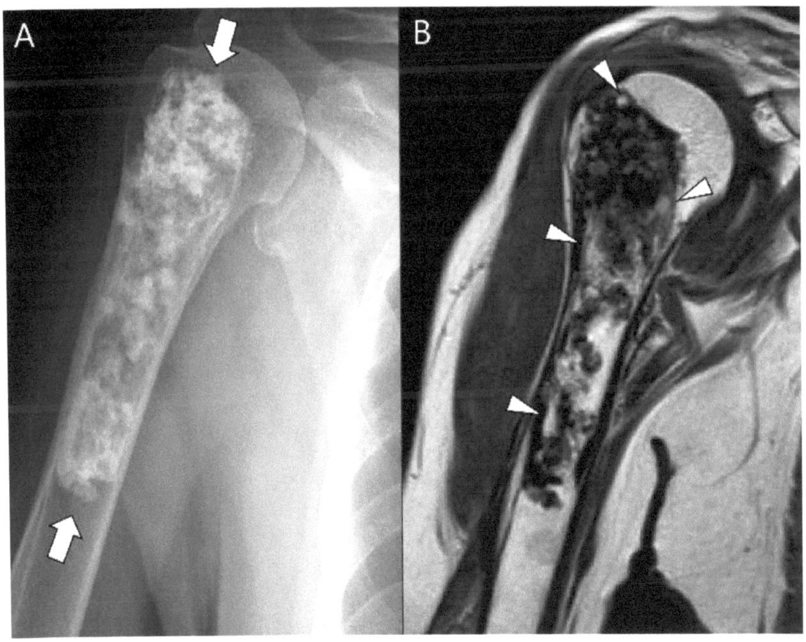

Figure 19. Atypical cartilaginous tumor of the humerus in an 81-year-old woman. (**A**) Plain radiograph demonstrates an intramedullary mass with prominent chondroid matrix mineralization (arrows) in the humerus. (**B**) Coronal T1-weighted image shows a lobulated intramedullary mass with areas of entrapped medullary fat (arrowheads).

Beyond CT and MRI, DCE-MRI can aid in the diagnosis of high-grade chondrosarcoma because it can reveal areas of fast enhancement due to richly vascularized intralesional septations [71,92]. However, DWI cannot differentiate low-grade lesions from high-grade chondrosarcomas [72]. Thus, novel tools for the objective grading of chondrosarcomas have recently been introduced, including texture analysis [73,93] and radiomics [94] with quantitative analysis. Deng et al. [93] reported that CT-based texture analysis showed potential for the grading of cartilaginous tumors in long bones. Gitto et al. [94] reported that their machine-learning approach showed satisfactory diagnostic performance for the classification of low-to-high-grade cartilaginous bone tumors based on radiomic features extracted from unenhanced MRI. One systemic review concluded that radiomics may allow the optimization of surgical decision making in chondrosarcoma despite weak evidence or insufficient study quality [95].

4. Current Treatments and Management

The therapeutic approach for chondrosarcomas is determined by the location and histologic grade. Surgical excision is the primary treatment for chondrosarcomas. Low-grade central chondrosarcoma can be treated with intralesional curettage, burring, and surgical adjuvant application such as hydrogen peroxide [96]. Tumors with extraosseous soft tissue extension, larger tumors, and axial skeleton tumors require wide excision. Wide en-bloc excision is the surgical approach of choice for intermediate or high-grade chondrosarcomas [97]. However, many patients show inoperable conditions at diagnosis or recur with metastatic disease, with more than 10% of recurrence cases showing a higher grade of malignancy than the first diagnosed grade [98].

Chemotherapy is usually ineffective in conventional and clear cell chondrosarcomas [97]. However, it may play a role in dedifferentiated chondrosarcomas containing high-grade spindle cell components [99]. A systematic review of 31 published studies suggested that adjuvant chemotherapy combined with surgical resection significantly improves disease-free survival in dedifferentiated chondrosarcomas compared to surgery alone [100]. In a non-randomized clinical cohort, adjuvant anthracyclin-based combination chemotherapy showed modest efficacy against mesenchymal chondrosarcomas [101].

Chondrogenic tumors are generally considered radioresistant because radiation-induced cytotoxicity requires actively dividing cells. Chondrogenic tumors are characterized by slow growth and a relatively low proportion of dividing cells [97]. However, radiation therapy can be administered after incomplete resection of high-grade conventional, dedifferentiated, or mesenchymal chondrosarcomas, with potential curative intent to maximize local control. Definitive radiation may also be indicated for palliative purposes [102].

5. Targets and Novel Treatment Options

Chondrosarcomas are poorly responsive to chemotherapy and radiation therapy, resulting in high morbidity and mortality [103]. Therefore, there is an urgent need to expand treatment options. Developing an efficient treatment strategy requires a better understanding of the molecular survival pathways involved in chondrosarcomas and their chemotherapy and radiation resistance mechanisms [104]. Chondrosarcoma subtypes differ at the molecular genetic level (Table 2) [105]. Recent studies have suggested several promising biomarkers and therapeutic targets for chondrosarcomas, with better understanding of chondrosarcoma genomic alterations and biology [103,105–110]. As shown in Table 2, the signaling pathways underpinning chondrosarcoma genesis such as IDH1/2 mutations, CDKN2A/B deletions, and TP53 mutations can be potential therapeutic targets [105]. The angiogenesis pathway is a potential effective target for preventing the growth and spread of chondrosarcoma [105]. Conventional chondrosarcomas are characterized by activation and/or overexpression of platelet-derived growth factor receptors PDGFR-alpha (PDGFRA) and PDGFR-beta (PDGFRB), and efforts to develop antiangiogenic therapies have produced many agents such as small molecule tyrosine kinase inhibitors and fully human monoclonal antibodies which affect angiogenesis [111]. Also, a multitargeted approach against multiple antiapoptotic proteins such as Bcl-2 (B-cell leukemia/lymphoma 2), Bcl-xL (Bcl-2 like 1), and XIAP (x-linked inhibitors or apoptosis) upregulated in chondrosarcomas can have a strong therapeutic potential to enhance the efficacy of radiation and chemotherapy [104]. These findings prompted research on the therapeutic efficacy of molecular-targeting therapies [103,112].

Table 2. Chondrosarcoma types and respective molecular features.

Chondrosarcoma (CS) Type	Molecular Features
Conventional central CS	IDH1/2 mutations COL2A1 mutations CDKN2A/B deletions
Conventional peripheral CS	EXT1/2 mutations
Conventional periosteal CS	Hedgehog pathway
Dedifferentiated CS	IDH1/2 mutations TP53 mutations PD-L1 expression
Mesenchymal CS	HEY1–NCOA2 fusion
Clear cell CS	No evidence of mutations

6. Conclusions

Chondrosarcomas are a heterogeneous group of malignant bone tumors that produce a chondroid (cartilaginous) matrix. Their clinical behaviors vary according to the histologic grade. The WHO defines these lesions as benign, intermediate, or malignant cartilaginous tumors. While most tumors are indolent, with a low potential for metastasis, some are aggressive, with a poor prognosis. Clinical management is guided by imaging findings, histopathological grading, and chondrosarcoma subtypes. Choosing the most appropriate diagnostic technique for grading chondroid tumors remains difficult because each modality has its own value; beyond CT and MRI, DCE-MRI supports chondrosarcoma grading, and new tools for quantitative analysis—including texture analysis and radiomics—have shown satisfactory diagnostic performance for chondrosarcoma classification. A limited range of treatment options exists for chondrosarcomas, including surgery and chemotherapy, and more therapeutic targets are needed. Multidisciplinary discussions of all modalities should be combined to determine the best treatment approach.

Author Contributions: Conceptualization: J.-H.K. and S.K.L. Writing Original Draft Preparation: J.-H.K. and S.K.L. Writing Review and Editing: S.K.L. Visualization: S.K.L. Supervision: S.K.L. All authors have read and agreed to the published version of the manuscript.

Funding: This research received no external funding.

Institutional Review Board Statement: Not applicable.

Informed Consent Statement: Not applicable.

Data Availability Statement: Not applicable.

Conflicts of Interest: The authors declare no conflict of interest.

References

1. Murphey, M.D.; Walker, E.A.; Wilson, A.J.; Kransdorf, M.J.; Temple, H.T.; Gannon, F.H. From the archives of the AFIP: Imaging of primary chondrosarcoma: Radiologic-pathologic correlation. *Radiographics* **2003**, *23*, 1245–1278. [CrossRef]
2. Marco, R.A.W.; Gitelis, S.; Brebach, G.T.; Healey, J.H. Cartilage tumors: Evaluation and treatment. *Am. Acad. Orthop. Surg.* **2000**, *8*, 292–304. [CrossRef]
3. Ahmed, A.R.; Tan, T.-S.; Unni, K.K.; Collins, M.S.; Wenger, D.E.; Sim, F.H. Secondary chondrosarcoma in osteochondroma: Report of 107 patients. *Clin. Orthop. Relat. Res.* **2003**, *411*, 193–206. [CrossRef]
4. Altay, M.; Bayrakci, K.; Yildiz, Y.; Erekul, S.; Saglik, Y. Secondary chondrosarcoma in cartilage bone tumors: Report of 32 patients. *J. Orthop. Sci.* **2007**, *12*, 415–423. [CrossRef]
5. Douis, H.; Saifuddin, A. The imaging of cartilaginous bone tumours. II. Chondrosarcoma. *Skelet. Radiol.* **2013**, *42*, 611–626. [CrossRef] [PubMed]

6. Aigner, T. Towards a new understanding and classification of chondrogenic neoplasias of the skeleton–biochemistry and cell biology of chondrosarcoma and its variants. *Virchows Arch.* **2002**, *441*, 219–230. [CrossRef] [PubMed]
7. Choi, J.H.; Ro, J.Y. The 2020 WHO Classification of Tumors of Bone: An Updated Review. *Adv. Anat. Pathol.* **2021**, *28*, 119–138. [CrossRef] [PubMed]
8. Douis, H.; Parry, M.; Vaiyapui, S.; Davies, A.M. What are the differentiating clinical and MRI-features of enchondromas from low-grade chondrosarcomas? *Eur. Radiol.* **2018**, *28*, 398–409. [CrossRef] [PubMed]
9. Wells, M.E.; Childs, B.R.; Eckhoff, M.D.; Rajani, R.; Potter, B.K.; Polfer, E.M. Atypical Cartilaginous Tumors: Trends in Management. *JAAOS Glob. Res. Rev.* **2021**, *5*, e21.00277. [CrossRef]
10. Bindiganavile, S.; Han, I.; Yun, J.Y.; Kim, H.-S. Long-term Outcome of Chondrosarcoma: A Single Institutional Experience. *Cancer Res. Treat.* **2015**, *47*, 897–903. [CrossRef]
11. Fromm, J.; Klein, A.; Baur-Melnyk, A.; Knösel, T.; Lindner, L.; Birkenmaier, C.; Roeder, F.; Jansson, V.; Dürr, H.R. Survival and prognostic factors in conventional central chondrosarcoma. *BMC Cancer* **2018**, *18*, 849. [CrossRef] [PubMed]
12. van Praag, V.M.; Rueten-Budde, A.J.; Ho, V.; Dijkstra, P.D.S.; Study group Bone and Soft tissue tumours (WeBot); Fiocco, M.; van de Sande, M.A.J. Incidence, outcomes and prognostic factors during 25 years of treatment of chondrosarcomas. *Surg. Oncol.* **2018**, *27*, 402–408. [CrossRef]
13. Engel, H.; Herget, G.W.; Füllgraf, H.; Sutter, R.; Benndorf, M.; Bamberg, F.; Jungmann, P.M. Chondrogenic Bone Tumors: The Importance of Imaging Characteristics. *Rofo* **2020**, *193*, 262–275. [CrossRef]
14. Kim, M.-J.; Cho, K.-J.; Ayala, A.G.; Ro, J.Y. Chondrosarcoma: With Updates on Molecular Genetics. *Sarcoma* **2011**, *2011*, 405437. [CrossRef]
15. Murphey, M.D.; Flemming, D.J.; Boyea, S.R.; Bojescul, J.A.; Sweet, D.E.; Temple, H.T. Enchondroma versus chondrosarcoma in the appendicular skeleton: Differentiating features. *Radiographics* **1998**, *18*, 1213–1237. [CrossRef] [PubMed]
16. Björnsson, J.; McLeod, R.A.; Unni, K.K.; Ilstrup, D.M.; Pritchard, D.J. Primary chondrosarcoma of long bones and limb girdles. *Cancer* **1998**, *83*, 2105–2119. [CrossRef]
17. Mavrogenis, A.F.; Gambarotti, M.; Angelini, A.; Palmerini, E.; Staals, E.L.; Ruggieri, P.; Papagelopoulos, P.J. Chondrosarcomas Revisited. *Orthopedics* **2012**, *35*, e379–e390. [CrossRef]
18. Brien, E.W.; Mirra, J.M.; Kerr, R. Benign and malignant cartilage tumors of bone and joint: Their anatomic and theoretical basis with an emphasis on radiology, pathology and clinical biology. I. The intramedullary cartilage tumors. *Skelet. Radiol.* **1997**, *26*, 325–353. [CrossRef]
19. Flemming, D.J.; Murphey, M.D. Enchondroma and Chondrosarcoma. *Semin. Musculoskelet. Radiol.* **2000**, *4*, 59–71. [CrossRef]
20. Bloem, J.L.; Reidsma, I.I. Bone and soft tissue tumors of hip and pelvis. *Eur. J. Radiol.* **2012**, *81*, 3793–3801. [CrossRef]
21. Hudson, T.M.; Manaster, B.J.; Springfield, D.S.; Spanier, S.S.; Enneking, W.F.; Hawkins, I.F. Radiology of medullary chondrosarcoma: Preoperative treatment planning. *Skelet. Radiol.* **1983**, *10*, 69–78. [CrossRef]
22. Reiter, F.B.; Ackerman, L.V.; Staple, T.W. Central Chondrosarcoma of the Appendicular Skeleton. *Radiology* **1972**, *105*, 525–530. [CrossRef] [PubMed]
23. Rosenthal, D.I.; Schiller, A.L.; Mankin, H.J. Chondrosarcoma: Correlation of radiological and histological grade. *Radiology* **1984**, *150*, 21–26. [CrossRef] [PubMed]
24. Logie, C.I.; Walker, E.A.; Forsberg, J.A.; Potter, B.K.; Murphey, M.D. Chondrosarcoma: A Diagnostic Imager's Guide to Decision Making and Patient Management. *Semin. Musculoskelet. Radiol.* **2013**, *17*, 101–115. [CrossRef]
25. Mayes, G.B.; Wallace, S.; Bernardino, M.E. Computed tomography of chondrosarcoma. *J. Comput. Tomogr.* **1981**, *5*, 345–348. [CrossRef]
26. Wuisman, P.I.J.M.; Jutte, P.C.; Ozaki, T. Secondary chondrosarcoma in osteochondromas Medullary extension in 15 of 45 cases. *Acta Orthop.* **1997**, *68*, 396–400. [CrossRef] [PubMed]
27. Pierz, K.A.; Stieber, J.R.; Kusumi, K.; Dormans, J.P. Hereditary multiple exostoses: One center's experience and review of etiology. *Clin. Orthop. Relat. Res.* **2002**, *401*, 49–59. [CrossRef] [PubMed]
28. Schwartz, H.S.; Zimmerman, N.B.; Simon, M.A.; Wroble, R.R.; Millar, E.A.; Bonfiglio, M. The malignant potential of enchondromatosis. *J. Bone Jt. Surg.* **1987**, *69*, 269–274. [CrossRef]
29. Liu, J.; Hudkins, P.G.; Swee, R.G.; Unni, K.K. Bone sarcomas associated with Ollier's disease. *Cancer* **1987**, *59*, 1376–1385. [CrossRef]
30. Mavrogenis, A.F.; Papagelopoulos, P.J.; Soucacos, P.N. Skeletal Osteochondromas Revisited. *Orthopedics* **2008**, *31*, 1018–1028. [CrossRef]
31. Lin, P.P.; Moussallem, C.D.; Deavers, M.T. Secondary Chondrosarcoma. *J. Am. Acad. Orthop. Surg.* **2010**, *18*, 608–615. [CrossRef]
32. Mandahl, N.; Gustafson, P.; Mertens, F.; Åkerman, M.; Baldetorp, B.; Gisselsson, D.; Knuutila, S.; Bauer, H.C.F.; Larsson, O. Cytogenetic aberrations and their prognostic impact in chondrosarcoma. *Genes Chromosom. Cancer* **2002**, *33*, 188–200. [CrossRef]
33. Murphey, M.D.; Choi, J.J.; Kransdorf, M.J.; Flemming, D.J.; Gannon, F.H. Imaging of Osteochondroma: Variants and Complications with Radiologic-Pathologic Correlation. *Radiographics* **2000**, *20*, 1407–1434. [CrossRef] [PubMed]
34. Lee, K.; Davies, A.; Cassar-Pullicino, V. Imaging the Complications of Osteochondromas. *Clin. Radiol.* **2002**, *57*, 18–28. [CrossRef]
35. Woertler, K.; Lindner, N.; Gosheger, G.; Brinkschmidt, C.; Heindel, W. Osteochondroma: MR imaging of tumor-related complications. *Eur. Radiol.* **2000**, *10*, 832–840. [CrossRef] [PubMed]
36. Norman, A.; Sissons, H.A. Radiographic hallmarks of peripheral chondrosarcoma. *Radiology* **1984**, *151*, 589–596. [CrossRef] [PubMed]
37. Bernard, S.A.; Murphey, M.D.; Flemming, D.J.; Kransdorf, M.J. Improved differentiation of benign osteochondromas from secondary chondrosarcomas with standardized measurement of cartilage cap at CT and MR imaging. *Radiology* **2010**, *255*, 857–865. [CrossRef]

38. Herget, G.W.; Strohm, P.; Rottenburger, C.; Kontny, U.; Krauß, T.; Bohm, J.; Sudkamp, N.; Uhl, M. Insights into Enchondroma, Enchondromatosis and the risk of secondary Chondrosarcoma. Review of the literature with an emphasis on the clinical behaviour, radiology, malignant transformation and the follow up. *Neoplasma* **2014**, *61*, 365–378. [CrossRef] [PubMed]
39. Verdegaal, S.H.; Bovée, J.V.; Pansuriya, T.C.; Grimer, R.J.; Ozger, H.; Jutte, P.C.; Julian, M.S.; Biau, D.J.; Geest, I.C.; Leithner, A.; et al. Incidence, Predictive Factors, and Prognosis of Chondrosarcoma in Patients with Ollier Disease and Maffucci Syndrome: An International Multicenter Study of 161 Patients. *Oncology* **2011**, *16*, 1771–1779. [CrossRef]
40. Robinson, P.; White, L.M.; Sundaram, M.; Kandel, R.; Wunder, J.; McDonald, D.J.; Janney, C.; Bell, R.S. Periosteal chondroid tumors: Radiologic evaluation with pathologic correlation. *AJR Am. J. Roentgenol.* **2001**, *177*, 1183–1188. [CrossRef]
41. Vanel, D.; De Paolis, M.; Monti, C.; Mercuri, M.; Picci, P. Radiological features of 24 periosteal chondrosarcomas. *Skelet. Radiol.* **2001**, *30*, 208–212. [CrossRef] [PubMed]
42. Papagelopoulos, P.J.; Galanis, E.C.; Mavrogenis, A.F.; Savvidou, O.D.; Bond, J.R.; Unni, K.K.; Sim, F.H. Survivorship Analysis in Patients with Periosteal Chondrosarcoma. *Clin. Orthop. Relat. Res.* **2006**, *448*, 199–207. [CrossRef] [PubMed]
43. Bertoni, F.; Boriani, S.; Laus, M.; Campanacci, M. Periosteal chondrosarcoma and periosteal osteosarcoma. Two distinct entities. *J. Bone Jt. Surg.* **1982**, *64*, 370–376. [CrossRef] [PubMed]
44. Nojima, T.M.; Unni, K.K.M.; McLeod, R.A.M.; Pritchard, D.J.M. Periosteal chondroma and periosteal chondrosarcoma. *Am. J. Surg. Pathol.* **1985**, *9*, 666–677. [CrossRef]
45. Collins, M.S.; Koyama, T.; Swee, R.G.; Inwards, C.Y. Clear cell chondrosarcoma: Radiographic, computed tomographic, and magnetic resonance findings in 34 patients with pathologic correlation. *Skelet. Radiol.* **2003**, *32*, 687–694. [CrossRef]
46. Kumar, R.; David, R.; Cierney, G., 3rd. Clear cell chondrosarcoma. *Radiology* **1985**, *154*, 45–48. [CrossRef]
47. Bagley, L.; Kneeland, J.B.; Dalinka, M.K.; Bullough, P.; Brooks, J. Unusual behavior of clear cell chondrosarcoma. *Skelet. Radiol.* **1993**, *22*, 279–282. [CrossRef]
48. PPresent, D.; Bacchini, P.; Pignatti, G.; Picci, P.; Bertoni, F.; Campanacci, M. Clear cell chondrosarcoma of bone. *Skelet. Radiol.* **1991**, *20*, 187–191. [CrossRef] [PubMed]
49. Swanson, P.E.; Lillemoe, T.J.; Manivel, J.C.; Wick, M.R. Mesenchymal chondrosarcoma. An immunohistochemical study. *Arch. Pathol. Lab. Med.* **1990**, *114*, 943–948.
50. Hoang, M.P.; Suarez, P.A.; Donner, L.R.; Ro, J.Y.; Ordñez, N.G.; Ayala, A.G.; Czerniak, B. Mesenchymal Chondrosarcoma: A Small Cell Neoplasm with Polyphenotypic Differentiation. *Int. J. Surg. Pathol.* **2000**, *8*, 291–301. [CrossRef]
51. Nakashima, Y.; Unni, K.K.; Shives, T.C.; Swee, R.G.; Dahlin, D.C. Mesenchymal chondrosarcoma of bone and soft tissue. A review of 111 cases. *Cancer* **1986**, *57*, 2444–2453. [CrossRef] [PubMed]
52. Ly, J.Q. Mesenchymal chondrosarcoma of the maxilla. *AJR Am. J. Roentgenol.* **2002**, *179*, 1077–1078. [CrossRef] [PubMed]
53. Shinaver, C.N.; Mafee, M.F.; Choi, K.H. MRI of mesenchymal chondrosarcoma of the orbit: Case report and review of the literature. *Neuroradiology* **1997**, *39*, 296–301. [CrossRef] [PubMed]
54. Capanna, R.; Bertoni, F.; Bettelli, G.; Picci, P.; Bacchini, P.; Present, D.; Giunti, A.; Campanacci, M. Dedifferentiated chondrosarcoma. *J. Bone Jt. Surg. Am.* **1988**, *70*, 60–69. [CrossRef]
55. Daly, P.J.; Sim, F.H.; Wold, L.E. Dedifferentiated chondrosarcoma of bone. *Orthopedics* **1989**, *12*, 763–767. [CrossRef] [PubMed]
56. Mercuri, M.; Picci, P.; Campanacci, L.; Rulli, E. Dedifferentiated chondrosarcoma. *Skelet. Radiol.* **1995**, *24*, 409–416. [CrossRef]
57. Grimer, R.J.; Gosheger, G.; Taminiau, A.; Biau, D.; Matejovsky, Z.; Kollender, Y.; San-Julian, M.; Gherlinzoni, F.; Ferrari, C. Dedifferentiated chondrosarcoma: Prognostic factors and outcome from a European group. *Eur. J. Cancer* **2007**, *43*, 2060–2065. [CrossRef]
58. Littrell, L.A.; Wenger, D.E.; Wold, L.E.; Bertoni, F.; Unni, K.K.; White, L.; Kandel, R.; Sundaram, M. Radiographic, CT, and MR Imaging Features of Dedifferentiated Chondrosarcomas: A Retrospective Review of 174 De Novo Cases. *Radiographics* **2004**, *24*, 1397–1409. [CrossRef]
59. Saifuddin, A.; Mann, B.; Mahroof, S.; Pringle, J.; Briggs, T.; Cannon, S. Dedifferentiated chondrosarcoma: Use of MRI to guide needle biopsy. *Clin. Radiol.* **2004**, *59*, 268–272. [CrossRef]
60. Kawaguchi, S.; Wada, T.; Nagoya, S.; Ikeda, T.; Isu, K.; Yamashiro, K.; Kawai, A.; Ishii, T.; Araki, N.; Myoui, A.; et al. Extraskeletal myxoid chondrosarcoma: A Multi-Institutional Study of 42 Cases in Japan. *Cancer* **2003**, *97*, 1285–1292. [CrossRef]
61. Drilon, A.D.; Popat, S.; Bhuchar, G.; D'Adamo, D.R.; Keohan, M.L.; Fisher, C.; Antonescu, C.R.; Singer, S.; Brennan, M.F.; Judson, I.; et al. Extraskeletal myxoid chondrosarcoma: A retrospective review from 2 referral centers emphasizing long-term outcomes with surgery and chemotherapy. *Cancer* **2008**, *113*, 3364–3371. [CrossRef]
62. Antonescu, C.R.; Argani, P.; Erlandson, R.A.; Healey, J.H.; Ladanyi, M.; Huvos, A.G. Skeletal and extraskeletal myxoid chondrosarcoma: A comparative clinicopathologic, ultrastructural, and molecular study. *Cancer* **1998**, *83*, 1504–1521. [CrossRef]
63. Aigner, T.; Oliveira, A.M.; Nascimento, A.G. Extraskeletal myxoid chondrosarcomas do not show a chondrocytic phenotype. *Mod. Pathol.* **2004**, *17*, 214–221. [CrossRef] [PubMed]
64. Rosenberg, A.E. WHO Classification of Soft Tissue and Bone, fourth edition: Summary and commentary. *Curr. Opin. Oncol.* **2013**, *25*, 571–573.
65. Gassert, F.G.; Breden, S.; Neumann, J.; Gassert, F.T.; Bollwein, C.; Knebel, C.; Lenze, U.; von Eisenhart-Rothe, R.; Mogler, C.; Makowski, M.R.; et al. Differentiating Enchondromas and Atypical Cartilaginous Tumors in Long Bones with Computed Tomography and Magnetic Resonance Imaging. *Diagnostics* **2022**, *12*, 2186. [CrossRef] [PubMed]

66. Choi, B.-B.; Jee, W.-H.; Sunwoo, H.-J.; Cho, J.-H.; Kim, J.-Y.; Chun, K.-A.; Hong, S.-J.; Chung, H.W.; Sung, M.-S.; Lee, Y.-S.; et al. MR differentiation of low-grade chondrosarcoma from enchondroma. *Clin. Imaging* **2013**, *37*, 542–547. [CrossRef]
67. Crim, J.; Schmidt, R.; Layfield, L.; Hanrehan, C.; Manaster, B.J. Can imaging criteria distinguish enchondroma from grade 1 chondrosarcoma? *Eur. J. Radiol.* **2015**, *84*, 2222–2230. [CrossRef]
68. Lisson, C.S.; Lisson, C.G.; Flosdorf, K.; Mayer-Steinacker, R.; Schulthesis, M.; von Baer, A.; Barth, T.F.E.; Beer, A.J.; Baumhauer, M.; Meier, R.; et al. Diagnostic value of MRI-based 3D texture analysis for tissue characterisation and discrimination of low-grade chondrosarcoma from enchondroma: A pilot study. *Eur. Radiol.* **2018**, *28*, 468–477. [CrossRef]
69. Skeletal Lesions Interobserver Correlation among Expert Diagnosticians (SLICED) Study Group. Reliability of histopathologic and radiologic grading of cartilaginous neoplasms in long bones. *J. Bone Jt. Surg. Am.* **2007**, *89*, 2113–2123. [CrossRef]
70. Eefting, D.; Schrage, Y.M.; Geirnaerdt, M.J.A.; Le Cessie, S.; Taminiau, A.H.M.; Bovée, J.V.M.G.; Hogendoorn, P.C.W. Assessment of Interobserver Variability and Histologic Parameters to Improve Reliability in Classification and Grading of Central Cartilaginous Tumors. *Am. J. Surg. Pathol.* **2009**, *33*, 50–57. [CrossRef]
71. De Coninck, T.; Jans, L.; Sys, G.; Verstraeten, T.; Forsyth, R.; Poffyn, B.; Verstraete, K. Dynamic contrast-enhanced MR imaging for differentiation between enchondroma and chondrosarcoma. *Eur. Radiol.* **2013**, *23*, 3140–3152. [CrossRef]
72. Douis, H.; Jeys, L.; Grimer, R.; Vaiyapuri, S.; Davies, A.M. Is there a role for diffusion-weighted MRI (DWI) in the diagnosis of central cartilage tumors? *Skelet. Radiol.* **2015**, *44*, 963–969. [CrossRef]
73. Fritz, B.; Muller, D.A.; Sutter, R.; Wurnig, M.C.; Wagner, M.W.; Pfirrmann, C.W.A.; Fischer, M.A. Magnetic Resonance Imaging-Based Grading of Cartilaginous Bone Tumors: Added Value of Quantitative Texture Analysis. *Investig. Radiol.* **2018**, *53*, 663–672. [CrossRef] [PubMed]
74. Pan, J.; Zhang, K.; Le, H.; Jiang, Y.; Li, W.; Geng, Y.; Li, S.; Hong, G. Radiomics Nomograms Based on Non-enhanced MRI and Clinical Risk Factors for the Differentiation of Chondrosarcoma from Enchondroma. *J. Magn. Reson. Imaging* **2021**, *54*, 1314–1323. [CrossRef] [PubMed]
75. Kumar, V.S.; Tyrrell, P.N.M.; Singh, J.; Gregory, J.; Cribb, G.L.; Cool, P. Surveillance of intramedullary cartilage tumours in long bones. *Bone Jt. J.* **2016**, *98-B*, 1542–1547. [CrossRef] [PubMed]
76. Deckers, C.; Schreuder, B.H.; Hannink, G.; de Rooy, J.W.; van der Geest, I. Radiologic follow-up of untreated enchondroma and atypical cartilaginous tumors in the long bones. *J. Surg. Oncol.* **2016**, *114*, 987–991. [CrossRef] [PubMed]
77. Parlier-Cuau, C.; Bousson, V.; Ogilvie, C.M.; Lackman, R.D.; Laredo, J.-D. When should we biopsy a solitary central cartilaginous tumor of long bones? Literature review and management proposal. *Eur. J. Radiol.* **2011**, *77*, 6–12. [CrossRef]
78. Vanel, D.; Ruggieri, P.; Ferrari, S.; Picci, P.; Gambarotti, M.; Staals, E.; Alberghini, M. The incidental skeletal lesion: Ignore or explore? *Cancer Imaging* **2009**, *9*, S38–S43. [CrossRef]
79. Patel, A.; Davies, A.; Botchu, R.; James, S. A pragmatic approach to the imaging and follow-up of solitary central cartilage tumours of the proximal humerus and knee. *Clin. Radiol.* **2019**, *74*, 517–526. [CrossRef] [PubMed]
80. Deckers, C.; de Rooy, J.W.J.; Flucke, U.; Schreuder, H.W.B.; Dierselhuis, E.F.; van der Geest, I.C.M. Midterm MRI Follow-Up of Untreated Enchondroma and Atypical Cartilaginous Tumors in the Long Bones. *Cancers* **2021**, *13*, 4093. [CrossRef]
81. van de Sande, M.A.J.; van der Wal, R.J.P.; Canete, A.N.; van Rijswijk, C.S.P.; Kroon, H.M.; Dijkstra, P.D.S.; Bloem, J.L.H. Radiologic differentiation of enchondromas, atypical cartilaginous tumors, and high-grade chondrosarcomas-Improving tumor-specific treatment: A paradigm in transit? *Cancer* **2019**, *125*, 3288–3291. [CrossRef]
82. Sullivan, C.W.; Kazley, J.M.; Murtaza, H.; Cooley, M.; Jones, D.; DiCaprio, M.R. Team Approach: Evaluation and Management of Low-Grade Cartilaginous Lesions. *JBJS Rev.* **2020**, *8*, e0054. [CrossRef]
83. Omlor, G.W.; Lohnherr, V.; Lange, J.; Gantz, S.; Mechtersheimer, G.; Merle, C.; Raiss, P.; Fellenberg, J.; Lehner, B. Outcome of conservative and surgical treatment of enchondromas and atypical cartilaginous tumors of the long bones: Retrospective analysis of 228 patients. *BMC Musculoskelet. Disord.* **2019**, *20*, 134. [CrossRef]
84. Akoh, C.C.; Craig, E.; Troester, A.M.; Miller, B.J. Radiographic Enchondroma Surveillance: Assessing Clinical Outcomes and Costs Effectiveness. *Iowa Orthop. J.* **2019**, *39*, 185–193. [PubMed]
85. Deckers, C.; Steyvers, M.J.; Hannink, G.; Schreuder, H.W.B.; de Rooy, J.W.J.; Van Der Geest, I.C.M. Can MRI differentiate between atypical cartilaginous tumors and high-grade chondrosarcoma? A systematic review. *Acta Orthop.* **2020**, *91*, 471–478. [CrossRef]
86. Laitinen, M.; Stevenson, J.D.; Parry, M.C.; Sumathi, V.; Grimer, R.J.; Jeys, L. The role of grade in local recurrence and the disease-specific survival in chondrosarcomas. *Bone Jt. J.* **2018**, *100-B*, 662–666. [CrossRef]
87. Jain, V.; Oliveira, I.; Chavda, A.; Khoo, M.; Saifuddin, A. MRI differentiation of low-grade and high-grade chondrosarcoma of the shoulder girdle, chest wall and pelvis. A pictorial review based on 111 consecutive cases. *Br. J. Radiol.* **2021**, *94*, 20201404. [CrossRef] [PubMed]
88. Douis, H.; Singh, L.; Saifuddin, A. MRI differentiation of low-grade from high-grade appendicular chondrosarcoma. *Eur. Radiol.* **2013**, *24*, 232–240. [CrossRef] [PubMed]
89. Alhumaid, S.M.; Iv, A.A.; Aljubair, H. Magnetic Resonance Imaging Role in the Differentiation Between Atypical Cartilaginous Tumors and High-Grade Chondrosarcoma: An Updated Systematic Review. *Cureus* **2020**, *12*, e11237. [CrossRef]
90. MacSweeney, F.; Darby, A.; Saifuddin, A. Dedifferentiated chondrosarcoma of the appendicular skeleton: MRI-pathological correlation. *Skelet. Radiol.* **2003**, *32*, 671–678. [CrossRef]
91. Yoo, H.J.; Hong, S.H.; Choi, J.-Y.; Moon, K.C.; Kim, H.-S.; Kang, H.S. Differentiating high-grade from low-grade chondrosarcoma with MR imaging. *Eur. Radiol.* **2009**, *19*, 3008–3014. [CrossRef]

92. Geirnaerdt, M.J.A.; Hogendoorn, P.; Bloem, J.L.; Taminiau, A.H.M.; Van Der Woude, H.-J. Cartilaginous Tumors: Fast Contrast-enhanced MR Imaging. *Radiology* **2000**, *214*, 539–546. [CrossRef]
93. Deng, X.-Y.; Chen, H.-Y.; Yu, J.-N.; Zhu, X.-L.; Chen, J.-Y.; Shao, G.-L.; Yu, R.-S. Diagnostic Value of CT- and MRI-Based Texture Analysis and Imaging Findings for Grading Cartilaginous Tumors in Long Bones. *Front. Oncol.* **2021**, *11*, 700204. [CrossRef]
94. Gitto, S.; Cuocolo, R.; Albano, D.; Chianca, V.; Messina, C.; Gambino, A.; Ugga, L.; Cortese, M.C.; Lazzara, A.; Ricci, D.; et al. MRI radiomics-based machine-learning classification of bone chondrosarcoma. *Eur. J. Radiol.* **2020**, *128*, 109043. [CrossRef]
95. Zhong, J.; Hu, Y.; Ge, X.; Xing, Y.; Ding, D.; Zhang, G.; Zhang, H.; Yang, Q.; Yao, W. A systematic review of radiomics in chondrosarcoma: Assessment of study quality and clinical value needs handy tools. *Eur. Radiol.* **2022**, *33*, 1433–1444. [CrossRef]
96. Leerapun, T.; Hugate, R.R.; Inwards, C.Y.; Scully, S.P.; Sim, F.H. Surgical Management of Conventional Grade I Chondrosarcoma of Long Bones. *Clin. Orthop. Relat. Res.* **2007**, *463*, 166–172. [CrossRef] [PubMed]
97. Gelderblom, H.; Hogendoorn, P.C.W.; Dijkstra, S.D.; van Rijswijk, C.S.; Krol, A.D.; Taminiau, A.H.M.; Bovee, J.V.M.G. The clinical approach towards chondrosarcoma. *Oncologist* **2008**, *13*, 320–329. [CrossRef] [PubMed]
98. Bovee, J.; Cleton-Jansen, A.-M.; Taminiau, A.H.; Hogendoorn, P. Emerging pathways in the development of chondrosarcoma of bone and implications for targeted treatment. *Lancet Oncol.* **2005**, *6*, 599–607. [CrossRef]
99. Dickey, I.D.; Rose, P.S.; Fuchs, B.; Wold, L.E.; Okuno, S.H.; Sim, F.H.; Scully, S.P. Dedifferentiated Chondrosarcoma: The Role of Chemotherapy with Updated Outcomes. *J. Bone Jt. Surg.* **2004**, *86*, 2412–2418. [CrossRef]
100. van Maldegem, A.M.; Bovée, J.V.; Gelderblom, H. Comprehensive analysis of published studies involving systemic treatment for chondrosarcoma of bone between 2000 and 2013. *Clin. Sarcoma Res.* **2014**, *4*, 11. [CrossRef] [PubMed]
101. Frezza, A.M.; Cesari, M.; Baumhoer, D.; Biau, D.; Bielack, S.; Campanacci, D.A.; Casanova, J.; Esler, C.; Ferrari, S.; Funovics, P.T.; et al. Mesenchymal chondrosarcoma: Prognostic factors and outcome in 113 patients. A European Musculoskeletal Oncology Society study. *Eur. J. Cancer* **2015**, *51*, 374–381. [CrossRef]
102. Suit, H.D.; Goitein, M.; Munzenrider, J.; Verhey, L.; Davis, K.R.; Koehler, A.; Linggood, R.; Ojemann, R.G. Definitive radiation therapy for chordoma and chondrosarcoma of base of skull and cervical spine. *J. Neurosurg.* **1982**, *56*, 377–385. [CrossRef]
103. MacDonald, I.J.; Lin, C.-Y.; Kuo, S.-J.; Su, C.-M.; Tang, C.-H. An update on current and future treatment options for chondrosarcoma. *Expert Rev. Anticancer. Ther.* **2019**, *19*, 773–786. [CrossRef] [PubMed]
104. Mery, B.; Espenel, S.; Guy, J.-B.; Rancoule, C.; Vallard, A.; Aloy, M.-T.; Rodriguez-Lafrasse, C.; Magné, N. Biological aspects of chondrosarcoma: Leaps and hurdles. *Crit. Rev. Oncol.* **2018**, *126*, 32–36. [CrossRef]
105. Tlemsani, C.; Larousserie, F.; De Percin, S.; Audard, V.; Hadjadj, D.; Chen, J.; Biau, D.; Anract, P.; Terris, B.; Goldwasser, F.; et al. Biology and Management of High-Grade Chondrosarcoma: An Update on Targets and Treatment Options. *Int. J. Mol. Sci.* **2023**, *24*, 1361. [CrossRef] [PubMed]
106. Nicolle, R.; Ayadi, M.; Gomez-Brouchet, A.; Armenoult, L.; Banneau, G.; Elarouci, N.; Tallegas, M.; Decouvelaere, A.-V.; Aubert, S.; Redini, F.; et al. Integrated molecular characterization of chondrosarcoma reveals critical determinants of disease progression. *Nat. Commun.* **2019**, *10*, 4622. [CrossRef]
107. Chen, J.-C.; Shih, H.-C.; Lin, C.-Y.; Guo, J.-H.; Huang, C.; Huang, H.-C.; Chong, Z.-Y.; Tang, C.-H. MicroRNA-631 Resensitizes Doxorubicin-Resistant Chondrosarcoma Cells by Targeting Apelin. *Int. J. Mol. Sci.* **2023**, *24*, 839. [CrossRef]
108. Miwa, S.; Yamamoto, N.; Hayashi, K.; Takeuchi, A.; Igarashi, K.; Tsuchiya, H. Therapeutic Targets and Emerging Treatments in Advanced Chondrosarcoma. *Int. J. Mol. Sci.* **2022**, *23*, 1096. [CrossRef] [PubMed]
109. Nazeri, E.; Savadkoohi, M.G.; Majidzadeh-A, K.; Esmaeili, R. Chondrosarcoma: An overview of clinical behavior, molecular mechanisms mediated drug resistance and potential therapeutic targets. *Crit. Rev. Oncol.* **2018**, *131*, 102–109. [CrossRef]
110. Thoenen, E.; Curl, A.; Iwakuma, T. TP53 in bone and soft tissue sarcomas. *Pharmacol. Ther.* **2019**, *202*, 149–164. [CrossRef]
111. Su, C.-M.; Fong, Y.-C.; Tang, C.-H. An overview of current and future treatment options for chondrosarcoma. *Expert Opin. Orphan. Drugs* **2014**, *2*, 217–227. [CrossRef]
112. Jeong, W.; Kim, H.-J. Biomarkers of chondrosarcoma. *J. Clin. Pathol.* **2018**, *71*, 579–583. [CrossRef] [PubMed]

cancers

MDPI

Review

Current Landscape of Immunotherapy for Advanced Sarcoma

Víctor Albarrán *, María Luisa Villamayor, Javier Pozas, Jesús Chamorro, Diana Isabel Rosero, María San Román, Patricia Guerrero, Patricia Pérez de Aguado, Juan Carlos Calvo, Coral García de Quevedo, Carlos González and María Ángeles Vaz

Medical Oncology Department, Ramon y Cajal University Hospital, 28034 Madrid, Spain
* Correspondence: vicalbarranfernandez@gmail.com

Simple Summary: The systemic treatment of advanced sarcoma remains challenging. Conventional chemotherapy and anti-angiogenic agents, even in the most responsive histologic subtypes, result in short responses and poor clinical outcomes. In a context where new therapeutic approaches are required, several strategies of immunotherapy have emerged as promising options, such as immune checkpoint inhibitors, vaccines, and adoptive cell therapy. In this review, we aim to summarize the current state and challenges of immunotherapy in patients with advanced bone and soft-tissue sarcomas.

Abstract: There is substantial heterogeneity between different subtypes of sarcoma regarding their biological behavior and microenvironment, which impacts their responsiveness to immunotherapy. Alveolar soft-part sarcoma, synovial sarcoma and undifferentiated pleomorphic sarcoma show higher immunogenicity and better responses to checkpoint inhibitors. Combination strategies adding immunotherapy to chemotherapy and/or tyrosine–kinase inhibitors globally seem superior to single-agent schemes. Therapeutic vaccines and different forms of adoptive cell therapy, mainly engineered TCRs, CAR-T cells and TIL therapy, are emerging as new forms of immunotherapy for advanced solid tumors. Tumor lymphocytic infiltration and other prognostic and predictive biomarkers are under research.

Keywords: bone sarcoma; soft-tissue sarcoma; immunotherapy; checkpoint inhibitors; TCR; TIL; vaccines; adoptive cell therapy

Citation: Albarrán, V.; Villamayor, M.L.; Pozas, J.; Chamorro, J.; Rosero, D.I.; San Román, M.; Guerrero, P.; Pérez de Aguado, P.; Calvo, J.C.; García de Quevedo, C.; et al. Current Landscape of Immunotherapy for Advanced Sarcoma. *Cancers* **2023**, *15*, 2287. https://doi.org/10.3390/cancers15082287

Academic Editors: Adam C. Berger and Catrin Sian Rutland

Received: 18 March 2023
Revised: 7 April 2023
Accepted: 12 April 2023
Published: 13 April 2023

1. Introduction

Sarcomas are a heterogeneous group of malignant tumors of mesenchymal origin. Their incidence in adults is low, comprising less than 1% of cancer diagnoses versus up to 15% of malignancies in the pediatric population [1]. More than 70 histologic subtypes have been identified, and they can be broadly classified into bone sarcomas (BS) and soft-tissue sarcomas (STS). The most frequent types of BS are osteosarcoma, chondrosarcoma (CS) and Ewing sarcoma (ES), whereas liposarcoma, leiomyosarcoma and undifferentiated pleomorphic sarcoma (UPS) are the most common subtypes of STS [2].

Conventional chemotherapy (CT) is still the standard treatment for unresectable or metastatic STS. Anthracyclines-based regimens, usually adriamycin plus ifosfamide, remain the upfront treatment [3], whereas other cytotoxic drugs are usually used in further lines (gemcitabine plus docetaxel [4], trabectedin [5,6], eribulin [7] or dacarbazine [8]). Several oral tyrosine–kinase inhibitors (TKI) have also demonstrated activity for STS, including multi-TKI pazopanib for non-adipocytic STS [9], anaplastic lymphoma kinase (ALK) inhibitors for myofibroblastic tumors with ALK fusions [10], and cediranib for alveolar soft part sarcoma (ASPS) [11].

In BS, multimodal treatment with CT, radiotherapy (RT) and radical surgery is recommended. For high-grade osteosarcoma, preoperative CT with a MAP regimen (doxorubicin, cisplatin, and high-dose methotrexate) is usually the front-line treatment for young patients [12]. In progressive disease, conventional CT with ifosfamide or cyclophosphamide plus carboplatin or etoposide is commonly used, with less evidence for other drugs, such as docetaxel and gemcitabine [13]. In ES, perioperative CT is indicated, usually with an interval VDC/IE scheme (vincristine, doxorubicin, cyclophosphamide, ifosfamide and etoposide) [14]. Topotecan plus cyclophosphamide and high-dose ifosfamide are the preferable options for further lines, followed by irinotecan plus temozolomide and docetaxel plus gemcitabine [15]. Several multi-TKI have also shown efficacy in advanced BS, mainly regorafenib, cabozantinib and apatinib for osteosarcoma [16].

Despite the recent incorporation of TKI and other drugs beyond conventional CT, the long-term prognosis of advanced sarcoma remains poor, with a median survival of 12–18 months for advanced STS [3], a 5-year survival rate <20% for osteosarcoma [17] and <40% for advanced ES [18]. The need for new therapeutic approaches, especially relevant given the predominance of these tumors in very young populations, explains the recurrent attempts to incorporate immunotherapy into the arsenal against advanced sarcoma.

The history of immunotherapy in sarcoma began with Coley's inoculations of erysipelas, inducing tumor regression in some patients [19], though its development was at a standstill for many decades. High-dose interleukin-2 (IL-2) therapy, approved for advanced melanoma and renal cell carcinoma in 1998 [20,21], demonstrated some activity in pretreated pediatric sarcoma [22], though its use was restricted due to the high incidence of severe toxicity (cytokine-induced capillary leak syndrome [23]).

Modern immunotherapy with immune checkpoint inhibitors (ICI) has revolutionized the treatment of solid tumors. Though the results of ICI in monotherapy are poorer in sarcoma than in other malignancies, their combination with other agents seems to have synergistic effects, and promising strategies, such as vaccines and adoptive cell therapy, are emerging. However, there is a wide clinical heterogeneity between different histologic subtypes, disease settings and treatment categories [24]. This review aims to summarize the biological basis, current state, and future challenges of immunotherapy in advanced sarcoma.

2. Immunogenicity of Sarcoma: Anti-Tumor Response and Biomarkers

2.1. Innate Immunity and Release of Neoantigens

In sarcoma, as in other solid tumors, the activation and migration of cytotoxic T lymphocytes play a key role in the anti-tumor immune response [25] (Figure 1). Cancer cells are initially attacked by macrophages and natural killer (NK) cells, the main components of innate immunity, leading to cell death. Some ligands on the membrane of sarcoma cells activate NK cells, mainly through NK cell group 2D receptors (NKG2D) [26], and facilitate apoptosis. However, the essential mechanism of cell death is necrosis, with the subsequent release of tumor-associated antigens (TAA) with damage-associated molecular patterns (DAMPs), including aberrant 'neoantigens' produced as a result of accumulative somatic mutations [27].

Tumors with a higher mutational burden (TMB) have an increased level of neoantigens and a higher immunogenicity [28]. Sarcoma is a heterogeneous disease with significant variability in TMB among different subtypes, as demonstrated by Chalmers et al. [29]. The median TMB exceeds 20 mutations per Mb of DNA in angiosarcoma, leiomyosarcoma and UPS but is lower than two mutations per Mb in myxofibrosarcoma, liposarcoma and synovial sarcoma. However, the neoantigens burden related to the TMB is not the only factor that determines tumor immunogenicity. The presence of certain chromosomal translocations in tumor cells give rise to fusion proteins that bind to major histocompatibility complex class I (MHC-I) molecules and activate cytotoxic CD8+ T cells, working as powerful neoantigens [30]. Worley et al. [31] showed that some subtypes of sarcoma, such as clear cell, synovial and desmoplastic round cell tumors, can be highly immunogenic due to these genetic alterations, despite their low median TMB.

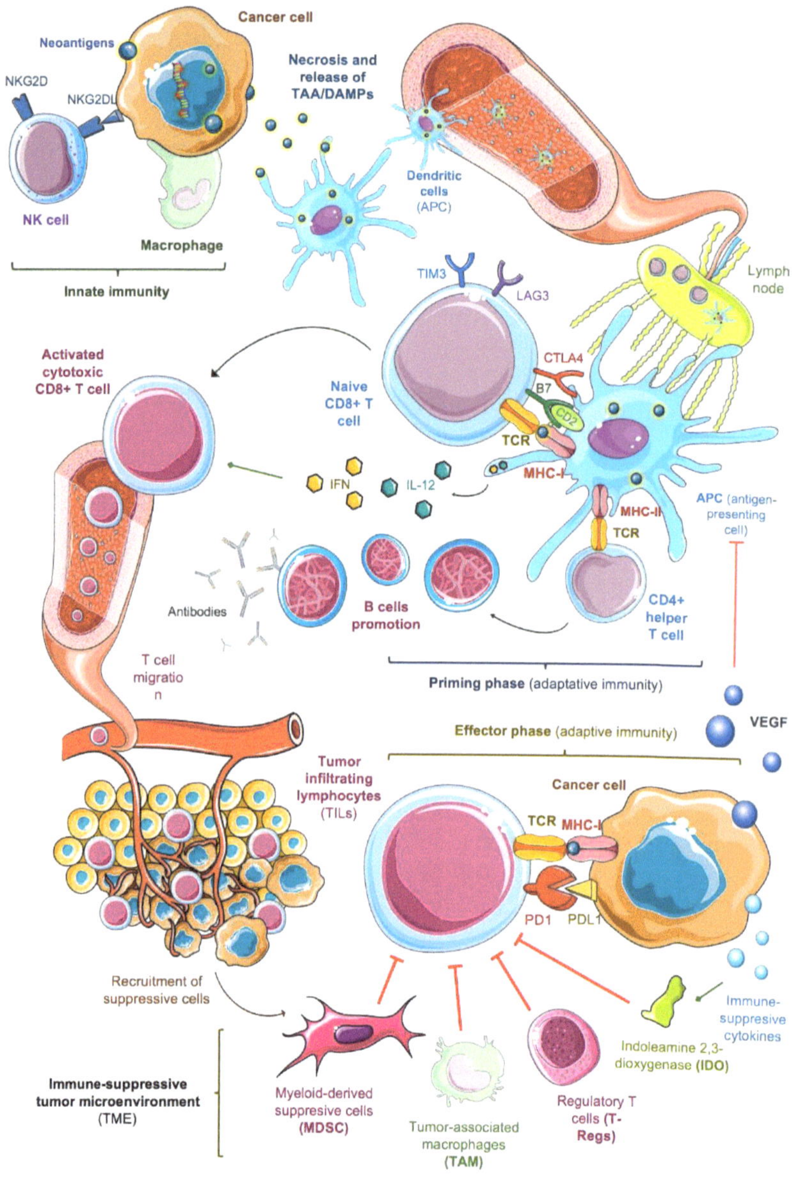

Figure 1. Mechanisms of anti-tumor response and potential immune biomarkers in solid tumors. NK: natural killer; TAA: tumor-associated antigen; DAMPs: damage-associated molecular patterns; TCR: T cell receptor; MHC: major histocompatibility complex; IL-12: interleukin-12; IFN: interferons; VEGF: vascular endothelial growth factor.

Kakimoto et al. [32] revealed that some high-grade sarcoma aberrantly express testis antigens (CTA), usually present in germ cells, such as melanoma-associated antigen (MAGE)-A4 and New York esophageal squamous cell carcinoma (NY-ESO)-1. MAGE-A4 was detected in 59% of synovial sarcoma and 56% of myxoid liposarcoma. NY-ESO-1 was found in 53% of synovial sarcomas.

The expression of NY-ESO-1 was associated with a better prognosis in high-grade sarcoma (5-year overall survival of 81% in the NY-ESO1+ group vs. 53% in NY-ESO-1-

group, $p < 0.05$), presumably due to the powerful immunogenicity conferred by these fusion proteins.

Tumors with a higher mutational burden (TMB) have an increased level of neoantigens and a higher immunogenicity [28]. Sarcoma is a heterogeneous disease with significant variability in TMB among different subtypes, as demonstrated by Chalmers et al. [29]. The median TMB exceeds 20 mutations per Mb of DNA in angiosarcoma, leiomyosarcoma and UPS but is lower than two mutations per Mb in myxofibrosarcoma, liposarcoma and synovial. However, the neoantigens burden related to the TMB is not the only factor that determines tumor immunogenicity. The presence of certain chromosomal translocations in tumor cells gives rise to fusion proteins that bind to major histocompatibility complex class I (MHC-I) molecules and activate cytotoxic CD8+ T cells, working as powerful neoantigens [30]. Worley et al. [31] showed that some subtypes of sarcoma, such as clear cell, synovial and desmoplastic round cell tumors, can be highly immunogenic due to these genetic alterations, despite their low median TMB.

2.2. Antigenic Presentation and Activation of T Cells

After necrosis, dendritic cells phagocytose the released DAMPs and migrate to lymph nodes, where they work as antigen-presenting cells (APC). APC are essential for unleashing adaptive immunity by activating naïve CD8+ T lymphocytes (priming phase). This process needs the interaction of T-cell receptors (TCR) with MHC-I molecules on the APC surface, through which the phagocytosed antigens are presented [33].

T-cell activation also requires co-stimulatory signals, such as the coupling between co-receptor B7 (CD80) and the ligand CD28 on the APC surface. Certain cytokines released by APCs, mainly type I interferons (IFN) and interleukin-12 (IL-12), promote the activity of cytotoxic T cells and contribute to the activation of CD4+ helper T cells -following the coupling of TCR and MHC class II-. This facilitates B cell promotion and antibody production. Zhou et al. [34] showed that IL-12 up-regulates the expression of *Fas* receptors in osteosarcoma and ES cells, increasing their sensitivity to *Fas*-induced apoptosis. Type I IFN (IFN-α/β) have antiangiogenic and antiproliferative properties, which have been studied in models of Kaposi sarcoma [35] and angiosarcoma [36].

On the other side, some competitive co-receptors on the T cell membrane, such as CTLA4, LAG3 and TIM3, work as co-inhibitory signals that control this process in negative feedback [37]. Dancsok et al. [38] studied 1072 sarcoma specimens, revealing LAG3 and TIM3 expression on the infiltrating T-cells of nearly 50% of them, reaching 80% in some subtypes, dedifferentiated liposarcoma, myxofibrosarcoma and UPS. These 'immune checkpoints' are pathologically stimulated by tumor cells as a mechanism of immune escape, which sets the rationale for the use of ICI.

2.3. Tumor Infiltration of Activated Lymphocytes

Activated cytotoxic T lymphocytes reach the tumor via blood vessels and kill malignant cells in the peripheral tissues (effector phase). The successful trafficking of T cells to the tumor site is a key component of an effective immune response [25]. Once cytotoxic T cells are primed, they undergo a shift in the expression of surface proteins, losing CD62L and CCR7, which mediate their access to lymph nodes, and gaining molecules that facilitate their migration to diseased tissues. These include selectins, which facilitate the rolling of T cells on the endothelium, and receptors for inflammatory chemokines (CXCL9, CXCL10) that mediate their extravasation [39].

Tumor-infiltrating lymphocytes (TILs) comprise different subtypes of lymphocytes with high immunogenicity against tumor cells CD4+, CD8+, CD20+ and FoxP3+ TILs. CD8+ TILs interact through their TCR with antigens presented by MHC class I molecules on the surface of cancer cells, unleashing the cytotoxic cascade that leads to necrosis. In this phase, as in the priming phase, the anti-tumor response is controlled by immune checkpoints and can be suppressed by the activation of inhibitory co-receptors of the lymphocyte, such as programmed cell death receptor PD1, due to the interaction with immune-suppressive proteins expressed by tumor cells and cells from the tumor microenvironment (TME), such as PD1 ligand PDL1- [40].

D'Angelo et al. [41] analyzed the variability in PDL1 expression in tumors, lymphocytes and macrophages among different subtypes of sarcoma. The expression of PDL1 was more frequent in lymphocytes and macrophages than in the tumor cells, where it was detected just in three histologic subtypes, gastrointestinal stromal tumor (GIST), radiation-associated pleomorphic sarcoma and spindle cell sarcoma. Globally, the PDL1 expression was positive in six samples among 50 (12%) and was significantly associated with a high density of CD8+ TILs.

A meta-analysis by Zheng et al. [42], containing 15 studies and 1451 patients with bone and soft-tissue sarcoma, concluded that high expression levels of PDL1 were associated with poorer overall survival (HR 1.27, $p < 0.001$) and events-free survival (HR 2.05, $p < 0.001$), confirming the negative prognostic role of PDL1 expression. This is consistent with other studies; Que et al. [43] showed that a positive PDL1 expression is associated with Foxp3+ T-regs infiltration and a poor clinical prognosis in STS.

Whereas the 'tumor killing' role mainly corresponds to cytotoxic CD8+ TILs, CD4+ cells contribute to their priming and proliferation [44]. A higher CD4+ and CD8+ TILs infiltration is related to better prognosis in several solid tumors [45]. On the contrary, a high density of TILs expressing transcription factor forkhead box protein 3 (FoxP3), a marker of immune-suppressive regulatory T cells (T-regs), has shown a positive correlation with poor clinical prognosis [46,47].

Globally, sarcoma has lower TILs infiltration than other solid tumors, with huge heterogeneity among histological subtypes. D'Angelo et al. [41] also analyzed the percentage of TILs subsets (CD3+, CD4+, CD8+ and FOXP3+) among different subtypes of sarcoma. A 'high density', defined as >5%- of CD3+ cells, was frequently found in GIST (41%), angiosarcoma (14%) and spindle cell sarcoma (14%). A high density of CD4+ cells was present in GIST (50%), angiosarcoma (25%) and pleomorphic rhabdomyosarcoma (25%), whereas a high density of CD8+ cells prevailed in GIST (27%) and spindle cell sarcoma (18%). The prognostic impact of TILs in sarcoma is not fully clear, though some studies suggest better survival rates in patients with higher infiltration levels of CD4+ TILs [48] and CD8+ TILs [49].

Though T cells have been the focus of anti-tumor immunity research, B cells are progressively gaining strong attention. The development of B cells in the TME depends on the maturity of tertiary lymphoid structures (TLS). TLS are ectopic lymphoid organs developed in tissues under chronic inflammation, including tumors. In immature TLS, B cells evolve as T-regs and release immune-suppressive cytokines, whereas, in mature TLS with a germinal center, B cells undergo affinity maturation and isotypic switching, resulting in plasmatic cells that secrete anti-tumor antibodies [50]. As a favorable lymphocytic infiltration, the presence of mature TLS has been associated with better clinical outcomes, and strategies to induce TLS neogenesis in immune-low tumors represent a promising pathway for cancer immunotherapy [51].

A favorable B population seems to have a key role in the response against sarcoma. Sorbye et al. [52] showed that a higher density of CD20+ TILs in STS is an independent positive prognostic factor. Petitprez et al. [53] proposed an immune-based classification of STS based on the TME composition, identifying an immune-high class E (SIC E) whose specimens were particularly rich in CD20+ TILs. They analyzed 47 patients with STS from the SARC028 trial [54] and found that a high infiltration of CD20+ cells determined the highest response rate and progression-free survival (PFS) to PD1 blockade, even in tumors with low CD8+ TILs infiltration.

2.4. Immune-Suppressive Tumor Microenvironment

The TME dynamics are affected by complex reciprocal interactions between immune-stimulatory and immune-suppressive cells. The recognition of tumor immunogenic epitopes by TILs promotes tumor regression by activating tumors into a T-cell-inflamed 'hot' state. On the contrary, immune-suppressive cells such as T-regs, myeloid-derived suppressor cells (MDSCs) and tumor-associated macrophages (TAMs) interfere with effector T cells and facilitate tumor immune escape [55]. The predominance of these immune-suppressive cells leads to non-T-cell inflamed or 'cold' tumors, promoting tumor progression and impoverishing clinical prognosis.

T-regs suppress the function of effector cells by direct contact through the interaction between granzymes and perforins with the CD8+ T cell membrane [56] but also by indirect mechanisms, such as the release of inhibitory cytokines growth factor β (TGF-β), interleukin-10 (IL-10), IL-35 and prostaglandin E2 (PGE2) [55] and the suppression of APCs through the downregulation of CD80 and other stimulatory coreceptors [57]. Smolle et al. [58] analyzed the infiltration of CD3+ FoxP3+ T-regs in 192 surgical samples of STS and found an increased risk of local recurrence in tumors with CD3+ FoxP3+ T-regs. An increasing prevalence of T cells with a regulatory phenotype (CD4+, FoxP3+) has been found in the advanced stages of Ewing [59] and Kaposi sarcoma [60].

MDSCs facilitate epithelial-to-mesenchymal transition [61], act as mediators of neo-angiogenesis through the release of vascular endothelial growth factor (VEGF), fibroblast growth factors (FGF) and matrix metalloproteinase 9 (MMP9) [62] and induce TME remodeling by establishing a 'pre-metastatic niche' [63]. The MDSCs trafficking to the tumor is mainly mediated by the CXCR2 receptor, which has been proposed as a potential target to alter the TME and attenuate tumor progression [64]. Sarcoma cells can produce CXCR2 ligands, such as CXCL8, that facilitate the arrival of MDSCs to the TME. Highfill et al. [65] showed that pediatric patients with advanced sarcoma display elevated serum levels of CXCL8, which are associated with poor survival rates. In murine models, the blockade of CXCR2 seems to suppress MDSCs trafficking and enhance the anti-tumor activity of PD1 blockade. These findings suggest that strategies to prevent the trafficking of MDSCs to the tumor bed may improve the efficacy of checkpoint inhibitors.

Monocyte-related MDSCs (M-MDSCs), together with circulating monocytes and tissue-resident macrophages, after being recruited to the tumor site in response to colony-stimulating factors (CSF) and several chemokines, can differentiate into TAMs [66]. TAMs can be polarized to M1-like (classically activated) or M2-like (alternatively activated) macrophages. TAMs with a M1-like phenotype display anti-tumor functions, whereas the M2-like phenotype is associated with pro-tumorigenic activity [67]. In fact, a high density of M2-like TAMs in the TME has been associated with poor clinical outcomes in many solid tumors [68].

The unfavorable prognostic role of M2-like TAMs has been established in STS. Higher levels of TAMs expressing M2-related markers (CD163+/CD204+) have been associated with poorer survival and higher disease stage in leiomyosarcoma [69], myxoid liposarcoma [70], synovial sarcoma [71] and UPS [72]. The prognostic significance of the M1/M2-phenotype in bone sarcoma is more controversial [73]. Some studies suggest a positive impact of polarized macrophages with an M1 phenotype [74], but others have found no clear correlation with survival [75]. Some studies have even reported longer survival rates in osteosarcoma [76] and ES [77] patients with a high density of M2-like TAMs.

Several cytokines released by malignant cells promote the production of indoleamine 2,3-dioxygenase (IDO1) and VEGF. IDO1 is an intracellular enzyme that reduces the activity of effector T cells through the suppression of the tryptophan pathway [25]. Some studies have suggested that the IDO1 pathway could contribute to the immune-suppressive phenotype of sarcoma cells and be a relevant mechanism of their primary resistance to PD1 blockade [78]. In fact, a high IDO1 expression may be used as a biomarker of poor response to anti-PD1 agents in sarcoma [79]. Around 39% of human sarcoma express IDO1, especially when the CD8+ TILs infiltration is high, setting a rationale for the dual blockade of IDO1 and immune checkpoints [80].

VEGF interferes with an antigenic presentation by inhibiting the maturation of dendritic cells [81], restricts the migration of lymphocytes into the tumor compartment [82], and favors the recruitment of T-regs, MDSCs and TAMs, contributing to a highly immune-suppressive TME [83]. The growth and dissemination of sarcoma strongly depend on angiogenesis, and VEGF circulating levels correlate with stage, grade, and risk of metastasis [84]. Overcoming these barriers of the TME remains a major challenge to move immunotherapy forward in advanced sarcoma.

3. Immunotherapy for Sarcoma: Clinical Results

3.1. Immune Checkpoint Inhibitors (ICIs)

The first studies with single-agent immunotherapy failed to demonstrate a significant anti-tumor activity (see Table 1). Anti-CTLA4 antibody ipilimumab in monotherapy showed negative results in recurrent synovial sarcoma [85] and pediatric sarcoma [86]. Anti-PD1 nivolumab was tried in the third line in 12 patients with advanced uterine leiomyosarcoma, with no objective responses [87].

Table 1. Published results of immunotherapy in sarcoma. AEs: adverse effects; irAEs: immune-related adverse effects; ORR: objective response rate (RECIST criteria); mPFS: median progression-free survival; mOS: median overall survival; m: months; w: weeks; PR: partial response; SD: stable disease; DCR: disease control rate (PR+SD); mDR: median duration of response; NA: not available; AST: aspartate aminotransferase; ALT: alanine aminotransferase; BS: bone sarcoma; STS: soft-tissue sarcoma; OST: osteosarcoma; ES: Ewing sarcoma; CS: chondrosarcoma; LMS: leiomyosarcoma; LPS: liposarcoma; SS: synovial sarcoma; UPS: undifferentiated pleomorphic sarcoma; KS: Kaposi sarcoma; ALP: alkaline phosphatase; CK: creatine kinase; mCP: metronomic cyclophosphamide; AS: angiosarcoma; ASPS: alveolar soft-part sarcoma; DDLPS: dedifferentiated liposarcoma; PPS: palmar-plantar syndrome; LDH: lactate dehydrogenase; MDSCs: myeloid-derived suppressor cells; MRCL: myxoid/round cell liposarcoma; NY-ESO1: New York esophageal squamous cell carcinoma 1; MAGE-A4: melanoma-associated antigen A4; DC: dendritic cells; MLS: myxoid liposarcoma; CRS: cytokine release syndrome; RMS: rhabdomyosarcoma; adj: adjuvant; maint: maintenance; mono: monotherapy.

Clinical Trial	Agent	Tumor	N	Age Range	Outcomes	Reported G3/G4 AEs
			Immune checkpoint inhibitors (ICI)			
Maki et al. (phase I/II) [85]	Ipilimumab	Synovial sarcoma	6	23–57	ORR 0% mPFS 1.5 m mOS 8.8 m	Nausea (50%), diarrhea (33.3%), lymphopenia (33.3%), hyperbilirubinemia (16.7%), thrombopenia (16.7%)
Merchant et al. (phase I) [86]	Ipilimumab	Pediatric sarcoma	17	2–17	ORR 0%; DCR 17.6% (3 SD) mPFS/mOS: NA	Diarrhea (9%), AST/ALT increase (6%), endocrinopathies (3%), other irAEs (9%)
Ben-Ami et al. (phase II) [87]	Nivolumab	Uterine leiomyo-sarcoma	12	29–73	ORR 0% mPFS 1.8 m; mOS NA	Lipase/amylase increase (8.3%), fatigue (8.3%), abdominal pain (8.3%)
Tawbi et al. (phase II) (SARC028) [54]	Pembrolizumab	BS cohort (22 OST, 13 ES, 5 CS)	40	16–70	mPFS 8 w; mOS 52 w OST: ORR 5%; DCR 32% (1 PR, 6 SD) ES: ORR 0%; DCR 15% (2 SD) CS: ORR 20%; DCR 40% (1 PR, 1 SD)	Interstitial nephritis (2%), infectious pneumonia (2%), bone pain (2%), pleural effusion (2%), hypoxia (2%)
		STS cohort (10 LMS, 10 LPS, 10 SS, 10 UPS)	40	18–81	mPFS 18 w; mOS 49 w LMS: ORR 0%; DCR 60% (6 SD) / LPS: ORR 20%; DCR 60% (2 PR, 4 SD)/SS: ORR 10%; DCR 30% (1 PR, 2 SD)/UPS: ORR 40%; DCR 70% (1 CR, 3 PR, 3 SD)	Pulmonary embolism (2%), adrenal insufficiency (2%), pneumonitis (2%)

Table 1. *Cont.*

Clinical Trial	Agent	Tumor	N	Age Range	Outcomes	Reported G3/G4 AEs
Blay et al. (phase II) (AcSé) [88]	Pembro-lizumab	Advanced rare sarcoma	98	>18 (NA)	ORR 15.3%; DCR 49% (1 CR, 14 PR, 33 SD); mDR 8.2 m; mPFS 2.8 m; mOS 19.7 m	NA
Delyon et al. (phase II) [89]	Pembrolizumab	Classic/ endemic KS	17	NA	ORR 70.1%; DCR 88% (1 CR, 10 PR, 4 SD)	Reversible acute cardiac decompensation (6%)
D'Angelo et al. (phase II) (Alliance A091401) [90]	Nivo/ipi vs. ipi	Advanced sarcoma (BS and STS)	85	21–81	ORR 16% vs. 5%; mDR 6.2 mmPFS 4.1 m vs. 1.7 m mOS 14.3 m vs. 10.7 m	Pain (7% vs. 5%), thrombopenia (0% vs. 2%), pulmonary edema (2% vs. 0%), respiratory failure (5% vs. 5%), skin infection (2% vs. 0%), intestinal obstruction (2% vs. 2%), spinal fracture (0% vs. 2%), thrombo-embolic event (2% vs. 2%), urinary tract infection (7% vs. 2%), urinary obstruction (0% vs. 5%), fistula (2% vs. 0%), vomiting (0% vs. 2%)
Somaiah et al. (phase II) [91]	Durva-lumab + tremeli-mumab	Advanced sarcoma (BS and STS)	57	35–59	mPFS 2.8 m; mOS: 21.6 m; PFS at 12 m (all): 28%; PFS at 12 m (ASPS): 80% ORR (irRECIST) (all): 12%; ORR (irRECIST) (ASPS): 40%	Lipase increase (7%), pneumonitis (6%), colitis (6%), myocarditis (4%), autoimmune disorders (4%), endocrine disorders (2%), diarrhea (2%), gastrointestinal disorders (2%), lung infection (2%), ALP increase (2%), amylase increase (2%), myositis (2%)
Immune checkpoint inhibitors (ICI) + conventional CT						
Livingston et al. (phase II) [92]	Pembro + doxorubicin	Anthracy-cline naïve STS	30	NA	ORR 36.7%; DCR 80% (1 CR, 10 PR, 13 SD) mPFS 5.7 m; mOS 17 m PFS at 6 m: 44% PFS at 12 m: 62%	Neutropenia (36.7%), anemia (26.7%), febrile neutropenia (16.7%), arthralgia (13.3%), lymphopenia (13.3%), nausea (13.3%), fatigue (10.0%), hyponatremia (10.0%), vomiting (10.0%), lung infection (10.0%), muscle weakness (10.0%)
Pollack et al. (phase I/II) [93]	Pembro + doxorubicin	Anthracy-cline naïve STS	37	25–80	ORR 19%; DCR 78% (7 PR, 22 SD); mPFS 8.1 m; mOS 27.6 m PFS at 12 m: 27%	Neutropenia (27.0%), oral mucositis (8.1%), anemia (5.4%), febrile neutropenia (5.4%), lymphopenia (5.4%), ejection fraction decrease (5.4%), anorexia (5.4%), diarrhea (2.7%), hypothyroidism (2.7%), nausea (2.7%), weight loss (2.7%)
Toulmonde et al. (phase II) [79]	Pembro + mCP	Advanced STS	50	18–84	ORR 2%; DCR 34% (1 PR, 16 SD); PFS at 6 m: 0% (LMS, UPS), 11.1% (GIST), 14.3% (others)	Anemia (7.0%), fatigue (3.5%), lymphopenia (3.5%), oral mucositis (3.5%)
Gordon et al. (phase I/II) (SAINT) [94]	Ipi/nivo + trabectedin (trab)	Advanced STS	79	NA	ORR 25.3%; DCR 87.3% (6 CR *, 14 PR, 49 SD) mPFS 6.7 m; mOS 24.6 m * One surgical CR	ALT increase (25%), fatigue (8.7%), AST increase (8.7%), decreased neutrophil count (5.4%), anemia (4.6%)
Pink et al. (phase II) (NITRA-SARC) [95]	Nivo + trab	Advanced STS	25	NA	ORR 8%; DCR 48% (2 PR, 10 SD); mPFS 4 m	Leukopenia (47.2%), neutropenia (41.7%), thrombopenia (33.3%), increased ALT (30.6%), anemia (27.8%)
Smrke et al. (phase I) [96]	Pembro + gemcitabine	LMS, UPS	13	40–67	LMS (11): DCR 73% (8 SD) UPS (2): DCR 100% (2 PR)	NA

Table 1. *Cont.*

Clinical Trial	Agent	Tumor	N	Age Range	Outcomes	Reported G3/G4 AEs
Nathenson et al. (phase II) [97]	Pembro + eribulin	LMS cohort	19	48–80	ORR 5.3%; DCR 26.3% (1 PR, 5 SD); mPFS 11 w	Most commonly, neutropenia, anemia, weight loss, diarrhea, lipase/ALP increase
Wagner et al. (phase I/II) [98]	Avelumab + trab	LMS, LPS	23	NA	ORR 13%; DCR 56% (3 PR, 10 SD); mPFS 8.3 m	NA
Toulmonde et al. (phase Ib) [99]	Durva + trab	Advanced STS cohort	16	NA	ORR 7%; PFS at 6 m: 28.6%	NA
Immune checkpoint inhibitors (ICI) + tyrosine–kinase inhibitors/antiangiogenic drugs						
Martin-Broto et al. (phase I/II) (IMMU-NOSARC)	Nivo + suni-tinib	BS cohort (17 OST, 14 CS, 8 ES, 1 UPS) [100]	40	21–74	ORR 5%; DCR 60% (1 CR, 1 PR, 22 SD) mPFS 3.7 m; mOS 14.2 m	Neutropenia (10%), anemia (10%), AST/ALT increase (7.5%), fatigue (5%), oral mucositis (5%), hemorrhage (2.5%), dysphagia (2.5%), thrombopenia (2.5%), malaise (2.5%), thromboembolism (2.5%), pneumonitis (2.5%)
		STS cohort [101]	43	19–77	ORR 9.3%; DCR 69.3% (1 CR, 3 PR, 26 SD); mPFS 5.9 m; mOS not reached (follow up 6.1 m)	AST increase (11.8%), ALT increase (9.8%), neutropenia (9.8%), fatigue (5.9%), thrombopenia (3.9%), diarrhea (3.9%), renal failure (3.9%)
Wilky et al. (phase II) [102]	Pembro + axitinib	12 ASPS, 6 LMS, 5 UPS, 2 DDLPS, 8 others (2 BS)	33	27–62	ORR 25%; DCR 53.1% (8 PR, 9 SD); mPFS (all): 4.7 m; mOS (all): 18.7 m; mPFS (ASPS): 12.4 m; mPFS (others): 3.0 m.	Hypertension (15%), autoimmune toxic effects (15%), nausea (6%), ALT/AST increase (3%), oral mucositis (3%), diarrhea (3%), abdominal pain (3%), hemoptysis (3%), hyperlipidemia (3%)
Xie et al. (phase II) [103]	Camre-lizumab + apatinib	CT-refractory OST	43	11–43	ORR 20.1% (9 PR) mDR 6.2 m mPFS 6.2 m; mOS 11.3 m	Wound dehiscence (14%), ALP increase (9.3%), AST/ALT increase (9.3%), blood bilirubin increase (9.3%), hypertriglyceridemia (7.0%), anorexia (7.0%), weight loss (7.0%), pneumothorax (7.0%), platelet count decrease (4.7%), diarrhea (4.7%), PPS (4.7%), limb pain (4.7%), leukopenia (4.7%), rash (4.7%), oral mucositis (4.7%), hypertension (4.7%), toothache (4.7%), nausea (4.7%), non-cardiac chest pain (4.7%), hypothyroidism (2.3%), LDH increase (2.3%), proteinuria (2.3%), cough (2.3%), hemorrhage (2.3%), fatigue (2.3%), peripheral neuroinflammation (2.3%)
Kim et al. (phase II) [104]	Durva + pazopanib	Advanced STS	47	NA	ORR 28.3% (1 CR, 12 PR) mPFS 8.6 m	NA
Cousin et al. (phase II) [105]	Avelumab + regorafenib	Advanced STS	43	NA	ORR 9.3%; DCR 48.8% (4 PR, 17 SD); mDR 7.8 m; mPFS 1.8 m; mOS 15.1 m	PPS (12.2%), fatigue (10.2%), diarrhea (10.2%)
Kelly et al. (phase II) [106]	Pembro + epacadostat	Advanced STS	29	24–78	ORR 3%; DCR 48% (1 PR, 13 SD); mPFS 8 w; PFS at 24 w 27.9%; mOS NA	AST increase (10%), ALT increase (3%), anemia (3%), hypophosphatemia (3%), lipase increase (3%)
Schöffski et al. (phase Ib) [107]	Pembro + olaratumab	Advanced STS	28	NA	ORR 21.4%; DCR 53.5%; mDR 16.2 m; mPFS 2.7 m; mOS 14.8 m	NA

Table 1. *Cont.*

Clinical Trial	Agent	Tumor	N	Age Range	Outcomes	Reported G3/G4 AEs
		Immune checkpoint inhibitors (ICI) + other agents				
Kelly et al. (phase II) [108]	Pembro + T-VEC	Advanced STS	20	24–90	ORR 35%; DCR 70% (7 PR, 7 SD); mPFS 17.1 w	Pneumonitis (5%), fever (5%), anemia (5%)
Chawla et al. (phase II) [109]	Trabectedin + nivo + T-VEC	Advanced sarcoma	36	NA	ORR 8.3%; DCR 86.1% (3 PR, 27 SD); mPFS 5.5 m; mOS 9.0 m; OS at 6 m 73%	Anemia (33.3%), ALT increase (22.2%), fatigue (11.1%), thrombopenia (11.1%), neutropenia (11.1%)
D'Angelo et al. (phase I) [110]	Nivo + bempegal-desleukin	Advanced STS	84	13–80	ORR 10.4% (PR: 3/8 in AS, 1/4 in ASPS, 2/10 in UPS, 1/10 in LMS, 1/10 in CS); mDR 9.3 m	Anemia (10%), lipase increase (10%), amylase increase (7%), hypertension (7%), pain (8%), thromboembolic events (5%)
Somaiah et al. (phase I) [111]	LV305	STS (13 SS, 6 MRCL)	24	25–72	ORR 4.2%; DCR 62.5% (1 PR, 14 SD); mPFS 4.6 m; mOS 33 m	No G3/G4 adverse events
Rosenbaum et al. (phase I) [112]	Avelumab + DCC-3014	Advanced STS (7 LMS)	13	32–71	DCR 23% (3 SD); decreased circulating MDSCs in 5/7 patients (median 26.9%)	ALT/AST increase (31%), CK increase (23%), amylase/lipase increase (16%), anemia (8%), hypertension (8%)
Chawla et al. (phase I) [113]	Avelumab + SNK01	Advanced sarcoma	15	20–75	ORR 13.3%; DCR 33.3% (2 PR, 3 SD); mPFS 11.1 w	No G3/G3 adverse events related to SNK01
		Therapeutic vaccines				
Kawaguchi et al. (phase I) [114]	SYT-SSX vaccine	SS	21	21–69	DCR 50% (6 SD out of 12 assessable patients)	NA
Takahashi et al. (phase II) [115]	Peptide vaccine	Advanced sarcoma	20	23–75	DCR 30% (6 SD); mOS 9.6 m	NA
Pipia et al. (phase I/II) [116]	DC vaccine	Advanced STS	74	NA	Cohort 1 (adj/maint): mOS 24.4 m; cohort 2 (mono): mOS 14.2 m	NA
Chawla et al. (phase II) [117]	Atezo +/− CMB305	SS, MLS	89	NA	mPFS 2.6 m vs. 1.6 m mOS 18 m in both groups	4 G3/G4 events in each group (not specified)
		Adoptive cell therapy				
Robbins et al. (phase II) [118]	NY-ESO1 TCR	SS	18	19–65	ORR 61% (1 CR, 10 PR); PFS 3–47 m; estimated 3-y OS: 38%	No toxicities attributed to the transferred cells *
D'Angelo et al. (phase II) [119]	NY-ESO1 TCR	SS	45	NA	ORR 33% (1 CR, 14 PR); mPFS 8.6–22.4 w	No toxicities attributed to the transferred cells *
Van Tine et al. (phase I) [120]	MAGE-A4 TCR	SS	8	NA	ORR 50%; DCR 87.5% (4 PR, 3 SD)	No toxicities attributed to the transferred cells *
D'Angelo et al. (phase II) [121]	MAGE-A4 TCR	STS (23 SS,2 MLS)	25	24–73	ORR 40%; DCR 84% (2 CR, 8 PR, 11 SD)	CRS (5%) *
Ahmed et al. (phase I/II) [122]	Her2-CAR T cells	Advanced sarcoma (16 OST, 3 others)	19	7–29	DCR 23.5% (4 SD); mOS 10.3 m	Anemia (5.3%), muscle weakness (5.3%), back pain (5.3%)
Navai et al. (phase I) [123]	Her2-CAR T cells	Advanced sarcoma (5 OST, 5 others)	10	4–54	ORR 20%; DCR 50% (2 CR, 3 SD) (CR in 1 OST and 1 RMS)	No toxicities attributed to the transferred cells *

* All patients experienced transient neutropenia and thrombopenia induced by the lymphodepleting CT with fludarabine plus cyclophosphamide and the transient toxicities associated with IL-2 infusion.

The first immunotherapy trial with positive results in sarcoma was phase II SARC028 with anti-PD1 pembrolizumab, including a cohort for BS and a cohort for STS [54]. In the BS cohort, there were only two objective responses (one osteosarcoma and one CS). In the STS cohort, with a total of 40 patients, there were seven responses (17.5%): four in UPS (including one complete response), two in dedifferentiated liposarcoma (DDLPS) and one in synovial sarcoma. Two expansion cohorts in UPS and DDLPS reported an objective response rate (ORR) of 23% and 10%, respectively [124]. Keung et al. [125] showed that patients from SARC028 who responded to pembrolizumab had higher densities of activated CD8+ TILs and an increased pre-treatment percentage of TAMs expressing PDL1 compared to non-responders.

Liu et al. [126] confirmed the activity of pembrolizumab in advanced STS in a real-world study, reporting an overall ORR of 19.4%.

Following the results of SARC028, a phase II randomized trial is currently studying neoadjuvant pembrolizumab combined with RT in high-risk UPS or DDLPS of the extremities (SUC2C-SARC032) [127]. Another phase II trial (STEREOSARC) is evaluating concomitant RT with atezolizumab in oligometastatic STS [128]. There is a rationale for combining ICIs and RT since RT induces the release of TAA following immunogenic cell death, which activates TILs and leads to the recruitment of more effector cells to the TME [129]. The primary endpoint is the progression-free survival rate at 6 months. If the results of these studies are favorable, ICIs may also gain ground in the context of early-stage disease.

ICIs have also been tested in less common histologic subtypes. The phase II trial AcSé evaluates pembrolizumab in different cohorts of patients with rare cancers. A total of 98 patients were enrolled in the sarcoma cohort [88], including 34 with chordoma, 14 with ASPS, 11 with SMARCA4-deficient malignant rhabdoid tumor (SMRT), 8 with desmoplastic small round cell tumor (DSRCT) and 31 with other histotypes. There were seven objective responses in ASPS (50%), three in SMRT (27%), one in DSCRT (12.5%), three in chordoma (8.8%) and one in other histotypes (3.2%). The greatest rates of PFS at 12 months were observed in ASPS (35.7%), chordoma (31.2%) and SMRT (18.2%).

SMARCA4-deficient thoracic sarcoma is a newly described entity of thoracic sarcomas that is associated with a poor prognosis. Partial responses to PD1/PDL1 blockade have been reported in PDL1-positive SMARCA4-deficient thoracic sarcomas with pembrolizumab [130–132], nivolumab [133] and atezolizumab plus bevacizumab combined with CT [134,135]. Marcrom et al. [136] reported a complete response of mediastinal clear cell sarcoma to pembrolizumab combined with RT. Yu et al. [137] reported a major response to pembrolizumab in an adult patient with an undifferentiated embryonal sarcoma of the liver. In metastatic myxofibrosarcoma, partial responses have been reported with pembrolizumab [138] and atezolizumab plus temozolomide [139]. These data, together with the data from AcSé trial, suggest that ICIs may be useful for certain subtypes of rare sarcoma, especially when selected by immune biomarkers such as the PDL1 expression.

A phase II trial with pembrolizumab in 17 patients with classic/endemic Kaposi sarcoma (KS) (71% of them pretreated with CT) showed 12 objective responses (ORR 70.1%), with 2 complete responses (CR) and 4 partial responses (PR) [89]. Interestingly, the lack of PDL1 expression on tumor and immune cells was associated with worse outcomes. Tabata et al. [140] reported a maintained response to ipilimumab plus nivolumab in a HIV-negative KS.

Nivolumab with and without ipilimumab was studied by D'Angelo et al. [90] in a phase II trial (Alliance A091401) in 85 patients with advanced sarcoma (BS and STS) after at least one previous line of treatment. The primary endpoint was the response rate. There were six objective responses in the combination group (16% vs. 5% in the monotherapy group), with a median duration of response of 6.2 months. There was a significant benefit for the combination group in terms of median PFS (4.1 m vs. 1.7 m) and median overall survival (OS) (14.3 m vs. 10.7 m). Responses occurred in UPS, LMS, myxofibrosarcoma, and angiosarcoma. Grade 3–4 toxicity was higher in the combined treatment (14 vs. 7%). Zhou et al. [141] reported two cases of PDL1-negative STS (DDLPS and myxofibrosarcoma) with long-term responses to ipilimumab plus nivolumab.

In the first-line setting, a retrospective study of nivolumab with or without ipilimumab in PDL1-positive STS found a significant benefit for the combination group in terms of ORR (13% vs. 7%), median PFS (4.1 m vs. 2.2 m) and median OS (12.2 m vs. 9.2 m) [142]. A similar ORR (15%) was observed in a retrospective study with ipilimumab plus nivolumab for advanced STS [143]. The combination ipilimumab plus nivolumab is also being evaluated by a phase II trial in patients with pre-treated classic KS (NCT03219671), with a promising ORR of 50% in an interim analysis [144].

A recent phase II trial has analyzed the combination of anti-CTLA4 tremelimumab plus anti-PDL1 durvalumab in 57 patients with advanced STS and BS after at least one line of systemic treatment [91]. The most represented subtypes were ASPS (18%), LPS (10%) and vascular tumors (18%). The median PFS and OS for all subtypes were 2.8 months and 21.6 months, respectively. The median PFS at 12 weeks (the primary endpoint) was 49% (95% CI 36–61). Global ORR was 12%, though there were significant differences among histological subtypes, with the greatest benefit observed in the ASPS subgroup (ORR 40%, including 2 CR). The authors concluded that tremelimumab and durvalumab is an active treatment for advanced sarcoma.

Lewin et al. have reported two ASPS with a durable response to durvalumab alone or combined with tremelimumab, confirming their activity in this subtype [145]. A randomized phase II trial (MEDI-SARC) is currently comparing durvalumab plus tremelimumab to CT (doxorubicin) in naïve-treatment STS [146]. Anti-PDL1 avelumab has been tried in recurrent osteosarcoma with negative results [147].

The combined blockade of PD1 and LAG3 is another encouraging strategy. A basket phase II study of anti-PD1 spartalizumab plus anti-LAG3 LAG525 in advanced solid tumors included a cohort of 10 patients with advanced sarcoma, reporting a disease control rate (DCR) of 40% at 24 weeks [148]. The combination nivolumab plus anti-LAG3 relatlimab, clearly superior to nivolumab alone in untreated melanoma according to the phase III trial RELATIVITY-047 [149], is being evaluated in metastatic STS by a phase II study (NCT04095208).

According to these results, there is not enough evidence to support the use of ICIs as monotherapy in the first-line setting of advanced sarcoma since the previous trials included mainly pre-treated patients. However, given the scarcity of therapeutic alternatives for CT-refractory patients, the authors consider that treatment with ICIs might be considered after progression to standard CT, especially in patients with immunogenic subtypes of STS (classic/endemic KS, UPS, synovial sarcoma and dedifferentiated liposarcoma). Dual blockade of CTLA4 and PD(L)1 with nivolumab plus ipilimumab or durvalumab plus tremelimumab seems to offer higher response rates and may be preferable to single-agent immunotherapy in fit patients. Although imperfect, predictive biomarkers such as high TMB, high PDL1 expression or dense lymphocytic infiltration may help clinicians decide to use ICIs in this subgroup of patients.

Several additional trials with monotherapy or a combination of ICIs are currently ongoing [25], such as anti-PDL1 atezolizumab alone or plus bevacizumab in ASPS (NCT03141684) and ipilimumab plus nivolumab in classical Kaposi sarcoma (NCT03219671).

3.2. Combination of ICIs and Conventional Chemotherapy

The combination of immunotherapy with CT is a promising approach to enhance antitumor activity in sarcoma. The DNA damage caused by cytotoxic drugs results in cell death, with the subsequent release of DAMPs and proteins that work as 'danger signals', upregulating PD1 and enhancing the activity of effector lymphocytes [150].

A phase II trial that studied pembrolizumab in combination with doxorubicin in 30 patients with unresectable STS, with no previous anthracycline therapy, showed interesting results, with a DCR of 80% and a global ORR of 36.7% [92]. The subtypes with the highest ORR were UPS (4/4 patients), epithelioid angiosarcoma (1/1 patient), leiomyosarcoma (4/10 patients) and liposarcoma (2/7 patients). The median PFS and OS were 5.7 months and 17 months, respectively. In this study, PDL1 expression was associated with improved ORR. The authors concluded that the combination of pembrolizumab and doxorubicin has manageable toxicity and promising activity in advanced STS.

Pollack et al. [93] performed another phase I/II trial with pembrolizumab plus doxorubicin in 37 anthracycline-naïve patients with advanced STS, reporting an ORR of 19%, a median PFS of 8.1 months and a median OS of 27.6 months. Similarly, durable partial responses were observed in two of three patients with UPS and two of four patients with DDLPS. A retrospective study of pembrolizumab plus doxorubicin in 21 patients with STS showed a similar DCR (71.4%) [151]. As in the clinical trials, patients with UPS, synovial sarcoma and angiosarcoma showed higher response rates.

The combination of pembrolizumab plus metronomic cyclophosphamide showed limited activity in the phase II trial PEMBROSARC [79], with just 1 partial response among 50 advanced STS. Interestingly, the response was observed in the only case, a solitary fibrous tumor, with a PDL1 expression >10% in immune cells.

Italiano et al. [152] demonstrated that the presence of tertiary lymphoid structures (TLS) was a powerful predictor of response among the patients from the PEMBROSARC trial, with an ORR of 40% in TLS+ and 26.7% in TLS- tumors. New strategies are needed to induce TLS neogenesis and sensitize TLS-negative tumors to immunotherapy. In fact, a cohort of 20 TLS- patients from the PEMBROSARC study were treated with pembrolizumab, low-dose cyclophosphamide and intra-tumoral injection of the toll-like receptor 4 (TLR4) agonist G100, which would potentially enhance the immune response against a TLS-negative TME [153]. G100 seemed to modulate the TME, increasing TILS infiltration, though the ratio CD8+/FoxP3+ decreased in 11 out of 14 assessable cases, suggesting a predominant recruitment of T-regs-, a finding that may explain the modest clinical results (PFS of 11.8% at 6 months).

A phase I/II clinical trial (SAINT) has analyzed the dual CTLA4/PD1 blockade with ipilimumab plus nivolumab added to trabectedin in advanced STS, with encouraging results. Among 79 patients enrolled in the phase II study, there was a DCR of 87.3% (6 CR, 14 PR, 49 SD), with a median PFS of 6.7 months and a median OS of 24.6 months [94]. A retrospective study of nivolumab alone combined with trabectedin in 28 pre-treated STS found a DCR of 72.7% (4 PR and 12 SD among 22 assessable patients), with a median PFS at 6 months of 68.2% [154].

The results of trabectedin combined with dual immune blockade (ipi/nivo) seem better than its combination with nivolumab alone, which obtained modest results in the phase II trial NITRASARC (2 PR among 25 evaluable patients, with an mPFS of 4 months) [95].

The phase I trial GEMMK [96] is studying the combination of pembrolizumab plus gemcitabine in STS; among 13 patients included, there were 11 leiomyosarcomas (LMS) (with 8 SD and 3 PD) and 2 UPS (with 2 PR). A phase II trial of pembrolizumab plus eribulin in STS (NCT 03899805) has reported preliminary data from the LMS cohort (19 patients), with limited efficacy (ORR 5.3% and DCR 26.3% after 12 weeks) [97].

A phase I/II study has evaluated anti-PDL1 avelumab combined with trabectedin in 33 patients with advanced liposarcoma and LMS, with an ORR of 13%, a DCR of 56% (3 PR, 10 SD) and a median PFS of 8.3 months [98]. Trabectedin has also been combined with durvalumab by a phase Ib study (TRAMUNE) in 16 patients with STS, with an ORR of 7% and a 6-month PFS of 28.6% [99].

The heterogeneity in the selection of patients may explain the significant differences in the results of these studies with combination strategies. The best results seem to be obtained with the early use of ICI+CT, particularly with dual immune blockade (CTLA4/PD1). As monotherapy with ICIs, the combination of ICI+CT seems to obtain better response rates in certain histologic subtypes of STS (such as liposarcoma and UPS). In our opinion, there are promising data supporting the combination of nivo/ipi plus trabectedin for advanced STS in anthracycline-refractory patients, as well as pembro plus doxorubicin in anthracycline-naïve patients. Phase III trials comparing these schemes to the standard treatment would be helpful to confirm these results and incorporate the ICI+CT combinations into the first-line of systemic treatment.

Several ongoing phase I/II clinical trials are studying combinations of different cytotoxic drugs with pembrolizumab (NCT 03899805, NCT 03123276, NCT 04332874), nivolumab (NCT 04535713, NCT 03590210) and durvalumab (NCT 03802071) [25]. Their results will hopefully help us identify proper ICI/CT synergistic combinations with manageable toxicity.

3.3. Combination of ICIs and Tyrosine–Kinase Inhibitors

Tyrosine–kinase inhibitors (TKIs) are progressively gaining ground in CT-refractory sarcoma. In addition to blocking the immune-suppressive effect of VEGF [83], multi-target TKIs, especially lenvatinib and cabozantinib, seem to decrease the arrival of MDSCs/TAMs to the TME, and increase the infiltration of dendritic cells, NK cells and CD8+ lymphocytes [155]. This favorable immune-modulating effect provides a rationale for their combination with ICIs.

The combination of nivolumab and sunitinib in advanced sarcoma has been studied by the phase I/II trial IMMUNOSARC, including 40 patients with BS and 50 patients with STS. In the BS cohort, there was 1 CR (2.5%), 1 PR (2.5%) and 22 SD (55%) among 40 assessable patients, with a median PFS of 3.7 months [100]. In the STS cohort, there was 1 CR (2.3%), 3 PR (7%) and 26 SD (60%) among 43 evaluable patients, with a median PFS of 5.9 months [101].

A phase II clinical trial [102] has evaluated pembrolizumab plus axitinib in 33 patients with advanced sarcoma, mainly STS (12 ASPS, 6 LMS, 5 UPS, 2 DDLPS and 8 other subtypes, including 2 BS). 51% of patients had received prior treatment with TKIs, and 15% of them with immunotherapy. Among 32 assessable patients, the ORR was 25% (8 PR) and the median PFS was 4.7 months in the intention-to-treat analysis. Six of the eight patients with a PR had ASPS, a subgroup in which this combination seems especially active (ORR 50%), as confirmed by a post-hoc analysis, median PFS of 12.4 months in the ASPS subgroup. Dorman et al. [156] reported a case of ASPS with a durable (15 months) response to pembrolizumab plus axitinib, in consonance with the previous data. The rarity of ASPS, which comprise just 1% of STS, and its unique biology, which seems to make it especially responsive to ICI/TKI combinations, have led some authors to warn about the potential skew of unselected STS studies that include a high proportion of this subtype.

The phase II trial APFAO [103] evaluated anti-PD1 camrelizumab in combination with apatinib in 43 CT-refractory osteosarcomas, with an ORR of 20.9% and a median PFS of 6.2 months. A PDL1 tumor proportion score >5% and lung metastasis correlated with longer PFS. A retrospective study in 33 STS (8 ASPS, 5 LMS, 3 UPS and 17 others) showed an interesting activity of this combination added to local therapy with radiofrequency ablation or transarterial chemoembolization [157].

Durvalumab plus pazopanib have also shown encouraging activity in advanced STS in a phase II trial with 47 patients, with an ORR of 28.3% (1 CR and 12 PR), a median duration of response of 11 months and a median PFS of 8.6 months [104]. Combinations of pazopanib with PD1 inhibitors have been explored by retrospective studies. Paoluzzi et al. [158] treated 28 patients (24 STS and 4 BS) with nivolumab (plus pazopanib in 18 cases), with a DCR of 50% and 3 PR (one CS, one epithelioid sarcoma and one maxillary osteosarcoma, last two patients on pazopanib). Arora et al. [159] reported a sustained response to pembrolizumab plus pazopanib in a patient with advanced UPS.

The combination of avelumab plus regorafenib has been tried by a phase II basket trial in advanced solid tumors (REGOMUNE), with an STS cohort including 49 patients (22 LMS, 9 synovial sarcomas, 4 liposarcomas, 4 UPS and 10 other subtypes) [105]. The results were similar to other combinations of ICI + TKIs, with 4 PR among 43 evaluable patients and a DCR of 48.8%.

Given the known immune-suppressive role of IDO1 in sarcoma, explained above in detail, there is a rationale for combining ICIs with TKI against IDO1. A phase II trial with pembrolizumab plus IDO1 inhibitor epacadostat showed modest efficacy in 29 pre-treated advanced sarcomas (ORR 3%, DCR 48%) [106]. A phase I trial has shown promising results of pembrolizumab plus olaratumab PDGFR inhibitor in 28 patients with advanced STS (ORR 21.4%, DCR 53.5%) [107].

In our opinion, the combination of ICI/TKI (nivolumab/sunitinib, pembrolizumab/axitinib, durvalumab/pazopanib or avelumab/regorafenib) should be considered in patients with advanced STS after progression to standard CT. The authors believe that further research is required to evaluate the benefits of these combinations compared to ICI or TKI monotherapy. It would also be interesting to evaluate this strategy in the first-line setting of patients *unfit* for anthracycline-based CT.

Several ongoing trials will assess the combination of ICIs with TKIs and other antiangiogenic drugs [25], including lenvatinib plus pembrolizumab (NCT 04784247), cabozantinib plus nivolumab and ipilimumab (NCT 04551430), anlotinib plus toripalimab (NCT 04172805), olaparib or cediranib plus durvalumab (NCT 03851614), epacadostat plus pembrolizumab (NCT 03414229), bevacizumab plus atezolizumab (NCT 03141684), tivozanib plus atezolizumab (NCT 05000294), dasatinib plus ipilimumab (NCT 01643278) and axitinib plus avelumab (NCT04258956).

3.4. Combination of ICIs and Other Agents

Various studies have explored the combination of ICIs with other immunomodulatory therapies. One of the most promising agents is talimogene laherparepvec (T-VEC), a modified immune-enhanced herpes simplex virus (HSV) type 1, engineered for intratumoral injection. It contains the coding sequence for granulocyte–macrophage colony-stimulating factor (GM-CSF) and proinflammatory cytokines, inducing an immune response against tumor cells. Preoperative RT plus T-VEC has been studied in 30 locally advanced STS, with modest results [160].

A phase II trial has evaluated T-VEC plus pembrolizumab in 20 cases of advanced sarcoma (after at least one standard therapy), with promising results (7 PR, with a median duration of response of 56 weeks, and 7 SD) and an acceptable safety profile. The most represented subtypes were LMS (25%), cutaneous angiosarcoma (15%) and UPS (10%) [108]. The mean TIL score was higher in responsive patients. A phase II trial has analyzed T-VEC plus nivolumab and trabectedin in 36 advanced sarcomas, including chordoma and desmoid tumor heavily pre-treated patients, with a median of 4 previous lines, with encouraging results: 3 PR, 27 SD (DCR 86.1%) and a median PFS of 5.5 months (being the median PFS of the immediately prior therapy 2.0 months) [109].

Bempegaldesleukin (NKTR-214), an interleukin-2 (IL-2) agonist, increases TIL infiltration and has shown encouraging activity in several refractory tumors [161], setting a rationale for its addition to PD1 inhibitors. A pilot study combining this agent with nivolumab in 84 patients showed positive results in angiosarcoma (3 PR out of 8), ASPS (1 PR out of 4) and UPS (2 PR out of 10) [110]. There was a correlation between the ORR, CD8+ TIL infiltration and PDL1 expression. Interestingly, the reduced expression of the Hedgehog signaling pathway was associated with better clinical outcomes, suggesting that the Hedgehog pathway enhances immune-suppressive mechanisms as it has been found in breast cancer [162] and skin basal cell carcinoma [163] and setting a rationale for combining immunotherapy with Hedgehog inhibitors.

Oleclumab is a monoclonal antibody that selectively binds and blocks the activity of CD73, a key enzyme for the extracellular generation of adenosine, which is a relevant biochemical component of the immunosuppressive TME [164]. An ongoing phase II study is evaluating the combination of durvalumab and oleclumab in specific sarcoma subtypes selected according to CD73 staining on the TME cells [165]. A basket phase II trial (CAIRE) is evaluating durvalumab plus tazemetostat, a molecule that inhibits EZH2, which leads to a functional alteration of Tregs and favors an effector-like profile in advanced solid tumors, including a cohort for STS [166].

Another approach is the modulation of immune cells in vivo. LV305 is a modified lentivirus-based vector designed to selectively transduce dendritic cells and promote the expression of NY-ESO1, unleashing an immune response against NY-ESO1-expressing cells. Somaiah et al. [111] tried LV305 in 24 patients (13 SS, 6 MRCL and 5 other subtypes), reporting 1 PR (SS) and 14 SD. NY-ESO1 expression was >75% in 67% of the subjects. A phase I study has evaluated avelumab plus SNK01, a therapy based on autologous modified NK cells, with enhanced cytotoxicity, in 15 patients with advanced sarcoma (6 LMS, 2 osteosarcoma and 7 other subtypes) [113], with 2 PR and 3 SD.

Other strategies to modulate the immune-suppressive TME of sarcoma have been explored in preclinical studies and early trials. Colony-stimulating factor 1 receptor (CSF1R) signaling regulates the infiltration of MDSCs and promotes their differentiation towards an M2 phenotype [167]. A phase Ib study has explored the combination of durvalumab with DCC-3014 (vimseltinib), a selective inhibitor of CSF1R, in 13 patients with advanced STS (7 leiomyosarcoma, 2 UPS, 2 DDLPS, 1 synovial sarcoma and 1 liposarcoma) [112]. There was a median decrease of 26.9% in circulating MDSCs, with disease stabilization in three patients.

3.5. Therapeutic Vaccines

Cancer vaccines are usually based on the exogenous administration of selected tumor antigens combined with adjuvants to induce the activity of APCs, mainly dendritic cells, aiming to stimulate the adaptive immune system against cancer cells [168].

Sato et al. [169] demonstrated that peptides derived from *SYT-SSX* fusion genes, resultant from chromosomal translocation t(X:18) specific to synovial sarcoma (SS), are recognized by circulating CD8+ T-cells in HLA-A24+ patients, and induce tumor-specific cytotoxic responses. They developed an *SYT-SSX* peptide vaccine and tried it on six patients with advanced SS, with no response [114]. However, they performed a second study adding interferon-alfa to the vaccine in 12 patients, observing 6 SD and a transient response in 1 patient [170].

Takahashi et al. [115] conducted a phase II clinical trial with a personalized vaccine in 20 patients with refractory BS and STS. A maximum of four HLA-matched peptides were selected, based on their high peptide-specific IgG responses in pre-vaccination plasma, and administered to each patient weekly for 6 weeks and each 2 weeks thereafter; six patients achieved SD, including one minor response in a malignant fibrous histiocytoma, and one durable SD (33 months) in a mediastinal SS. The authors concluded that this treatment could be feasible for the vast majority of refractory sarcoma patients, with high rates of immunological responses.

Neoantigens such as cancer–testis antigens (CTA), aberrantly expressed in a high percentage of high-grade sarcoma, mainly MAGE and NY-ESO, as explained above in detail, might also be used to design personalized vaccines [32]. A phase I/II trial has evaluated an autologous dendritic-cell vaccine based on CTA (CaTeVac) in 74 patients with advanced STS, with a cohort receiving the vaccine as adjuvant or maintenance treatment after the first or second line of systemic therapy and another cohort receiving it in monotherapy after at least one CT line [116]. Median OS was 24.1 months in the first cohort and 14.2 months in the second one, suggesting a positive impact on overall survival. A phase I study tried a dendritic cell vaccine-targeting MAGE-A1, MAGE-A3 and NY-ESO1 combined with decitabine in 10 children with neuroblastoma, Ewing's sarcoma, osteosarcoma and rhabdomyosarcoma. Six of nine patients developed a response to MAGE-A1, MAGE-A3 or NY-ESO-1 peptides post-vaccine, concluding that the chemoimmunotherapy approach using DAC/DC-CT vaccine is feasible [171].

More recently, Chawla et al. [117] have published a phase II randomized study of CMB305 and atezolizumab compared with atezolizumab alone in STS expressing NY-ESO1. CMB305 is a vaccination regimen designed to prime the CD8+ T-cell population specific for NY-ESO1 and then use a TLR4 agonist to unleash the antitumor immune response. The study recruited 89 patients with SS and myxoid liposarcoma (MLS). Though the PFS increase was not significant, the patients treated with the combination acquired a higher rate of NY-ESO1-specific T-cells, which was associated with longer OS in a post hoc analysis.

In addition to vaccines based on specific peptides, a potential approach to induce tumor recognition is the production of vaccines derived from whole tumor cells combined with immune-enhancing adjuvants (such as IFN-γ and GM-CSF), with interesting preliminary data in a cohort of STS [172].

3.6. Adoptive Cell Therapy

Adoptive cell therapy (ACT) encompasses several strategies to improve the activity of autologous T cells, which are obtained through leukapheresis and genetically engineered to overcome tumor immune evasion. Modified T cells are reinfused into the patient, usually in combination with adjuvant IL-2, after lymphodepleting CT (commonly fludarabine plus cyclophosphamide). ACT includes chimeric antigen receptor (CAR) T cell therapy, engineered TCRs and TIL therapy [173].

3.6.1. Engineered TCR

In sarcoma, the aberrant expression of CTA not present in normal cells makes TCR therapy particularly attractive since T cells can be equipped with newly engineered TCRs transduced through retroviral vectors that enable them to target these specific neoantigens.

Robbins et al. [118] conducted a phase II trial with letetresgene autoleucel (Lete-cel), a therapy based on T cells transduced with a TCR against NY-ESO1, plus adjuvant IL-2- in HLA-A02 patients with metastatic melanoma and synovial sarcoma. Among 18 patients with NY-ESO1+ SS, 11 objective responses were documented (61%), including 1 CR and 10 PR (ranging from 3 to 47 months). The estimated overall 3-year OS rate was 38%. The authors concluded that the adoptive transfer of autologous T cells transduced with a retrovirus encoding a TCR against an HLA-A*0201 restricted NY-ESO-1 epitope can be an effective therapy for synovial cell sarcomas. Considering that all the patients were heavily pre-treated and refractory to standard CT, the authors consider it improbable that the lymphodepleting therapy could significantly contribute to the clinical responses.

More recently, Lete-cel has been evaluated in 45 patients with advanced SS after standard first-line CT, enrolled in 4 cohorts according to NY-ESO1 tumor expression [119]. Objective responses were documented in all cohorts, with a total of 1 CR and 14 PR (ORR 20–50%). The median PFS ranged from 8.6 to 22.4 weeks between cohorts. A post-hoc analysis revealed that responders had higher pre-infusion levels of IL-15, which may be used as a predictive biomarker [174]. NY-ESO1-targeted TCR therapy is also being evaluated in myxoid/round cell liposarcoma (MRCL) [175].

A first-in-human study with T cells equipped with afamitresgene autoleucel (Afami-cel), a genetically engineered autologous specific peptide enhanced affinity receptor (SPEAR) targeting MAGE-A4, reported 4 PR and 3 SD among eight assessed patients with advanced SS [120]. Following these results, Afami-cel was evaluated by a phase II trial (SPEARHEAD-1) in 32 eligible patients (HLA-A02) with MAGE-A4-expressing STS (87.5% SS, 12,5% MLS) [121]. Among 25 evaluable subjects (23 SS and 2 MLS), there were 2 CR, 8 PR and 11 SD (DCR 84%). The safety profile was favorable, with mainly low-grade cytokine release syndrome (CRS) and reversible hematologic toxicities due to lymphodepleting CT. A pooled efficacy analysis of phase I and II trials with Afami-cel revealed that the baseline tumor burden, prior systemic treatment history and MAGE-04 expression levels are potential predictors of response [176].

3.6.2. CAR T Cells

The use of CAR T cells has also been explored in sarcoma, with the advantage over TCR therapy of not being restricted to HLA-A02 carriers. In contrast to hematologic malignancies, the efficacy of CAR T cells in solid tumors is limited by their intense antigenic heterogeneity derived from their polyclonal expansion and accumulative mutations, which makes it hard to find homogeneously expressed targets, particularly without unacceptable off-tumor toxicity [177].

Human epidermal growth factor receptor 2 (Her2) is the most studied target for CAR T cells in sarcoma. A phase I/II trial evaluated Her2-CAR T cells in 19 Her2+ sarcoma (16 of them osteosarcoma), with modest activity (4 SD ranging from 12 weeks to 14 months). Her2-CAR T cells persisted for at least 6 weeks in seven of the nine evaluable patients. Three of these patients had their tumors removed, with one showing $\geq$90% necrosis. The median overall survival was 10.3 months (ranging 5.1 to 29.1 months) [122].

In a phase I trial with Her2-CAR T cells in 10 Her2+ sarcoma, there were 2 long-term CR, 1 in osteosarcoma and 1 in rhabdomyosarcoma, added to 3 SD [123]. A phase I study is currently using the combination of Her2-CAR T therapy with PD1 blockade pembrolizumab or nivolumab in advanced Her2+ sarcoma (NCT 049955003).

Other phase I trials are currently evaluating CAR T cells targeting epidermal growth factor receptor (EGFR) (NCT 03618381) and tumor antigen GD2 (NCT 04539366, NCT 03635632, NCT 02107963, NCT 03721068) [177]. Preclinical studies have also suggested a promising activity of CAR T cells targeting the type I insulin-like growth factor receptor (IGF1R) and the tyrosine–kinase-like orphan receptor 1 (ROR1), which are highly expressed in sarcoma cell lines [178]. Other potential targets for CAR T therapy in sarcoma are CD44v6 [179] and NK cell activating receptor group 2-member D ligand (NKG2DL) [180,181].

3.6.3. TIL Therapy

TIL therapy is based on the extraction of tumor-infiltrating lymphocytes for ex vivo expansion and reinfusion to the patient, after lymphodepleting conditioning, in combination with immune-enhancing adjuvants such as IL-2. It has an exciting potential, being the only ACT-using cells with multiple TCR clones able to cover the antigenic heterogeneity of solid tumors in contrast to engineered TCRs and CAR T cells, which target specific antigens [182]. Up to date, TIL therapy has been mainly developed in melanoma with a positive phase III trial versus ipilimumab in the first-line setting [183] and has promising results in cervical cancer [184–186], non-small cell lung cancer [187], ovarian cancer [188], colorectal cancer [189], breast cancer [190] and cholangiocarcinoma [191].

Mullinax et al. [192] demonstrated the feasibility of TIL therapy in STS. They successfully propagated TILs from 70 STS surgical specimens, cocultured them with tumor cells, and expanded them using a rapid expansion protocol, observing that nearly all specimens generated TILs, mainly CD3+, which were responsive to the autologous tumor. Ko et al. [193] published another preclinical study that suggests that certain sarcoma subtypes can potentially yield an appropriate number of cells for TIL therapy.

Zhou et al. [194] conducted a retrospective study of 60 patients with CT-resistant metastatic osteosarcoma who were treated with TIL therapy plus nivolumab. Of the patients, 83.3% had lung metastasis, and 83.3% had presented a poor response to neoadjuvant CT. The results were encouraging, with an ORR of 36.7% (2 CR and 20 PR), a DCR of 80% and a median PFS of 5.8 months. An infusion of $\geq 5 \times 10$ [9] TIL cells, a percentage of CD8+ TIL $\geq 60\%$ and a percentage of CD4+/FoxP3+ TIL <20% were significantly associated with response to treatment. Overall, OS was 23.7 months in responders versus 8.7 months in non-responders ($p < 0.0001$).

Despite these interesting preclinical and retrospective data, the use of TIL therapy in sarcoma is not endorsed by any clinical trial to date. A phase I study of TILs in STS (NCT 04052334) and two phase II basket trials, one with a cohort for carcinosarcoma (NCT 03610490) and another with a cohort for STS (NCT 03935893), are ongoing [177]. Further research is required to understand how to overcome the challenges still faced by TIL therapy, such as the negative impact of the immunosuppressive TME or its application in 'cold' tumors, including many sarcoma subtypes.

4. Conclusions

Immunotherapy is progressively acquiring a role in the treatment of advanced sarcoma, though the biological heterogeneity among histologic subtypes impacts their clinical response. Globally, soft tissue sarcomas (STS) are more responsive than bone sarcomas (BS). Undifferentiated pleomorphic (UPS), alveolar soft part (ASPS), synovial and Kaposi sarcomas seem especially immunogenic, followed by some liposarcoma and leiomyosarcoma. The presence of tertiary lymphoid structures (TLS), a high density of infiltrating CD8+ T cells and a high PDL1 expression in cells of the TME have been identified as prognostic and predictive biomarkers.

Immune checkpoint inhibitors (ICI) have poor results in monotherapy, except for certain subtypes. The SARC028 trial is the most relevant study with an ICI alone (pembrolizumab), with an ORR of 40% in UPS and 20% in liposarcoma and chondrosarcoma but <10% in the other subtypes. Kaposi and other rare forms of sarcoma, such as chordoma and SMARCA4-deficient rhabdoid tumors, might also benefit from PD1 inhibitors. Dual blockade with ipilimumab plus nivolumab showed superiority compared to single immunotherapy (Alliance A091401), and durvalumab plus tremelimumab seems especially active in ASPS.

Multiple studies have evaluated ICI combined with conventional CT, trying to achieve a synergistic effect. Pembrolizumab plus doxorubicin has positive results in anthracycline-naïve STS, and some other combinations, ipi/nivo plus trabectedin, pembro plus gemcitabine and avelumab plus trabectedin, have promising results in CT-refractory STS. The IMMUNOSARC trial is the main study combining ICI with tyrosine–kinase inhibitors (nivolumab plus sunitinib), reaching a DCR >60% both in STS and BS, though other combinations have shown similar outcomes in phase I and II trials. ICI has also been combined with immunomodulatory agents, with promising results of oncolytic viral therapy (T-VEC) added to PD1 blockade.

Going beyond checkpoint inhibitors, therapeutic vaccines and adoptive cell therapy are incorporated into the arsenal of immunotherapy, with encouraging results. Engineered TCR targeting aberrant neoantigens NY-ESO1 (Lete-cel) and MAGE-A4 (Afami-cel) have shown ORR 40–60% in synovial sarcoma, with a favorable safety profile. Her2-targeted CAR-T cells and TIL therapy have promising preclinical and retrospective data in advanced sarcoma, with the potential benefit of their combination with PD1 inhibitors. Further research is still needed to overcome the theoretical and practical obstacles faced by adoptive cell therapy, especially in such aggressive tumors with a powerful immunosuppressive microenvironment.

Author Contributions: V.A.—conception of the work, writing and review of literature; J.P., J.C., D.I.R., M.S.R., P.G., P.P.d.A., J.C.C., C.G.d.Q. and C.G.—contribution to writing and review of literature; M.L.V., M.Á.V.—provision of study material and supervision of the work. All authors have read and agreed to the published version of the manuscript.

Funding: This research received no external funding.

Conflicts of Interest: The authors declare no conflict of interest.

References

1. Siegel, R.L.; Miller, K.D.; Fuchs, H.E.; Jemal, A. Cancer Statistics, 2021. *CA Cancer J. Clin.* **2021**, *71*, 7–33. [CrossRef]
2. Brennan, M.F.; Antonescu, C.R.; Moraco, N.; Singer, S. Lessons learned from the study of 10,000 patients with soft tissue sarcoma. *Ann. Surg.* **2014**, *260*, 416–421. [CrossRef]
3. Judson, I.; Verweij, J.; Gelderblom, H.; Hartmann, J.T.; Schöffski, P.; Blay, J.-Y.; Kerst, J.M.; Sufliarsky, J.; Whelan, J.; Hohenberger, P.; et al. Doxorubicin alone versus intensified doxorubicin plus ifosfamide for first-line treatment of advanced or metastatic soft-tissue sarcoma: A randomised controlled phase 3 trial. *Lancet Oncol.* **2014**, *15*, 415–423. [CrossRef] [PubMed]
4. Seddon, B.; Strauss, S.J.; Whelan, J.; Leahy, M.; Woll, P.J.; Cowie, F.; Rothermundt, C.; Wood, Z.; Benson, C.; Ali, N.; et al. Gemcitabine and docetaxel versus doxorubicin as first-line treatment in previously untreated advanced unresectable or metastatic soft-tissue sarcomas (GeDDiS): A randomised controlled phase 3 trial. *Lancet Oncol.* **2017**, *18*, 1397–1410. [CrossRef] [PubMed]
5. Demetri, G.D.; von Mehren, M.; Jones, R.L.; Hensley, M.L.; Schuetze, S.M.; Staddon, A.; Milhem, M.; Elias, A.; Ganjoo, K.; Tawbi, H.; et al. Efficacy and Safety of Trabectedin or Dacarbazine for Metastatic Liposarcoma or Leiomyosarcoma After Failure of Conventional Chemotherapy: Results of a Phase III Randomized Multicenter Clinical Trial. *J. Clin. Oncol.* **2016**, *34*, 786–793. [CrossRef] [PubMed]
6. Kawai, A.; Araki, N.; Sugiura, H.; Ueda, T.; Yonemoto, T.; Takahashi, M.; Morioka, H.; Hiraga, H.; Hiruma, T.; Kunisada, T.; et al. Trabectedin monotherapy after standard chemotherapy versus best supportive care in patients with advanced, translocation-related sarcoma: A randomised, open-label, phase 2 study. *Lancet Oncol.* **2015**, *16*, 406–416. [CrossRef] [PubMed]
7. Schöffski, P.; Chawla, S.; Maki, R.G.; Italiano, A.; Gelderblom, H.; Choy, E.; Grignani, G.; Camargo, V.; Bauer, S.; Rha, S.Y.; et al. Eribulin versus dacarbazine in previously treated patients with advanced liposarcoma or leiomyosarcoma: A randomised, open-label, multicentre, phase 3 trial. *Lancet* **2016**, *387*, 1629–1637. [CrossRef]

8. García-Del-Muro, X.; López-Pousa, A.L.; Maurel, J.; Martín, J.; Martínez-Trufero, J.; Casado, A.; Gómez-España, A.; Fra, J.; Cruz, J.; Poveda, A.; et al. Randomized Phase II Study Comparing Gemcitabine Plus Dacarbazine Versus Dacarbazine Alone in Patients With Previously Treated Soft Tissue Sarcoma: A Spanish Group for Research on Sarcomas Study. *J. Clin. Oncol.* **2011**, *29*, 2528–2533. [CrossRef]
9. Van Der Graaf, W.T.; Blay, J.Y.; Chawla, S.P.; Kim, D.W.; Bui-Nguyen, B.; Casali, P.G.; Aglietta, M.; Staddon, A.P.; Beppu, Y.; Cesne, A.L.; et al. Faculty Opinions recommendation of Pazopanib for metastatic soft-tissue sarcoma (PALETTE): A randomised, double-blind, placebo-controlled phase 3 trial. *Lancet Lond. E* **2012**, *379*, 1879–1886. [CrossRef]
10. Butrynski, J.E.; D'Adamo, D.R.; Hornick, J.L.; Cin, P.D.; Antonescu, C.R.; Jhanwar, S.C.; Ladanyi, M.; Capelletti, M.; Rodig, S.J.; Ramaiya, N.; et al. Crizotinib in *ALK*-Rearranged Inflammatory Myofibroblastic Tumor. *N. Engl. J. Med.* **2010**, *363*, 1727–1733. [CrossRef]
11. Judson, I.; Morden, J.P.; Kilburn, L.; Leahy, M.; Benson, C.; Bhadri, V.; Campbell-Hewson, Q.; Cubedo, R.; Dangoor, A.; Fox, L.; et al. Cediranib in patients with alveolar soft-part sarcoma (CASPS): A double-blind, placebo-controlled, randomised, phase 2 trial. *Lancet Oncol.* **2019**, *20*, 1023–1034. [CrossRef] [PubMed]
12. Marina, N.M.; Smeland, S.; Bielack, S.S.; Bernstein, M.; Jovic, G.; Krailo, M.D.; Hook, J.M.; Arndt, C.; van den Berg, H.; Brennan, B.; et al. Comparison of MAPIE versus MAP in patients with a poor response to preoperative chemotherapy for newly diagnosed high-grade osteosarcoma (EURAMOS-1): An open-label, international, randomised controlled trial. *Lancet Oncol.* **2016**, *17*, 1396–1408. [CrossRef] [PubMed]
13. Palmerini, E.; Torricelli, E.; Cascinu, S.; Pierini, M.; De Paolis, M.; Donati, D.; Ferrari, S. Is there a role for chemotherapy after local relapse in high-grade osteosarcoma? *Pediatr. Blood Cancer* **2019**, *66*, e27792. [CrossRef] [PubMed]
14. Womer, R.B.; West, D.C.; Krailo, M.D.; Dickman, P.S.; Pawel, B.R.; Grier, H.E.; Marcus, K.; Sailer, S.; Healey, J.H.; Dormans, J.P.; et al. Randomized Controlled Trial of Interval-Compressed Chemotherapy for the Treatment of Localized Ewing Sarcoma: A Report From the Children's Oncology Group. *J. Clin. Oncol.* **2012**, *30*, 4148–4154. [CrossRef]
15. Felix, A.; Berlanga, P.; Toulmonde, M.; Landman-Parker, J.; Dumont, S.; Vassal, G.; Gaspar, N. Systematic review of phase-I/II trials enrolling refractory and recurrent Ewing sarcoma: Actual knowledge and future directions to optimize the research. *Cancer Med.* **2021**, *10*, 1589–1604. [CrossRef]
16. Albarrán, V.; Villamayor, M.L.; Chamorro, J.; Rosero, D.I.; Pozas, J.; Román, M.S.; Calvo, J.C.; de Aguado, P.P.; Moreno, J.; Guerrero, P.; et al. Receptor Tyrosine Kinase Inhibitors for the Treatment of Recurrent and Unresectable Bone Sarcomas. *Int. J. Mol. Sci.* **2022**, *23*, 13784. [CrossRef]
17. Ferrari, S.; Briccoli, A.; Mercuri, M.; Bertoni, F.; Picci, P.; Tienghi, A.; Del Prever, A.B.; Fagioli, F.; Comandone, A.; Bacci, G. Postrelapse Survival in Osteosarcoma of the Extremities: Prognostic Factors for Long-Term Survival. *J. Clin. Oncol.* **2003**, *21*, 710–715. [CrossRef]
18. Cotterill, S.; Ahrens, S.; Paulussen, M.; Jürgens, H.; Voûte, P.; Gadner, H.; Craft, A. Prognostic Factors in Ewing's Tumor of Bone: Analysis of 975 Patients From the European Intergroup Cooperative Ewing's Sarcoma Study Group. *J. Clin. Oncol.* **2000**, *18*, 3108–3114. [CrossRef]
19. Coley, W.B. II Contribution to the Knowledge of Sarcoma. *Ann. Surg.* **1891**, *14*, 199–220. [CrossRef]
20. Rosenberg, S.A. IL-2: The First Effective Immunotherapy for Human Cancer. *J. Immunol.* **2014**, *192*, 5451–5458. [CrossRef]
21. Fyfe, G.A.; Fisher, R.I.; Rosenberg, S.A.; Sznol, M.; Parkinson, D.R.; Louie, A.C. Long-term response data for 255 patients with met-astatic renal cell carcinoma treated with high-dose recombinant interleukin-2 therapy. *J. Clin. Oncol. Oncol. Oncol.* **1996**, *14*, 2410–2411. [CrossRef] [PubMed]
22. Schwinger, W.; Klass, V.; Benesch, M.; Lackner, H.; Dornbusch, H.J.; Sovinz, P.; Moser, A.; Schwantzer, G.; Urban, C. Feasibility of high-dose interleukin-2 in heavily pretreated pediatric cancer patients. *Ann. Oncol.* **2005**, *16*, 1199–1206. [CrossRef] [PubMed]
23. Baluna, R.; Vitetta, E.S. Vascular leak syndrome: A side effect of immunotherapy. *Immunopharmacology* **1997**, *37*, 117–132. [CrossRef] [PubMed]
24. Saerens, M.; Brusselaers, N.; Rottey, S.; Decruyenaere, A.; Creytens, D.; Lapeire, L. Immune checkpoint inhibitors in treatment of soft-tissue sarcoma: A systematic review and meta-analysis. *Eur. J. Cancer* **2021**, *152*, 165–182. [CrossRef]
25. Dajsakdipon, T.; Siripoon, T.; Ngamphaiboon, N.; Ativitavas, T.; Dejthevaporn, T. Immunotherapy and Biomarkers in Sarcoma. *Curr. Treat. Options Oncol.* **2022**, *23*, 415–438. [CrossRef]
26. Chen, D.S.; Mellman, I. Oncology Meets Immunology: The Cancer-Immunity Cycle. *Immunity* **2013**, *39*, 1–10. [CrossRef]
27. Rosin, D.L.; Okusa, M.D. Dangers within: DAMP responses to damage and cell death in kidney disease. *J. Am. Soc. Nephrol. JASN* **2011**, *22*, 416–425. [CrossRef]
28. Fumet, J.-D.; Truntzer, C.; Yarchoan, M.; Ghiringhelli, F. Tumour mutational burden as a biomarker for immunotherapy: Current data and emerging concepts. *Eur. J. Cancer* **2020**, *131*, 40–50. [CrossRef]
29. Chalmers, Z.R.; Connelly, C.F.; Fabrizio, D.; Gay, L.; Ali, S.M.; Ennis, R.; Schrock, A.; Campbell, B.; Shlien, A.; Chmielecki, J.; et al. Analysis of 100,000 human cancer genomes reveals the landscape of tumor mutational burden. *Genome Med.* **2017**, *9*, 34. [CrossRef]
30. Bocchia, M.; Wentworth, P.A.; Southwood, S.; Sidney, J.; McGraw, K.; Scheinberg, D.A.; Sette, A. Specific binding of leukemia oncogene fusion protein peptides to HLA class I molecules. *Blood* **1995**, *85*, 2680–2684. [CrossRef]

31. Worley, B.S.; Broeke, L.T.V.D.; Goletz, T.J.; Pendleton, C.D.; Daschbach, E.M.; Thomas, E.K.; Marincola, F.M.; Helman, L.J.; Berzofsky, J.A. Antigenicity of fusion proteins from sarcoma-associated chromosomal translocations. *Cancer Res.* **2001**, *61*, 6868–6875. [PubMed]
32. Kakimoto, T.; Matsumine, A.; Kageyama, S.; Asanuma, K.; Matsubara, T.; Nakamura, T.; Sudo, A. Immunohistochemical expression and clinicopathological assessment of the cancer testis antigens NY-ESO-1 and MAGE-A4 in high-grade soft-tissue sarcoma. *Oncol. Lett.* **2019**, *17*, 3937–3943. [CrossRef] [PubMed]
33. Dustin, M.L. The immunological synapse. *Cancer Immunol. Res.* **2014**, *2*, 1023–1033. [CrossRef] [PubMed]
34. Zhou, Z.; Lafleur, E.A.; Koshkina, N.V.; Worth, L.L.; Lester, M.S.; Kleinerman, E.S. Interleukin-12 Up-Regulates Fas Expression in Human Osteosarcoma and Ewing's Sarcoma Cells by Enhancing Its Promoter Activity. *Mol. Cancer Res.* **2005**, *3*, 685–692. [CrossRef] [PubMed]
35. Marchisone, C.; Benelli, R.; Albini, A.; Santi, L.; Noonan, D.M. Inhibition of Angiogenesis by Type I Interferons in Models of Kaposi'S Sarcoma. *Int. J. Biol. Markers* **1999**, *14*, 257–262. [CrossRef]
36. Taylor, K.L.; Oates, R.K.; Grane, R.; Leaman, D.W.; Borden, E.C.; Lindner, D.J. IFN-alpha1,8 inhibits tumor-induced angiogenesis in murine angiosarcomas. *J. Interferon Cytokine Res.* **2006**, *26*, 353–361. [CrossRef]
37. Chen, L.; Flies, D.B. Molecular mechanisms of T cell co-stimulation and co-inhibition. *Nat. Rev. Immunol.* **2013**, *13*, 227–242. [CrossRef]
38. Dancsok, A.R.; Setsu, N.; Gao, D.; Blay, J.-Y.; Thomas, D.; Maki, R.G.; Nielsen, T.O.; Demicco, E.G. Expression of lymphocyte immunoregulatory biomarkers in bone and soft-tissue sarcomas. *Mod. Pathol.* **2019**, *32*, 1772–1785. [CrossRef]
39. Slaney, C.Y.; Kershaw, M.H.; Darcy, P.K. Trafficking of T Cells into Tumors. *Cancer Res.* **2014**, *74*, 7168–7174. [CrossRef]
40. Kornepati, A.V.R.; Vadlamudi, R.K.; Curiel, T.J. Programmed death ligand 1 signals in cancer cells. *Nat. Rev. Cancer* **2022**, *22*, 174–189. [CrossRef]
41. D'Angelo, S.P.; Shoushtari, A.N.; Agaram, N.P.; Kuk, D.; Qin, L.-X.; Carvajal, R.D.; Dickson, M.A.; Gounder, M.; Keohan, M.L.; Schwartz, G.K.; et al. Prevalence of tumor-infiltrating lymphocytes and PD-L1 expression in the soft tissue sarcoma microenvironment. *Hum. Pathol.* **2014**, *46*, 357–365. [CrossRef] [PubMed]
42. Zheng, C.; You, W.; Wan, P.; Jiang, X.; Chen, J.; Zheng, Y.; Li, W.; Tan, J.; Zhang, S. Clinicopathological and prognostic significance of PD-L1 expression in sarcoma: A systematic review and meta-analysis. *Medicine* **2018**, *97*, e11004. [CrossRef] [PubMed]
43. Que, Y.; Xiao, W.; Guan, Y.-X.; Liang, Y.; Yan, S.-M.; Chen, H.-Y.; Li, Q.-Q.; Xu, B.-S.; Zhou, Z.-W.; Zhang, X. PD-L1 Expression Is Associated with FOXP3+ Regulatory T-Cell Infiltration of Soft Tissue Sarcoma and Poor Patient Prognosis. *J. Cancer* **2017**, *8*, 2018–2025. [CrossRef] [PubMed]
44. Bremnes, R.M.; Busund, L.-T.; Kilvær, T.L.; Andersen, S.; Richardsen, E.; Paulsen, E.E.; Hald, S.; Khanehkenari, M.R.; Cooper, W.A.; Kao, S.C.; et al. The Role of Tumor-Infiltrating Lymphocytes in Development, Progression, and Prognosis of Non-Small Cell Lung Cancer. *J. Thorac. Oncol.* **2016**, *11*, 789–800. [CrossRef] [PubMed]
45. Hadrup, S.; Donia, M.; Straten, P.T. Effector CD4 and CD8 T Cells and Their Role in the Tumor Microenvironment. *Cancer Microenviron.* **2012**, *6*, 123–133. [CrossRef]
46. Saleh, R.; Elkord, E. FoxP3+ T regulatory cells in cancer: Prognostic biomarkers and therapeutic targets. *Cancer Lett.* **2020**, *490*, 174–185. [CrossRef]
47. Chen, B.; Li, H.; Liu, C.; Xiang, X.; Wang, S.; Wu, A.; Shen, Y.; Li, G. Prognostic value of the common tumour-infiltrating lymphocyte subtypes for patients with non-small cell lung cancer: A meta-analysis. *PLoS ONE* **2020**, *15*, e0242173. [CrossRef]
48. Bi, Q.; Liu, Y.; Yuan, T.; Wang, H.; Li, B.; Jiang, Y.; Mo, X.; Lei, Y.; Xiao, Y.; Dong, S.; et al. Predicted CD4$^+$ T cell infiltration levels could indicate better overall survival in sarcoma patients. *J. Int. Med. Res.* **2021**, *49*, 300060520981539. [CrossRef]
49. Fujii, H.; Arakawa, A.; Utsumi, D.; Sumiyoshi, S.; Yamamoto, Y.; Kitoh, A.; Ono, M.; Matsumura, Y.; Kato, M.; Konishi, K.; et al. CD8$^+$tumor-infiltrating lymphocytes at primary sites as a possible prognostic factor of cutaneous angiosarcoma. *Int. J. Cancer* **2013**, *134*, 2393–2402. [CrossRef]
50. Fridman, W.H.; Sibéril, S.; Pupier, G.; Soussan, S.; Sautès-Fridman, C. Activation of B cells in Tertiary Lymphoid Structures in cancer: Anti-tumor or anti-self? *Semin. Immun.* **2023**, *65*, 101703. [CrossRef]
51. Sautès-Fridman, C.; Petitprez, F.; Calderaro, J.; Fridman, W.H. Tertiary lymphoid structures in the era of cancer immunotherapy. *Nat. Rev. Cancer* **2019**, *19*, 307–325. [CrossRef] [PubMed]
52. Sorbye, S.W.; Kilvaer, T.; Valkov, A.; Donnem, T.; Smeland, E.; Al-Shibli, K.; Bremnes, R.M.; Busund, L.-T. Prognostic Impact of Lymphocytes in Soft Tissue Sarcomas. *PLoS ONE* **2011**, *6*, e14611. [CrossRef] [PubMed]
53. Petitprez, F.; de Reyniès, A.; Keung, E.Z.; Chen, T.W.-W.; Sun, C.-M.; Calderaro, J.; Jeng, Y.-M.; Hsiao, L.-P.; Lacroix, L.; Bougoüin, A.; et al. B cells are associated with survival and immunotherapy response in sarcoma. *Nature* **2020**, *577*, 556–560. [CrossRef] [PubMed]
54. Tawbi, H.A.; Burgess, M.; Bolejack, V.; Van Tine, B.A.; Schuetze, S.M.; Hu, J.; D'Angelo, S.; Attia, S.; Riedel, R.F.; Priebat, D.A.; et al. Pembrolizumab in advanced soft-tissue sarcoma and bone sarcoma (SARC028): A multicentre, two-cohort, single-arm, open-label, phase 2 trial. *Lancet Oncol.* **2017**, *18*, 1493–1501. [CrossRef]
55. Shimizu, K.; Iyoda, T.; Okada, M.; Yamasaki, S.; Fujii, S. Immune suppression and reversal of the suppressive tumor microenvironment. *Int. Immunol.* **2018**, *30*, 445–455. [CrossRef]
56. Cao, X.; Cai, S.F.; Fehniger, T.A.; Song, J.; Collins, L.I.; Piwnica-Worms, D.R.; Ley, T.J. Granzyme B and Perforin Are Important for Regulatory T Cell-Mediated Suppression of Tumor Clearance. *Immunity* **2007**, *27*, 635–646. [CrossRef]

57. Mahnke, K.; Ring, S.; Johnson, T.S.; Schallenberg, S.; Schönfeld, K.; Storn, V.; Bedke, T.; Enk, A.H. Induction of immunosuppressive functions of dendritic cellsin vivo by CD4+CD25+ regulatory T cells: Role of B7-H3 expression and antigen presentation. *Eur. J. Immunol.* **2007**, *37*, 2117–2126. [CrossRef]
58. Smolle, M.A.; Herbsthofer, L.; Granegger, B.; Goda, M.; Brcic, I.; Bergovec, M.; Scheipl, S.; Prietl, B.; Pichler, M.; Gerger, A.; et al. T-regulatory cells predict clinical outcome in soft tissue sarcoma patients: A clinico-pathological study. *Br. J. Cancer* **2021**, *125*, 717–724. [CrossRef]
59. Brinkrolf, P.; Landmeier, S.; Altvater, B.; Chen, C.; Pscherer, S.; Rosemann, A.; Ranft, A.; Dirksen, U.; Juergens, H.; Rossig, C. A high proportion of bone marrow T cells with regulatory phenotype (CD4+CD25^{hi}FoxP3+) in Ewing sarcoma patients is associated with metastatic disease. *Int. J. Cancer* **2009**, *125*, 879–886. [CrossRef]
60. Camarillo, D.; Leslie, K.; Unemori, P.; Yashi, A.K.; McCune-Smith, K.; Maurer, T. Regulatory T cells are present in Kaposi's sarcoma and increasingly frequent in advanced disease. *Infect. Agents Cancer* **2009**, *4*, P12–1. [CrossRef]
61. Colomiere, M.; Ward, A.C.; Riley, C.; Trenerry, M.K.; Cameron-Smith, D.; Findlay, J.; Ackland, L.; Ahmed, N. Cross talk of signals between EGFR and IL-6R through JAK2/STAT3 mediate epi-thelial-mesenchymal transition in ovarian carcinomas. *Br. J. Cancer* **2009**, *100*, 134–144. [CrossRef] [PubMed]
62. Tartour, E.; Pere, H.; Maillere, B.; Terme, M.; Merillon, N.; Taieb, J.; Sandoval, F.; Quintin-Colonna, F.; Lacerda, K.; Karadimou, A.; et al. Angiogenesis and immunity: A bidirectional link potentially relevant for the monitoring of antiangiogenic therapy and the development of novel therapeutic combination with immunotherapy. *Cancer Metastasis Rev.* **2011**, *30*, 83–95. [CrossRef] [PubMed]
63. Wang, Y.; Ding, Y.; Guo, N.; Wang, S. MDSCs: Key Criminals of Tumor Pre-metastatic Niche Formation. *Front. Immunol.* **2019**, *10*, 172. [CrossRef]
64. Han, X.; Shi, H.; Sun, Y.; Shang, C.; Luan, T.; Wang, D.; Ba, X.; Zeng, X. CXCR2 expression on granulocyte and macrophage progenitors under tumor conditions con-tributes to mo-MDSC generation via SAP18/ERK/STAT3. *Cell Death Dis.* **2019**, *10*, 598. [CrossRef] [PubMed]
65. Highfill, S.L.; Cui, Y.; Giles, A.J.; Smith, J.P.; Zhang, H.; Morse, E.; Kaplan, R.N.; Mackall, C.L. Disruption of CXCR2-Mediated MDSC Tumor Trafficking Enhances Anti-PD1 Efficacy. *Sci. Transl. Med.* **2014**, *6*, 237ra67. [CrossRef]
66. Mantovani, A.; Marchesi, F.; Malesci, A.; Laghi, L.; Allavena, P. Tumour-associated macrophages as treatment targets in oncology. *Nat. Rev. Clin. Oncol.* **2017**, *14*, 399–416. [CrossRef]
67. Malfitano, A.M.; Pisanti, S.; Napolitano, F.; Di Somma, S.; Martinelli, R.; Portella, G. Tumor-Associated Macrophage Status in Cancer Treatment. *Cancers* **2020**, *12*, 1987. [CrossRef]
68. Jung, K.Y.; Cho, S.W.; Kim, Y.; Kim, D.; Oh, B.-C.; Park, D.J.; Park, Y.J. Cancers with Higher Density of Tumor-Associated Macrophages Were Associated with Poor Survival Rates. *J. Pathol. Transl. Med.* **2015**, *49*, 318–324. [CrossRef]
69. Lee, C.-H.; Espinosa, I.; Vrijaldenhoven, S.; Subramanian, S.; Montgomery, K.D.; Zhu, S.; Marinelli, R.J.; Peterse, J.L.; Poulin, N.; Nielsen, T.O.; et al. Prognostic Significance of Macrophage Infiltration in Leiomyosarcomas. *Clin. Cancer Res.* **2008**, *14*, 1423–1430. [CrossRef]
70. Nabeshima, A.; Matsumoto, Y.; Fukushi, J.; Iura, K.; Matsunobu, T.; Endo, M.; Fujiwara, T.; Iida, K.; Hatano, M.; Yokoyama, N.; et al. Tumour-associated macrophages correlate with poor prognosis in myxoid liposarcoma and promote cell motility and invasion via the HB-EGF-EGFR-PI3K/Akt pathways. *Br. J. Cancer* **2015**, *112*, 547–555. [CrossRef]
71. Oike, N.; Kawashima, H.; Ogose, A.; Hotta, T.; Hatano, H.; Ariizumi, T.; Sasaki, T.; Yamagishi, T.; Umezu, H.; Endo, N. Prognostic impact of the tumor immune microenvironment in synovial sarcoma. *Cancer Sci.* **2018**, *109*, 3043–3054. [CrossRef]
72. Komohara, Y.; Takeya, H.; Wakigami, N.; Kusada, N.; Bekki, H.; Ishihara, S.; Takeya, M.; Nakashima, Y.; Oda, Y. Positive correlation between the density of macrophages and T-cells in undif-ferentiated sarcoma. *Med. Mol. Morphol.* **2019**, *52*, 44–51. [CrossRef] [PubMed]
73. Fujiwara, T.; Healey, J.; Ogura, K.; Yoshida, A.; Kondo, H.; Hata, T.; Kure, M.; Tazawa, H.; Nakata, E.; Kunisada, T.; et al. Role of Tumor-Associated Macrophages in Sarcomas. *Cancers* **2021**, *13*, 1086. [CrossRef] [PubMed]
74. Dumars, C.; Ngyuen, J.-M.; Gaultier, A.; Lanel, R.; Corradini, N.; Gouin, F.; Heymann, D.; Heymann, M.-F. Dysregulation of macrophage polarization is associated with the metastatic process in osteosarcoma. *Oncotarget* **2016**, *7*, 78343–78354. [CrossRef]
75. Buddingh, E.P.; Kuijjer, M.L.; Duim, R.A.; Bürger, H.; Agelopoulos, K.; Myklebost, O.; Serra, M.; Mertens, F.; Hogendoorn, P.C.; Lankester, A.C.; et al. Tumor-Infiltrating Macrophages Are Associated with Metastasis Suppression in High-Grade Osteosarcoma: A Rationale for Treatment with Macrophage Activating Agents. *Clin. Cancer Res.* **2011**, *17*, 2110–2119. [CrossRef] [PubMed]
76. Gomez-Brouchet, A.; Illac, C.; Gilhodes, J.; Bouvier, C.; Aubert, S.; Guinebretiere, J.M.; Marie, B.; Larousserie, F.; Entz-Werlé, N.; de Pinieux, G.; et al. CD163-positive tumor-associated macrophages and CD8-positive cytotoxic lymphocytes are powerful diagnostic markers for the therapeutic stratification of osteosarcoma patients: An immuno-histochemical analysis of the biopsies fromthe French OS2006 phase 3 trial. *OncoImmunology* **2017**, *6*, e1331193.
77. Handl, M.; Hermanova, M.; Hotarkova, S.; Jarkovsky, J.; Mudry, P.; Shatokhina, T.; Vesela, M.; Sterba, J.; Zambo, I. Clinicopatho-logical correlation of tumor-associated macrophages in Ewing sarcoma. *Biomed. Pap.* **2018**, *162*, 54–60. [CrossRef]
78. Lussier, D.M.; O'Neill, L.; Nieves, L.M.; McAfee, M.S.; Holechek, S.A.; Collins, A.W.; Dickman, P.; Jacobsen, J.; Hingorani, P.; Blattman, J.N. Enhanced T-Cell Immunity to Osteosarcoma Through Antibody Blockade of PD-1/PD-L1 Interactions. *J. Immun.* **2015**, *38*, 96–106. [CrossRef]

79. Toulmonde, M.; Penel, N.; Adam, J.; Chevreau, C.; Blay, J.Y.; Le Cesne, A.; Bompas, E.; Piperno-Neumann, S.; Cousin, S.; Grellety, T.; et al. Use of PD-1 Targeting, Macrophage Infiltration, and IDO Pathway Activation in Sarcomas: A Phase 2 Clinical Trial. *JAMA Oncol.* **2018**, *4*, 93–97. [CrossRef]
80. Nafia, I.; Toulmonde, M.; Bortolotto, D.; Chaibi, A.; Bodet, D.; Rey, C.; Velasco, V.; Larmonier, C.B.; Cerf, L.; Adam, J.; et al. IDO Targeting in Sarcoma: Biological and Clinical Implications. *Front. Immunol.* **2020**, *11*, 274. [CrossRef]
81. Mimura, K.; Kono, K.; Takahashi, A.; Kawaguchi, Y.; Fujii, H. Vascular endothelial growth factor inhibits the function of human mature dendritic cells mediated by VEGF receptor-2. *Cancer Immunol. Immunother.* **2006**, *56*, 761–770. [CrossRef]
82. Oelkrug, C.; Ramage, J.M. Enhancement of T cell recruitment and infiltration into tumours. *Clin. Exp. Immunol.* **2014**, *178*, 1–8. [CrossRef] [PubMed]
83. Yang, J.; Yan, J.; Liu, B. Targeting VEGF/VEGFR to Modulate Antitumor Immunity. *Front. Immunol.* **2018**, *9*, 978. [CrossRef]
84. Dubois, S.; Demetri, G. Markers of angiogenesis and clinical features in patients with sarcoma. *Cancer* **2007**, *109*, 813–819. [CrossRef] [PubMed]
85. Maki, R.G.; Jungbluth, A.A.; Gnjatic, S.; Schwartz, G.K.; D'Adamo, D.R.; Keohan, M.L.; Wagner, M.J.; Scheu, K.; Chiu, R.; Ritter, E.; et al. A Pilot Study of Anti-CTLA4 Antibody Ipilimumab in Patients with Synovial Sarcoma. *Sarcoma* **2013**, *2013*, 1–8. [CrossRef]
86. Merchant, M.S.; Wright, M.; Baird, K.; Wexler, L.H.; Rodriguez-Galindo, C.; Bernstein, D.; Delbrook, C.; Lodish, M.; Bishop, R.; Wolchok, J.D.; et al. Phase I Clinical Trial of Ipilimumab in Pediatric Patients with Advanced Solid Tumors. *Clin. Cancer Res.* **2016**, *22*, 1364–1370. [CrossRef]
87. Ben-Ami, E.; Barysauskas, C.M.; Solomon, S.; Tahlil, K.; Malley, R.; Hohos, M.; Polson, K.; Loucks, M.; Severgnini, M.; Patel, T.; et al. Immunotherapy with single agent nivolumab for advanced leiomyosarcoma of the uterus: Results of a phase 2 study: Nivolumab for Uterine Leiomyosarcoma. *Cancer* **2017**, *123*, 3285–3290. [CrossRef]
88. Blay, J.-Y.; Penel, N.; Ray-Coquard, I.L.; Cousin, S.; Bertucci, F.; Bompas, E.; Eymard, J.C.; Saada-Bouzid, E.; Soulie, P.; Boudou-Rouquette, P.; et al. High clinical activity of pembrolizumab in chordoma, alveolar soft part sarcoma (ASPS) and other rare sarcoma histotypes: The French AcSé pembrolizumab study from Unicancer. *J. Clin. Oncol.* **2021**, *39* (Suppl. 15), 11520. [CrossRef]
89. Delyon, J.; Resche-Rigon, M.; Renaud, M.; Le Goff, J.; Dalle, S.; Heidelberger, V.; Da Meda, L.; Allain, V.; Toullec, L.; Carcelain, G.; et al. 1077MO PD1 blockade with pembrolizumab in classic and endemic Kaposi sarcoma: A multicenter phase II study. *Ann. Oncol.* **2020**, *31*, S732. [CrossRef]
90. D'Angelo, S.P.; Mahoney, M.R.; Van Tine, B.A.; Atkins, J.; Milhem, M.M.; Jahagirdar, B.N.; Antonescu, C.R.; Horvath, E.; Tap, W.D.; Schwartz, G.K.; et al. Nivolumab with or without ipilimumab treatment for metastatic sarcoma (Alliance A091401): Two open-label, non-comparative, randomised, phase 2 trials. *Lancet Oncol.* **2018**, *19*, 416–426. [CrossRef]
91. Somaiah, N.; Conley, A.P.; Parra, E.R.; Lin, H.; Amini, B.; Soto, L.S.; Salazar, R.; Barreto, C.; Chen, H.; Gite, S.; et al. Durvalumab plus tremelimumab in advanced or metastatic soft tissue and bone sarcomas: A single-centre phase 2 trial. *Lancet Oncol.* **2022**, *23*, 1156–1166. [CrossRef]
92. Livingston, M.B.; Jagosky, M.H.; Robinson, M.M.; Ahrens, W.A.; Benbow, J.H.; Farhangfar, C.J.; Foureau, D.M.; Maxwell, D.M.; Baldrige, E.A.; Begic, X.; et al. Phase II Study of Pembrolizumab in Combination with Doxorubicin in Metastatic and Unresectable Soft-Tissue Sarcoma. *Clin. Cancer Res.* **2021**, *27*, 6424–6431. [CrossRef]
93. Pollack, S.M.; Redman, M.W.; Baker, K.K.; Wagner, M.J.; Schroeder, B.A.; Loggers, E.T.; Trieselmann, K.; Copeland, V.C.; Zhang, S.; Black, G.; et al. Assessment of Doxorubicin and Pembrolizumab in Patients With Advanced Anthracycline-Naive Sarcoma: A Phase 1/2 Nonrandomized Clinical Trial. *JAMA Oncol.* **2020**, *6*, 1778. [CrossRef]
94. Gordon, E.M.; Chawla, S.P.; Tellez, W.A.; Younesi, E.; Thomas, S.; Chua-Alcala, V.S.; Chomoyan, H.; Valencia, C.; Brigham, D.A.; Moradkhani, A.; et al. SAINT: A Phase I/Expanded Phase II Study Using Safe Amounts of Ipilimumab, Nivolumab and Trabectedin as First-Line Treatment of Advanced Soft Tissue Sarcoma. *Cancers* **2023**, *15*, 906. [CrossRef]
95. Pink, D.; Andreou, D.; Flörcken, A.; Golf, A.; Richter, S.; Kessler, T.; Kortüm, M.; Schmidt, C.A.; Kasper, B.; Wardelmann, E.; et al. Efficacy and safety of nivolumab and trabectedin in pretreated patients with advanced soft tissue sarcomas (STS): Preliminary results of a phase II trial of the German Interdisciplinary Sarcoma Group (GISG-15, NitraSarc) for the non-L sarcoma cohort. *J. Clin. Oncol.* **2021**, *39*, 11545. [CrossRef]
96. Smrke, A.; Ostler, A.; Napolitano, A.; Vergnano, M.; Asare, B.; Fotiadis, N.; Thway, K.; Zaidi, S.; Miah, A.; van der Graaf, W.; et al. 1526MO GEMMK: A phase I study of gemcitabine (gem) and pembrolizumab (pem) in patients (pts) with leiomyosarcoma (LMS) and undifferentiated pleomorphic sarcoma UPS). *Ann. Oncol.* **2021**, *32*, S1114. [CrossRef]
97. Nathenson, M.; Choy, E.; Carr, N.D.; Hibbard, H.D.; Mazzola, E.; Catalano, P.J.; Thornton, K.A.; Morgan, J.A.; Cote, G.M.; Merriam, P.; et al. Phase II study of eribulin and pembrolizumab in patients (pts) with metastatic soft tissue sarcomas (STS): Report of LMS cohort. *J. Clin. Oncol.* **2020**, *38*, 11559. [CrossRef]
98. Wagner, M.J.; Zhang, Y.; Cranmer, L.D.; Loggers, E.T.; Black, G.; McDonnell, S.; Maxwell, S.; Johnson, R.; Moore, R.; Hermida de Viveiros, P.; et al. A Phase 1/2 Trial Combining Avelumab and Trabectedin for Advanced Lipo-sarcoma and Leiomyosarcoma. *Clin. Cancer Res.* **2022**, *28*, 2306–2312. [CrossRef]
99. Toulmonde, M.; Brahmi, M.; Giraud, A.; Chakiba, C.; Bessede, A.; Kind, M.; Toulza, E.; Pulido, M.; Albert, S.; Guégan, J.-P.; et al. Trabectedin plus Durvalumab in Patients with Advanced Pretreated Soft Tissue Sarcoma and Ovarian Carcinoma (TRAMUNE): An Open-Label, Multicenter Phase Ib Study. *Clin. Cancer Res.* **2021**, *28*, 1765–1772. [CrossRef]
100. Palmerini, E.; Lopez-Pousa, A.; Grignani, G.; Redondo, A.; Hindi, N.; Stacchiotti, S.; Sebio, A.; Lopez-Martin, J.A.; Morales, C.M.V.; Martinez-Trufero, J.; et al. IMMUNOSARC: A collaborative Spanish (GEIS) and Italian (ISG) sarcoma groups phase I/II

trial of sunitinib and nivolumab in advanced soft tissue and bone sarcoma: Results from the phase II part, bone sarcoma cohort. *J. Clin. Oncol.* **2020**, *38*, 11522. [CrossRef]
101. Broto, J.M.; Hindi, N.; Grignani, G.; Trufero, J.M.; Redondo, A.; Valverde, C.; Pousa, A.L.; Stacchiotti, S.; Palmerini, E.; De Alava, E.; et al. IMMUNOSARC: A collaborative Spanish (GEIS) and Italian (ISG) sarcoma groups phase I/II trial of sunitinib plus nivolumab in advanced soft tissue and bone sarcomas: Results of the phase II- soft-tissue sarcoma cohort. *Ann. Oncol.* **2019**, *30*, v684. [CrossRef]
102. Wilky, B.A.; Trucco, M.M.; Subhawong, T.K.; Florou, V.; Park, W.; Kwon, D.; Wieder, E.D.; Kolonias, D.; Rosenberg, A.E.; Kerr, D.A.; et al. Axitinib plus pembrolizumab in patients with advanced sarcomas including alveolar soft-part sarcoma: A single-centre, single-arm, phase 2 trial. *Lancet Oncol.* **2019**, *20*, 837–848. [CrossRef]
103. Xie, L.; Xu, J.; Sun, X.; Guo, W.; Gu, J.; Liu, K.; Zheng, B.; Ren, T.; Huang, Y.; Tang, X.; et al. Apatinib plus camrelizumab (anti-PD1 therapy, SHR-1210) for advanced osteosarcoma (APFAO) progressing after chemotherapy: A single-arm, open-label, phase 2 trial. *J. Immunother. Cancer* **2019**, *8*, e000798. [CrossRef] [PubMed]
104. Kim, H.S.; Cho, H.J.; Yun, K.-H.; Lee, Y.H.; Kim, S.H.; Baek, W.; Jeon, M.K. Durvalumab and pazopanib in patients with advanced soft tissue sarcoma: A single-center, single-arm, phase 2 trial. *J. Clin. Oncol.* **2021**, *39*, 11551. [CrossRef]
105. Cousin, S.; Bellera, C.; Guegan, J.-P.; Valentin, T.; Bahleda, R.; Metges, J.-P.; Cassier, P.; Cantarel, C.; Ceruso, M.S.; Kind, M.; et al. 1494P Regomune-a phase II study of regorafenib + avelumab in solid tumors: Results of the soft tissue sarcoma (STS) cohort. *Ann. Oncol.* **2022**, *33*, S1230. [CrossRef]
106. Kelly, C.M.; Chi, P.; Dickson, M.A.; Gounder, M.M.; Keohan, M.L.; Qin, L.-X.; Adamson, T.; Condy, M.M.; Biniakewitz, M.; Phelan, H.; et al. A phase II study of epacadostat and pembrolizumab in patients with advanced sarcoma. *J. Clin. Oncol.* **2019**, *37*, 11049. [CrossRef]
107. Schöffski, P.; Bahleda, R.; Wagner, A.; Burgess, M.; Junker, N.; Chisamore, M.; Peterson, P.; Ceccarelli, M.; William, T. 154P Results of an open-label, phase Ia/Ib study of olaratumab plus pembrolizumab in patients with unresectable, locally advanced or metastatic soft tissue sarcoma. *Ann. Oncol.* **2021**, *32*, S1447. [CrossRef]
108. Kelly, C.M.; Antonescu, C.R.; Bowler, T.; Munhoz, R.; Chi, P.; Dickson, M.A.; Gounder, M.M.; Keohan, M.L.; Movva, S.; Dholakia, R.; et al. Objective Response Rate Among Patients With Locally Advanced or Metastatic Sarcoma Treated With Talimogene Laherparepvec in Combination With Pembrolizumab: A Phase 2 Clinical Trial. *JAMA Oncol.* **2020**, *6*, 402. [CrossRef]
109. Chawla, N.S.; Kim, T.; Sherman, T.; Dang, J.; Chua, V.S.; Moradkhani, A.; Bhuiyan, I.; Krkyan, N.; Fernando, M.; Colletti, E.; et al. A phase 2 study of talimogene laherparepvec, nivolumab, and trabectedin (TNT) in advanced sarcoma. *J. Clin. Oncol.* **2021**, *39*, 11567. [CrossRef]
110. D'Angelo, S.P.; Richards, A.L.; Conley, A.P.; Woo, H.J.; Dickson, M.A.; Gounder, M.; Kelly, C.; Keohan, M.L.; Movva, S.; Thornton, K.; et al. Pilot study of bempegaldesleukin in combination with nivolumab in patients with metastatic sarcoma. *Nat. Commun.* **2022**, *13*, 1–11. [CrossRef]
111. Somaiah, N.; Block, M.S.; Kim, J.W.; Shapiro, G.I.; Do, K.T.; Hwu, P.; Eder, J.P.; Jones, R.L.; Lu, H.; ter Meulen, J.H.; et al. First-in-Class, First-in-Human Study Evaluating LV305, a Dendritic-Cell Tropic Lentiviral Vector, in Sarcoma and Other Solid Tumors Expressing NY-ESO-1. *Clin. Cancer Res.* **2019**, *25*, 5808–5817. [CrossRef]
112. Rosenbaum, E.; Movva, S.; Kelly, C.M.; Dickson, M.A.; Keohan, M.L.; Gounder, M.M.; Thornton, K.A.; Chi, P.; Chan, J.E.; Nacev, B.; et al. A phase 1b study of avelumab plus DCC-3014, a potent and selective inhibitor of colony stimulating factor 1 receptor (CSF1R), in patients with advanced high-grade sarcoma. *J. Clin. Oncol.* **2021**, *39*, 11549. [CrossRef]
113. Chawla, S.P.; Chua-Alcala, V.S.; Gordon, E.M.; Kim, T.T.; Feske, W.; Gibson, B.L.; Chang, P.Y.; Robinson, D.; Song, P.Y. Interim analysis of a phase I study of SNK01 (Autologous Nongenetically Modified Natural Killer Cells with Enhanced Cytotoxicity) and avelumab in advanced refractory sarcoma. *J. Clin. Oncol.* **2022**, *40*, 11517. [CrossRef]
114. Kawaguchi, S.; Wada, T.; Ida, K.; Sato, Y.; Nagoya, S.; Tsukahara, T.; Kimura, S.; Sahara, H.; Ikeda, H.; Shimozawa, K.; et al. Phase I vaccination trial of SYT-SSX junction peptide in patients with disseminated synovial sarcoma. *J. Transl. Med.* **2005**, *3*, 1. [CrossRef]
115. Takahashi, R.; Ishibashi, Y.; Hiraoka, K.; Matsueda, S.; Kawano, K.; Kawahara, A.; Kage, M.; Ohshima, K.; Yamanaka, R.; Shichijo, S.; et al. Phase II study of personalized peptide vaccination for refractory bone and soft tissue sarcoma patients. *Cancer Sci.* **2013**, *104*, 1285–1294. [CrossRef]
116. Pipia, N.; Baldueva, I.; Nekhaeva, T.; Novik, A.; Danilova, A.; Avdonkina, N.; Protsenko, S.; Komarov, Y.; Girdyuk, D.; Oganesyan, A.; et al. Autologous dendritic-cell vaccine based on cancer-testis antigens "CaTeVac" in the treatment of soft tissue sarcoma. *Ann. Oncol.* **2018**, *29*, viii410–viii411. [CrossRef]
117. Chawla, S.P.; Van Tine, B.A.; Pollack, S.M.; Ganjoo, K.N.; Elias, A.D.; Riedel, R.F.; Attia, S.; Choy, E.; Okuno, S.H.; Agulnik, M.; et al. Phase II Randomized Study of CMB305 and Atezolizumab Compared With Atezolizumab Alone in Soft-Tissue Sarcomas Expressing NY-ESO-1. *J. Clin. Oncol.* **2022**, *40*, 1291–1300. [CrossRef]
118. Robbins, P.F.; Kassim, S.H.; Tran, T.L.N.; Crystal, J.S.; Morgan, R.A.; Feldman, S.A.; Yang, J.C.; Dudley, M.E.; Wunderlich, J.R.; Sherry, R.M.; et al. A Pilot Trial Using Lymphocytes Genetically Engineered with an NY-ESO-1–Reactive T-cell Receptor: Long-term Follow-up and Correlates with Response. *Clin. Cancer Res.* **2015**, *21*, 1019–1027. [CrossRef] [PubMed]
119. D'Angelo, S.; Demetri, G.; Tine, B.V.; Druta, M.; Glod, J.; Chow, W.; Araujo, D. 298 Final analysis of the phase 1 trial of NY-ESO-1–specific T-cell receptor (TCR) T-cell therapy (letetresgene autoleucel; GSK3377794) in patients with advanced synovial sarcoma (SS) [Internet]. In *Regular and Young Investigator Award Abstracts*; BMJ Publishing Group Ltd.: London, UK, 2020; p. A182–3. Available online: https://jitc.bmj.com/lookup/doi/10.1136/jitc-2020-SITC2020.0298 (accessed on 13 April 2023).

120. Van Tine, B.A.; Butler, M.O.; Araujo, D.; Johnson, M.L.; Clarke, J.; Liebner, D.; Odunsi, K.; Olszanski, A.J.; Basu, S.; Brophy, F.; et al. ADP-A2M4 (MAGE-A4) in patients with synovial sarcoma. *Ann. Oncol.* **2019**, *30*, v684–5. [CrossRef]
121. D'Angelo, S.P.; Van Tine, B.A.; Attia, S.; Blay, J.-Y.; Strauss, S.J.; Morales, C.M.V.; Razak, A.R.A.; Van Winkle, E.; Trivedi, T.; Biswas, S.; et al. SPEARHEAD-1: A phase 2 trial of afamitresgene autoleucel (Formerly ADP-A2M4) in patients with advanced synovial sarcoma or myxoid/round cell liposarcoma. *J. Clin. Oncol.* **2021**, *39*, 11504. [CrossRef]
122. Ahmed, N.; Brawley, V.S.; Hegde, M.; Robertson, C.; Ghazi, A.; Gerken, C.; Liu, E.; Dakhova, O.; Ashoori, A.; Corder, A.; et al. Human Epidermal Growth Factor Receptor 2 (HER2) –Specific Chimeric Antigen Receptor–Modified T Cells for the Immunotherapy of HER2-Positive Sarcoma. *J. Clin. Oncol.* **2015**, *33*, 1688–1696. [CrossRef] [PubMed]
123. Navai, S.A.; Derenzo, C.; Joseph, S.; Sanber, K.; Byrd, T.; Zhang, H.; Mata, M.; Gerken, C.; Shree, A.; Mathew, P.R.; et al. Abstract LB-147: Administration of HER2-CAR T cells after lymphodepletion safely improves T cell expansion and induces clinical responses in patients with advanced sarcomas. *Cancer Res.* **2019**, *79* (Suppl. 13), LB-147. [CrossRef]
124. Burgess, M.A.; Bolejack, V.; Schuetze, S.; Van Tine, B.A.; Attia, S.; Riedel, R.F.; Hu, J.S.; Davis, L.E.; Okuno, S.H.; Priebat, D.A.; et al. Clinical activity of pembrolizumab (P) in undifferentiated pleomorphic sarcoma (UPS) and dedifferentiated/pleomorphic liposarcoma (LPS): Final results of SARC028 expansion cohorts. *J. Clin. Oncol.* **2019**, *37*, 11015. [CrossRef]
125. Keung, E.Z.; Burgess, M.; Salazar, R.; Parra, E.R.; Rodrigues-Canales, J.; Bolejack, V.; Tawbi, H.A. Correlative Analyses of the SARC028 Trial Reveal an Association Between Sar-coma-Associated Immune Infiltrate and Response to Pembrolizumab. *Clin. Cancer Res.* **2020**, *26*, 1258–1266. [CrossRef]
126. Liu, J.; Fan, Z.; Bai, C.; Li, S.; Xue, R.; Gao, T.; Zhang, L.; Tan, Z.; Fang, Z. Real-world experience with pembrolizumab in patients with advanced soft tissue sarcoma. *Ann. Transl. Med.* **2021**, *9*, 339. [CrossRef]
127. Saif, A.; Verbus, E.A.; Sarvestani, A.L.; Teke, M.E.; Lambdin, J.; Hernandez, J.M.; Kirsch, D.G. A Randomized Trial of Pembrolizumab & Radiotherapy Versus Radiotherapy in High-Risk Soft Tissue Sarcoma of the Extremity (SU2C-SARC032). *Ann. Surg. Oncol.* **2022**, *30*, 683–685. [CrossRef]
128. le Guevelou, J.; Debaigt, C.; Saada-Bouzid, E.; Viotti, J.; Khalladi, N.; Thibouw, D.; Thariat, J. Phase II study of concomitant radiotherapy with atezolizumab in oli-gometastatic soft tissue sarcomas: STEREOSARC trial protocol. *BMJ Open* **2020**, *10*, e038391. [CrossRef]
129. Keung, E.Z.; Tsai, J.-W.; Ali, A.M.; Cormier, J.N.; Bishop, A.J.; Guadagnolo, B.A.; Roland, C.L. Analysis of the immune infiltrate in undifferentiated pleomorphic sarcoma of the extremity and trunk in response to radiotherapy: Rationale for combination neoadjuvant immune checkpoint inhibition and radio-therapy. *OncoImmunology* **2018**, *7*, e1385689. [CrossRef]
130. Takada, K.; Sugita, S.; Murase, K.; Kikuchi, T.; Oomori, G.; Ito, R.; Hayasaka, N.; Miyanishi, K.; Iyama, S.; Ikeda, H.; et al. Exceptionally rapid response to pembrolizumab in a SMARCA4-deficient thoracic sarcoma overexpressing PD-L1: A case report. *Thorac. Cancer* **2019**, *10*, 2312–2315. [CrossRef]
131. Anžič, N.; Krasniqi, F.; Eberhardt, A.-L.; Tzankov, A.; Haslbauer, J.D. Ipilimumab and Pembrolizumab Mixed Response in a 41-Year-Old Patient with SMARCA4-Deficient Thoracic Sarcoma: An Interdisciplinary Case Study. *Case Rep. Oncol.* **2021**, *14*, 706–715. [CrossRef]
132. Henon, C.; Blay, J.-Y.; Massard, C.; Mir, O.; Bahleda, R.; Dumont, S.; Postel-Vinay, S.; Adam, J.; Soria, J.-C.; Le Cesne, A. Long lasting major response to pembrolizumab in a thoracic malignant rhabdoid-like SMARCA4-deficient tumor. *Ann. Oncol.* **2019**, *30*, 1401–1403. [CrossRef] [PubMed]
133. Iijima, Y.; Sakakibara, R.; Ishizuka, M.; Honda, T.; Shirai, T.; Okamoto, T.; Tateishi, T.; Sakashita, H.; Tamaoka, M.; Takemoto, A.; et al. Notable response to nivolumab during the treatment of SMARCA4-deficient thoracic sarcoma: A case report. *Immunotherapy* **2020**, *12*, 563–569. [CrossRef] [PubMed]
134. Kawachi, H.; Kunimasa, K.; Kukita, Y.; Nakamura, H.; Honma, K.; Kawamura, T.; Inoue, T.; Tamiya, M.; Kuhara, H.; Nishino, K.; et al. Atezolizumab with bevacizumab, paclitaxel and carboplatin was effective for patients with SMARCA4-deficient thoracic sarcoma. *Immunotherapy* **2021**, *13*, 799–806. [CrossRef] [PubMed]
135. Kunimasa, K.; Okami, J.; Takenaka, S.; Honma, K.; Kukita, Y.; Nagata, S.; Kawamura, T.; Inoue, T.; Tamiya, M.; Kuhara, H.; et al. Conversion Surgery for Advanced Thoracic SMARCA4-Deficient Undifferentiated Tumor With Atezolizumab in Combination With Bevacizumab, Paclitaxel, and Carboplatin Treatment: A Case Report. *JTO Clin. Res. Rep.* **2021**, *2*, 100235. [CrossRef]
136. Marcrom, S.; De Los Santos, J.F.; Conry, R.M. Complete response of mediastinal clear cell sarcoma to pembrolizumab with ra-diotherapy. *Clin. Sarcoma Res.* **2017**, *7*, 14. [CrossRef]
137. Yu, X.-H.; Huang, J.; Ge, N.-J.; Yang, Y.-F.; Zhao, J.-Y. Recurrent undifferentiated embryonal sarcoma of the liver in adult patient treated by pembrolizumab: A case report. *World J. Clin. Cases* **2021**, *9*, 2281–2288. [CrossRef]
138. Song, H.-N.; Kang, M.G.; Park, J.R.; Hwang, J.-Y.; Kang, J.H.; Lee, W.S.; Lee, G.-W. Pembrolizumab for Refractory Metastatic Myxofibrosarcoma: A Case Report. *Cancer Res. Treat.* **2018**, *50*, 1458–1461. [CrossRef]
139. Lambden, J.P.; Kelsten, M.F.; Schulte, B.C.; Pa, S.A.; Hayes, J.P.; Villaflor, V.; Agulnik, M. Metastatic Myxofibrosarcoma with Durable Response to Temozolomide Followed by Atezolizumab: A Case Report. *Oncologist* **2021**, *26*, 549–553. [CrossRef]
140. Tabata, M.M.; Novoa, R.A.; Bui, N.Q.; Zaba, L.C. Successful treatment of HIV-negative Kaposi sarcoma with ipilimumab and nivolumab and concurrent management of baseline psoriasis and bullous pemphigoid. *JAAD Case Rep.* **2020**, *6*, 447–449. [CrossRef]

141. Zhou, M.; Bui, N.; Lohman, M.; van de Rjin, M.; Hwang, G.; Ganjoo, K. Long-Term Remission with Ipilimumab/Nivolumab in Two Patients with Different Soft Tissue Sarcoma Subtypes and No PD-L1 Expression. *Case Rep. Oncol.* **2021**, *14*, 459–465. [CrossRef]
142. Chen, Y.; Liu, X.; Liu, J.; Liang, D.; Zhao, M.; Yu, W.; Chen, P. Nivolumab plus ipilimumab versus nivolumab in individuals with treatment-naive programmed death-ligand 1 positive metastatic soft tissue sarcomas: A multicentre retrospective study. *BMC Cancer* **2021**, *21*, 1–8. [CrossRef] [PubMed]
143. Zhou, M.; Bui, N.; Bolleddu, S.; Lohman, M.; Becker, H.-C.; Ganjoo, K. Nivolumab plus ipilimumab for soft tissue sarcoma: A single institution retrospective review. *Immunotherapy* **2020**, *12*, 1303–1312. [CrossRef] [PubMed]
144. Zer, A.; Icht, O.; Joseph, L.; Avram, D.; Jacobi, O.; Fenig, E.; Shamai, S.; Frommer, R.S.; Bernstine, H.; Weitzen, R.; et al. A phase II single-arm study of nivolumab and ipilimumab (Nivo/Ipi) in previously treated Classic Kaposi sarcoma (CKS). *J. Clin. Oncol.* **2019**, *37*, 11064. [CrossRef]
145. Lewin, J.; Davidson, S.; Anderson, N.D.; Lau, B.Y.; Kelly, J.; Tabori, U.; Salah, S.; Butler, M.O.; Aung, K.L.; Shlien, A.; et al. Response to Immune Checkpoint Inhibition in Two Patients with Alveolar Soft-Part Sarcoma. *Cancer Immunol. Res.* **2018**, *6*, 1001–1007. [CrossRef] [PubMed]
146. Grünwald, V.; Bauer, S.; Hermes, B.; Ivanyi, P.; Lindner, L.H.; Pink, D.; Reichardt, P.; Richter, S.; Tuchscherer, A. A randomized phase II study of durvalumab and tremelimumab compared to dox-orubicin in patients with advanced or metastatic soft tissue sarcoma (MEDISARC, AIO-STS 0415). *J. Clin. Oncol.* **2019**, *37* (Suppl. 15), TPS11075. [CrossRef]
147. Bishop, M.W.; Kaste, S.C.; Sykes, A.; Pan, H.; Cruz, F.S.D.; Whittle, S.; Mascarenhas, L.; Thomas, P.G.; Youngblood, B.; Harman, J.L.; et al. OSTPDL1: A phase II study of avelumab, a monoclonal antibody targeting programmed death-ligand 1 (PD-L1) in adolescent and young adult patients with recurrent or progressive osteosarcoma. *J. Clin. Oncol.* **2020**, *38*, 10521. [CrossRef]
148. Uboha, N.V.; Milhem, M.M.; Kovacs, C.; Amin, A.; Magley, A.; Das Purkayastha, D.; Piha-Paul, S.A. Phase II study of spartalizumab (PDR001) and LAG525 in advanced solid tumors and hematologic malignancies. *J. Clin. Oncol.* **2019**, *37*, 2553. [CrossRef]
149. Tawbi, H.A.; Schadendorf, D.; Lipson, E.J.; Ascierto, P.A.; Matamala, L.; Castillo Gutiérrez, E.; Long, G.V. Relatlimab and Nivolumab versus Nivolumab in Untreated Advanced Mela-noma. *N. Engl. J. Med.* **2022**, *386*, 24–34. [CrossRef]
150. Rivera Vargas, T.; Apetoh, L. Danger signals: Chemotherapy enhancers? *Immunol. Rev.* **2017**, *280*, 175–193. [CrossRef]
151. Tian, Z.; Yang, Y.; Yang, J.; Zhang, P.; Zhang, F.; Du, X.; Li, C.; Wang, J. Safety and Efficacy of PD-1 Inhibitors Plus Chemotherapy in Advanced Soft Tissue Sarcomas: A Retrospective Study. *Cancer Manag. Res.* **2020**, *12*, 1339–1346. [CrossRef]
152. Italiano, A.; Bessede, A.; Bompas, E.; Piperno-Neumann, S.; Chevreau, C.; Penel, N.; Bertucci, F.; Toulmonde, M.; Bellera, C.A.; Guegan, J.-P.; et al. PD1 inhibition in soft-tissue sarcomas with tertiary lymphoid structures: A multicenter phase II trial. *J. Clin. Oncol.* **2021**, *39*, 11507. [CrossRef]
153. Spalato-Ceruso, M.; Bouteiller, F.; Guegan, J.-P.; Toulmonde, M.; Bessede, A.; Kind, M.; Cousin, S.; Buy, X.; Palussiere, J.; Le Loarer, F.; et al. Pembrolizumab combined with low-dose cyclophosphamide and in-tra-tumoral injection of the toll-like receptor 4 agonist G100 in patients with advanced pretreated soft tissue sarcoma: Results from the PEMBROSARC basket study. *J. Hematol. Oncol. J. Hematol. Oncol.* **2022**, *15*, 157. [CrossRef] [PubMed]
154. Chawla, S.P.; Sankhala, K.K.; Ravicz, J.; Kang, G.; Liu, S.; Stumpf, N.; Leong, B.; Kim, S.; Arasheben, S.; Tseng, W.W.; et al. Clinical experience with combination chemo-/immunotherapy using trabectedin and nivolumab for advanced soft tissue sarcoma. *J. Clin Oncol.* **2018**, *36* (Suppl. 15), e23568. [CrossRef]
155. Fleuren, E.D.G.; Terry, R.L.; Meyran, D.; Omer, N.; Trapani, J.A.; Haber, M.; Neeson, P.J.; Ekert, P.G. Enhancing the Potential of Immunotherapy in Paediatric Sarcomas: Breaking the Immunosuppressive Barrier with Receptor Tyrosine Kinase Inhibitors. *Biomedicines* **2021**, *9*, 1798. [CrossRef] [PubMed]
156. Dorman, K.; Burkhard-Meier, A.; Di Gioia, D.; Kunz, W.G.; Knösel, T.; Holch, J.W.; Lindner, L.H. Treatment of metastatic alveolar soft part sarcoma with axitinib and pembrolizumab in an 80-year-old patient with a history of autoimmune disorders. *Anti-Cancer Drugs* **2022**, *34*, 311–316. [CrossRef]
157. Li, H.-L.; Li, Y.; Hu, H.-T.; Shao, S.-S.; Chen, C.-S.; Guo, C.-Y.; Zhao, Y.; Yao, Q.-J. Clinical observation of local intervention combined with camrelizumab and apatinib in the treatment of metastatic soft-tissue sarcoma. *J. Cancer Res. Ther.* **2021**, *17*, 1718. [CrossRef]
158. Paoluzzi, L.; Cacavio, A.; Ghesani, M.; Karambelkar, A.; Rapkiewicz, A.; Weber, J.; Rosen, G. Response to anti-PD1 therapy with nivolumab in metastatic sarcomas. *Clin. Sarcoma Res.* **2016**, *6*, 1–7. [CrossRef]
159. Arora, S.; Rastogi, S.; Shamim, S.A.; Barwad, A.; Sethi, M. Good and sustained response to pembrolizumab and pazopanib in advanced undifferentiated pleomorphic sarcoma: A case report. *Clin. Sarcoma Res.* **2020**, *10*, 1–6. [CrossRef]
160. Monga, V.; Miller, B.J.; Tanas, M.; Boukhar, S.; Allen, B.; Anderson, C.; Stephens, L.; Hartwig, S.; Varga, S.; Houtman, J.; et al. Intratumoral talimogene laherparepvec injection with concurrent preoperative radiation in patients with locally advanced soft-tissue sarcoma of the trunk and extremities: Phase IB/II trial. *J. Immunother. Cancer* **2021**, *9*, e003119. [CrossRef]
161. Diab, A.; Tannir, N.M.; Bentebibel, S.-E.; Hwu, P.; Papadimitrakopoulou, V.; Haymaker, C.; Kluger, H.M.; Gettinger, S.N.; Sznol, M.; Tykodi, S.S.; et al. Bempegaldesleukin (NKTR-214) plus Nivolumab in Patients with Advanced Solid Tumors: Phase I Dose-Escalation Study of Safety, Efficacy, and Immune Activation (PIVOT-02). *Cancer Discov.* **2020**, *10*, 1158–1173. [CrossRef]

162. Hanna, A.; Metge, B.J.; Bailey, S.K.; Chen, D.; Chandrashekar, D.S.; Varambally, S.; Samant, R.S.; Shevde, L.A. Inhibition of Hedgehog signaling reprograms the dysfunctional immune microenvironment in breast cancer. *Oncoimmunology* **2019**, *8*, 1548241. [CrossRef] [PubMed]
163. Grund-Gröschke, S.; Ortner, D.; Szenes-Nagy, A.B.; Zaborsky, N.; Weiss, R.; Neureiter, D.; Wipplinger, M.; Risch, A.; Hammerl, P.; Greil, R. Epidermal activation of Hedgehog signaling establishes an immuno-suppressive microenvironment in basal cell carcinoma by modulating skin immunity. *Mol. Oncol.* **2020**, *14*, 1930–1946. [CrossRef] [PubMed]
164. Yang, H.; Yao, F.; Davis, P.F.; Tan, S.T.; Hall, S.R.R. CD73, Tumor Plasticity and Immune Evasion in Solid Cancers. *Cancers* **2021**, *13*, 177. [CrossRef] [PubMed]
165. Somaiah, N.; Livingston, J.A.A.; Ravi, V.; Lin, H.Y.; Amini, B.; Solis, L.M.; Conley, A.P.; Zarzour, M.A.; Ludwig, J.A.; Ratan, R.; et al. A phase II multi-arm study to test the efficacy of oleclumab and durvalumab in specific sarcoma subtypes. *J. Clin. Oncol.* **2022**, *40* (Suppl. 16), TPS11594. [CrossRef]
166. Italiano, A.; Isambert, N.; Metges, J.-P.; Toulmonde, M.; Cousin, S.; Pernot, S.; Spalato, M.; Grellety, T.; Auzanneau, C.; Lortal, B.; et al. CAIRE: A basket multicenter open-label phase 2 study evaluating the EZH2 inhibitor tazemetostat in combination with durvalumab in patients with advanced solid tumors. *J. Clin. Oncol.* **2022**, *40* (Suppl. 16), TPS2703. [CrossRef]
167. Lin, C.-C. Clinical Development of Colony-Stimulating Factor 1 Receptor (CSF1R) Inhibitors. *J. Immunother. Precis. Oncol.* **2021**, *4*, 105–114. [CrossRef]
168. Saxena, M.; van der Burg, S.H.; Melief, C.J.M.; Bhardwaj, N. Therapeutic cancer vaccines. *Nat. Rev. Cancer* **2021**, *21*, 360–378. [CrossRef]
169. Sato, Y.; Nabeta, Y.; Tsukahara, T.; Hirohashi, Y.; Syunsui, R.; Maeda, A.; Sahara, H.; Ikeda, H.; Torigoe, T.; Ichimiya, S.; et al. Detection and Induction of CTLs Specific for SYT-SSX-Derived Peptides in HLA-A24+ Patients with Synovial Sarcoma. *J. Immunol.* **2002**, *169*, 1611–1618. [CrossRef]
170. Kawaguchi, S.; Tsukahara, T.; Ida, K.; Kimura, S.; Murase, M.; Kano, M.; Emori, M.; Nagoya, S.; Kaya, M.; Torigoe, T.; et al. SYT-SSX breakpoint peptide vaccines in patients with synovial sarcoma: A study from the Japanese Musculoskeletal Oncology Group. *Cancer Sci.* **2012**, *103*, 1625–1630. [CrossRef]
171. Krishnadas, D.K.; Shusterman, S.; Bai, F.; Diller, L.; Sullivan, J.E.; Cheerva, A.C.; George, R.E.; Lucas, K.G. A phase I trial combining decitabine/dendritic cell vaccine targeting MAGE-A1, MAGE-A3 and NY-ESO-1 for children with relapsed or therapy-refractory neuroblastoma and sarcoma. *Cancer Immunol. Immunother.* **2015**, *64*, 1251–1260. [CrossRef]
172. Dillman, R.O.; Wiemann, M.; Nayak, S.K.; DeLeon, C.; Hood, K.; DePriest, C. Interferon-gamma or granulocyte-macrophage col-ony-stimulating factor administered as adjuvants with a vaccine of irradiated autologous tumor cells from short-term cell line cultures: A randomized phase 2 trial of the cancer biotherapy research group. *J. Immun. Hagerstown Md.* **2003**, *26*, 367–373.
173. Kershaw, M.; Westwood, J.A.; Darcy, P. Gene-engineered T cells for cancer therapy. *Nat. Rev. Cancer* **2013**, *13*, 525–541. [CrossRef] [PubMed]
174. Gyurdieva, A.; Zajic, S.; Chang, Y.-F.; Houseman, E.A.; Zhong, S.; Kim, J.; Nathenson, M.; Faitg, T.; Woessner, M.; Turner, D.C.; et al. Biomarker correlates with response to NY-ESO-1 TCR T cells in patients with synovial sarcoma. *Nat. Commun.* **2022**, *13*, 1–18. [CrossRef] [PubMed]
175. D'Angelo, S.P.; Druta, M.; Liebner, D.A.; Schuetze, S.; Somaiah, N.; Van Tine, B.A.; Tap, W.D.; Pulham, T.; Chagin, K.; Norry, E.; et al. Pilot study of NY-ESO-1^{c259} T cells in advanced myxoid/round cell liposarcoma. *J. Clin. Oncol.* **2018**, *36*, 3005. [CrossRef]
176. D'Angelo, S.P.; Attia, S.; Blay, J.-Y.; Strauss, S.J.; Morales, C.M.V.; Razak, A.R.A.; Van Tine, B.A. Identification of response stratification factors from pooled efficacy analyses of afami-tresgene autoleucel ("Afami-cel" [Formerly ADP-A2M4]) in metastatic synovial sarcoma and myxoid/round cell liposarcoma phase 1 and phase 2 trials. *J. Clin. Oncol.* **2022**, *40* (Suppl. 16), 11562.
177. Moreno Tellez, C.; Leyfman, Y.; D'Angelo, S.P.; Wilky, B.A.; Dufresne, A. Immunotherapy in Sarcoma: Where Do Things Stand? *Surg. Oncol. Clin. N. Am.* **2022**, *31*, 381–397. [CrossRef]
178. Huang, X.; Park, H.; Greene, J.; Pao, J.; Mulvey, E.; Zhou, S.X.; Albert, C.M.; Moy, F.; Sachdev, D.; Yee, D.; et al. IGF1R-and ROR1-Specific CAR T Cells as a Potential Therapy for High Risk Sarcomas. *PLoS ONE* **2015**, *10*, e0133152. [CrossRef]
179. Leuci, V.; Casucci, G.M.; Grignani, G.; Rotolo, R.; Rossotti, U.; Vigna, E.; Gammaitoni, L.; Mesiano, G.; Fiorino, E.; Donini, C.; et al. CD44v6 as innovative sarcoma target for CAR-redirected CIK cells. *Oncoimmunology* **2018**, *7*, e1423167. [CrossRef]
180. Lehner, M.; Götz, G.; Proff, J.; Schaft, N.; Dörrie, J.; Full, F.; Ensser, A.; Muller, Y.; Cerwenka, A.; Abken, H.; et al. Redirecting T Cells to Ewing's Sarcoma Family of Tumors by a Chimeric NKG2D Receptor Expressed by Lentiviral Transduction or mRNA Transfection. *PLoS ONE* **2012**, *7*, e31210. [CrossRef]
181. Fernández, L.; Metais, J.-Y.; Escudero, A.; Vela, M.; Valentín, J.; Vallcorba, I.; Leivas, A.; Torres, J.; Valeri, A.; Patiño-García, A.; et al. Memory T Cells Expressing an NKG2D-CAR Efficiently Target Osteosarcoma Cells. *Clin. Cancer Res.* **2017**, *23*, 5824–5835. [CrossRef]
182. Zhao, Y.; Deng, J.; Rao, S.; Guo, S.; Shen, J.; Du, F.; Wu, X.; Chen, Y.; Li, M.; Chen, M.; et al. Tumor Infiltrating Lymphocyte (TIL) Therapy for Solid Tumor Treatment: Progressions and Challenges. *Cancers* **2022**, *14*, 4160. [CrossRef] [PubMed]
183. Rohaan, M.W.; Borch, T.H.; van den Berg, J.H.; Met, Ö.; Kessels, R.; Foppen, M.H.G.; Granhøj, J.S.; Nuijen, B.; Nijenhuis, C.; Jedema, I.; et al. Tumor-Infiltrating Lymphocyte Therapy or Ipilimumab in Advanced Melanoma. *N. Engl. J. Med.* **2022**, *387*, 2113–2125. [CrossRef] [PubMed]

184. Huang, H.; Nie, C.-P.; Liu, X.-F.; Song, B.; Yue, J.-H.; Xu, J.-X.; He, J.; Li, K.; Feng, Y.-L.; Wan, T.; et al. Phase I study of adjuvant immunotherapy with autologous tumor-infiltrating lymphocytes in locally advanced cervical cancer. *J. Clin. Investig.* **2022**, *132*, e157726. [CrossRef]
185. Stevanović, S.; Helman, S.R.; Wunderlich, J.R.; Langhan, M.M.; Doran, S.L.; Kwong, M.L.M.; Somerville, R.P.; Klebanoff, C.A.; Kammula, U.S.; Sherry, R.M.; et al. A Phase II Study of Tumor-infiltrating Lymphocyte Therapy for Human Papillomavirus–associated Epithelial Cancers. *Clin. Cancer Res.* **2019**, *25*, 1486–1493. [CrossRef] [PubMed]
186. Jazaeri, A.A.; Zsiros, E.; Amaria, R.N.; Artz, A.S.; Edwards, R.P.; Wenham, R.M.; Slomovitz, B.M.; Walther, A.; Thomas, S.S.; Chesney, J.A.; et al. Safety and efficacy of adoptive cell transfer using autologous tumor infiltrating lymphocytes (LN-145) for treatment of recurrent, metastatic, or persistent cervical carcinoma. *J. Clin. Oncol.* **2019**, *37*, 2538. [CrossRef]
187. Creelan, B.; Wang, C.; Teer, J.; Toloza, E.; Yao, J.; Kim, S.; Landin, A.; Mullinax, J.; Saller, J.; Saltos, A.; et al. Tumor-infiltrating lymphocyte treatment for anti-PD-1-resistant metastatic lung cancer: A phase 1 trial. *Immune ACCESS* **2021**, *27*, 1410–1418. [CrossRef]
188. Pedersen, M.; Westergaard, M.; Milne, K.; Nielsen, M.; Borch, T.H.; Poulsen, L.G.; Hendel, H.W.; Kennedy, M.; Briggs, G.; Ledoux, S.; et al. Adoptive cell therapy with tumor-infiltrating lymphocytes in patients with metastatic ovarian cancer: A pilot study. *Oncoimmunology* **2018**, *7*, e1502905. [CrossRef]
189. Tran, E.; Robbins, P.F.; Lu, Y.-C.; Prickett, T.D.; Gartner, J.J.; Jia, L.; Pasetto, A.; Zheng, Z.; Ray, S.; Groh, E.M.; et al. T-Cell Transfer Therapy Targeting Mutant KRAS in Cancer. *N. Engl. J. Med.* **2016**, *375*, 2255–2262. [CrossRef]
190. Zacharakis, N.; Chinnasamy, H.; Black, M.; Xu, H.; Lu, Y.C.; Zheng, Z.; Feldman, S.A. Immune recognition of somatic mutations leading to complete durable re-gression in metastatic breast cancer. *Nat. Med.* **2018**, *24*, 724–730. [CrossRef]
191. Tran, E.; Turcotte, S.; Gros, A.; Robbins, P.F.; Lu, Y.C.; Dudley, M.E.; Wunderlich, J.R.; Somerville, R.P.; Hogan, K.; Hinrichs, C.S.; et al. Cancer Immunotherapy Based on Mutation-Specific CD4+ T Cells in a Patient with Epi-thelial Cancer. *Science* **2014**, *344*, 641–645. [CrossRef]
192. Mullinax, J.E.; Hall, M.; Beatty, M.; Weber, A.M.; Sannasardo, Z.; Svrdlin, T.; Hensel, J.; Bui, M.; Richards, A.; Gonzalez, R.J.; et al. Expanded Tumor-infiltrating Lymphocytes From Soft Tissue Sarcoma Have Tu-mor-specific Function. *J. Immunother Hagerstown Md.* **2021**, *44*, 63–70. [CrossRef] [PubMed]
193. Ko, A.; Coward, V.S.; Gokgoz, N.; Dickson, B.C.; Tsoi, K.; Wunder, J.S.; Andrulis, I.L. Investigating the Potential of Isolating and Expanding Tumour-Infiltrating Lymphocytes from Adult Sarcoma. *Cancers* **2022**, *14*, 548. [CrossRef] [PubMed]
194. Zhou, X.; Wu, J.; Duan, C.; Liu, Y. Retrospective Analysis of Adoptive TIL Therapy plus Anti-PD1 Therapy in Patients with Chemotherapy-Resistant Metastatic Osteosarcoma. *J. Immunol. Res.* **2020**, *2020*, 1–12. [CrossRef] [PubMed]

cancers

MDPI

Review

Gender Differences in Soft Tissue and Bone Sarcoma: A Narrative Review

Ilaria Cosci [1], Paolo Del Fiore [2,*], Simone Mocellin [2,3,†] and Alberto Ferlin [4,5,†]

1 Veneto Institute of Oncology IOV-IRCCS, 35128 Padova, Italy; ilaria.cosci@iov.veneto.it
2 Soft-Tissue, Peritoneum and Melanoma Surgical Oncology Unit, Veneto Institute of Oncology IOV-IRCCS, 35128 Padua, Italy; simone.mocellin@iov.veneto.it
3 Department of Surgical, Oncological and Gastroenterological Sciences (DISCOG), University of Padua, 35128 Padova, Italy
4 Unit of Andrology and Reproductive Medicine, University Hospital of Padova, 35128 Padova, Italy; alberto.ferlin@unipd.it
5 Department of Medicine, University of Padova, 35128 Padova, Italy
* Correspondence: paolo.delfiore@iov.veneto.it
† These authors contributed equally to this work.

Simple Summary: This review focusing on gender differences in the incidence of soft tissue and bone sarcomas. Sarcomas are rare cancers arising from mesenchymal tissues, which are different from the epithelial tissues and originate from the embryonic mesodermal layer. These cancers can be classified into bone or soft tissue sarcomas. Most sarcomas occur without known causes; however, certain genetic syndromes and environmental factors are known to be associated with these malignancies. Studies have indicated a male predominance in sarcoma incidence, which is also seen in other cancers like colorectal and lung cancers. Notably, childhood sarcomas exhibit significant gender differences, with a stronger association with the male sex, particularly in soft tissue sarcomas. The biological reasons for these sex differences are not well understood, and this review seeks to shed light on these underlying factors to aid in prevention and treatment strategies.

Abstract: Sarcomas, uncommon malignancies, stem from mesenchymal tissues, distinct from epithelial tissues, originating in the embryonic mesodermal layer. These sarcomas have been categorized as either bone or soft tissue sarcomas, depending on their originating tissue. The majority of sarcomas occur sporadically with their etiology being unknown, but there are several, well-established genetic predisposition syndromes and some environmental exposures associated with specific sarcomas. Recently, many studies have shown that sarcomas, in analogy with colorectal, skin, head and neck, esophageal, lung, and liver carcinomas, also have a male sex predilection. Significant gender differences have already been observed in childhood sarcomas. Among the tumors strongly associated with the male sex, childhood sarcomas have been identified as being particularly sensitive to the biological differences between the sexes, with special regard to soft tissue sarcomas. As the biological mechanisms underlying the sex differences in the incidence of soft tissue sarcomas remain largely unexplored, this review aims to highlight the factors underlying these differences to inform prevention and treatment.

Keywords: gender differences; sarcoma; soft tissue cancer

Citation: Cosci, I.; Del Fiore, P.; Mocellin, S.; Ferlin, A. Gender Differences in Soft Tissue and Bone Sarcoma: A Narrative Review. *Cancers* **2024**, *16*, 201. https://doi.org/10.3390/cancers16010201

Academic Editors: Matthew T. Houdek and Catrin Sian Rutland

Received: 17 November 2023
Revised: 28 December 2023
Accepted: 28 December 2023
Published: 31 December 2023

1. Introduction

Sarcomas are tumors of mesenchymal origin, accounting for approximately 15% of all cancers in children and 1% of all malignancies in adults [1]. According to histopathological criteria and the primary site of occurrence within tissues, the World Health Organization has delineated over 70 distinctive histological subtypes of soft tissue sarcomas (the most common of which are shown in Figure 1) and about 10 subtypes of bone sarcomas (the most common of which are shown in Figure 2) [2].

Type of Soft Tissue Sarcoma	Characteristics and Common Locations	Typical Age Groups Affected
Adult Fibrosarcoma	Affects fibrous tissue in legs, arms or trunk.	Most common in ages 20–60, but can occur at any age.
Alveolar Soft-Part Sarcoma	Rare cancer that often starts in the legs.	Primarily affects young adults.
Clear Cell Sarcoma	Often starts in tendons of the arms legs with features resembling malignant melanoma.	Occurs in various age groups.
Epithelioid Sarcoma	Most often starts in tissues under the skin of hands, forearms, feet, or lower legs.	Teens and young adults are often affected.
Desmoplastic Small Round Cell Tumor	Rare sarcoma often found in the abdomen.	Primarily affects teens and young adults.
Fibromyxoid Sarcoma Low-Grade	Slow-growing cancer; often starting in the trunk or limbs.	More common in young to middle-aged adults.
Gastrointestinal Stromal Tumor (GIST)	Starts in the digestive tract.	Occurs in various age groups.
Angiosarcoma	Starts in blood vessels (hemangiosarcomas) or lymph vessels (lymphangiosarcomas).	Can occur after radiation therapy, in the breast, and limbs with lymphedema.
Kaposi Sarcoma	Starts in cells lining lymph or blood vessels.	Common in individuals with compromised immune systems.

Type of Soft Tissue Sarcoma	Characteristics and Common Locations	Typical Age Groups Affected
Leiomyo Sarcoma	Starts in smooth muscle tissue, often in the abdomen or other body parts.	Occurs in various age groups.
Liposarcomas	Malignant tumors of fat tissue, can occur in various body locations.	Most common in adults between 50 and 65 years old.
Malignant Mesenchymoma	Rare type showing features of fibrosarcoma and at least two other sarcoma types.	Occurs in various age groups.
Malignant Peripheral Nerve Sheath Tumors	Start in cells surrounding nerves.	Occurs in various age groups.
Myxofibro Sarcomas Low-Grade	Found in arms and legs, often in elderly people, primarily just under the skin.	Most common in elderly individuals.
Rhabdomyo Sarcoma	Most common in children, typically found in various body locations.	Primarily affects children.
Synovial Sarcoma	Malignant tumor around joints, common in the hip, knee, ankle, and shoulder.	More common in children and young adults but can occur in older people.
Undifferentiated Pleomorphic Sarcoma (UPS)	Often found in the arms or legs, may start inside the abdomen.	Most common in older adults.

Figure 1. The most common soft tissue sarcomas.

Type of Bone Sarcoma	Characteristics and Common Location	Typical Age Group Affected
Osteosarcoma	Highly aggressive, affects long bones (often around knee), commonly seen in teenagers and young adults.	Adolescents (10–19 years) and Young Adults (20–29 years) and Older Adults (60+years)
Ewing's Sarcoma	Aggressive, affects bones and soft tissues, commonly found in pelvis, legs, arms, chest wall.	Children (0–14 years) and Young Adults (15–29 years)
Chondrosarcoma	Arises from cartilage, usually in the pelvis, hip, shoulder, or ribs, tends to grow slowly.	Middle Aged Adults (30–49 years) and Older Adults (50+ years)
Chordoma	Arises from remnants of the notochord, typically found in the base of the skull or lower spine.	Middle –Aged Adults (30–49 years) and Older Adults (50+ years)

Figure 2. The most common bone sarcomas.

Sarcomas are rare tumors (their annual incidence is lower than six new cases per 100,000 people) that can occur in virtually any part of the body. Their prognosis varies greatly across different histological subtypes and depends also on the stage the tumor is diagnosed. Due to their rarity and heterogeneity, sarcomas show better outcomes when treated within the frame of specialized centers, with inappropriate medical management being reported in more than 70% of patients treated outside dedicated centers [3,4].

Although most sarcomas typically occur without a known cause, some types are attributable to various confirmed genetic predisposition syndromes and specific environmental factors. However, although the etiology of sarcomas is not yet clear, numerous studies agree that sarcomas, like other types of cancer known so far, arise predominantly in men. This gender disparity has already been observed in childhood sarcomas, where there is a strong association between the male sex and soft tissue sarcomas. Because the precise biological mechanisms responsible for these sex-based differences in sarcoma incidence remain largely unexplored, this review seeks to shed light on the factors that underlie these distinctions, improving our understanding for prevention and treatment.

1.1. *Environmental Factors and Sarcomas*

Although the etiology of sarcomas remains poorly understood, several environmental and genetic factors have been identified that are responsible for their development.

Previous studies have shown that exposure to radiation for the treatment of other cancers, such as lymphoma, breast, testicular, ovarian, prostate, and lung, increases the risk of developing sarcomas. While radiation-associated sarcomas (RAS) are typically infrequent, affecting less than 1% of individuals undergoing radiation therapy, their occurrence is anticipated to rise due to the expanded utilization of radiation therapy for specific tumor types and the overall enhancement in cancer patient survival rates.

Radiation-associated sarcoma is a rare iatrogenic malignancy that occurs after radiotherapy for a high malignant grade and accounts for approximately 0.5–5.5% of all sarcomas. This event appears to be due to the ability that ionizing radiation has to damage the DNA triggering a spectrum of molecular changes, ranging from minor single base substitutions to severe genomic disorders. The use of whole genome sequencing in a limited subgroup of radiation-related neoplasms (carcinomas and sarcomas) has allowed the identification of particular traits highlighting a greater prevalence of small deletions (with a higher ratio to insertions) compared to non-radiation-induced tumors.

Furthermore, these neoplasms also show balanced inversions responsible for chromotripsy and other structural abnormalities damage [5].

The 3-year survival rate for individuals with radiation-associated sarcoma varies and is generally considered poor due to the aggressive nature of the disease and its resistance to chemotherapy. A study by Xi et al. reported that the 3-year overall survival rate was 32.4% among treated patients with RAS, with the median survival being 21.2 months. Complete surgical resection was identified as a major prognostic factor for survival. Another study by Wei et al. reported a 3-year overall survival rate of 19.1%, with a 3-year survival rate with no disease at 11.1% [6,7].

In a recent retrospective study focusing on radiation-induced sarcomas, it was found that the overall median survival for RAS patients at 3 years was 36 months.

This study, however, noted no significant survival differences when stratifying patients by various factors such as age at radiation therapy, latency time, and age at RAS occurrence. This indicates that these factors may not have a straightforward impact on the survival of RAS patients [5].

While the 5-year survival rate for individuals with RAS ranges from 17% to 58%, it is considerably lower than the 54% to 76% rate observed in the cases of sporadic sarcomas [8].

A study comparing radiation-associated sarcoma of the pelvis/sacrum (RASB) and primary osteosarcoma/spindle cell sarcoma of pelvis/sacrum (POPS) found that older RASB patients were less likely to receive chemotherapy and more likely to have higher perioperative mortality and worse 5-year disease-specific survival. No difference was noted in local recurrence or metastasis-free survival at 5 years. Overall, RASB shows poorer outcomes compared to POPS [9].

Diagnostic protocols for RAS continue to follow guidelines established in the 1940s, which have undergone few changes over time. These guidelines are based on three fundamental principles: the documented history of radiation exposure preceding the development of the sarcoma, the onset of the sarcoma in or near the radiation-exposed area, and the confirmation of the sarcoma through a histological diagnosis and histological diagnosis of sarcoma other than cancer in the first instance [10].

Although ionizing radiation remains the most relevant environmental factor associated with the development of sarcoma, other environmental elements have been investigated. In particular, increased incidence and mortality for some types of cancer, particularly soft tissue sarcomas (STS), have been observed in individuals exposed to both agricultural and non-agricultural chemicals. The link between these chemicals and STS was initially highlighted by Hardell et al. in 1977 [11]. This evidence was supported in 1995 by the same authors in a Swedish case–control study conducted in the 1970s and 1980s, which confirmed a significantly increased risk of STS with exposure to these compounds, with an odds ratio (OR) of 2.7 (95% confidence interval (CI) 1.9, 4.7) and OR of 3.3 (95% CI: 1.8, 6.1) obtained for phenoxy acetic acids and chlorophenols, respectively [12].

A recent systematic review that included a total of 50 publications and 35 meta-analyses highlighted, in 16 studies involving 2254 participants, a combined odds ratio (OR) for sarcoma of 1.85 (95% CI: 1, 22, 2,82) about exposure to phenoxy herbicides and chlorophenols, with a pooled standardized mortality ratio based on four cohort studies with 59,289 participants of 40.93 (95% CI: 2.19, 765.90) [13]. Furthermore, exposure to vinyl chloride monomers resulted in pooled hazard ratios of 19.23 (95% CI: 2.03, 182.46) for hepatic angiosarcoma and 2.23 (95% CI: 1.55, 3, 22) for other STS, respectively, in three cohort studies involving 12,816 participants [14], while, in four cohort studies including 30,797 participants, an association between dioxin exposure and increased STS mortality was observed, showing a combined standardized mortality ratio of 2.56 (95% CI: 1.60, 4.10). Finally, woodworking occupations also represent an increased risk of developing STS, presenting an aggregate odds ratio of 2.16 (95% CI: 1.39, 3.36) [15].

1.2. Genetic Susceptibility and Sarcoma

Genetically, sarcomas are a type of tumor that can be divided into two groups. One group is characterized by simple karyotype defects consisting of disease-specific chromosomal translocations leading to abnormal gene (and protein) function, such as Ewing sarcoma, alveolar rhabdomyosarcoma, and synovial sarcoma. Furthermore, there are instances where complex karyotypes are observed, signifying a notable disturbance in genomic stability, seen for example in leiomyosarcoma, liposarcoma, undifferentiated pleomorphic sarcoma, osteosarcoma, angiosarcoma, and malignant peripheral nerve sheath tumor [16].

In addition, it is now known that individuals with certain genetic syndromes are predisposed to developing sarcoma. In particular, many inherited syndromes such as Familial gastrointestinal stromal tumor syndrome (GIST), Li–Fraumeni syndrome (LFS), neurofibromatosis (NF1), and retinoblastoma (Rb), Bloom syndrome (BS), fumarate hydratase (FH), Rothmund–Thompson syndrome (RTS), and Werner syndrome (WS) are responsible for an increased risk of developing this type of cancer [17].

Familial gastrointestinal stromal tumor syndrome (GIST)

Familial gastrointestinal stromal tumor (GIST) syndrome is a condition associated with sarcoma development, originating from the interstitial cells of Cajal in the gastrointestinal tract. Recognized recently as distinct, GISTs were previously classified as leiomyomas or leiomyosarcomas. Most GISTs (75% to 80%) exhibit c-kit gene mutations, especially in exon 11, which lead to increased function of a tyrosine kinase receptor, fostering cell proliferation. Additionally, mutations in the PDGFRA gene account for 5% to 15% of GIST cases [18]. Moreover, the syndrome is characterized by activating mutations in the KIT or PDGFRA genes, which include inherited mutations in specific exons of these genes. These mutations lead to continuous kinase activation, promoting tumorigenesis. Individuals with these mutations may develop multiple GISTs in the stomach or bowel [18].

Li–Fraumeni syndrome (LFS)

Li–Fraumeni syndrome (LFS) was initially identified and documented by Li and Fraumeni, along with their colleagues, after examining four family cases wherein either siblings or first cousins manifested pediatric sarcoma, while a parent had experienced early-onset cancer.

The Li–Fraumeni syndrome definition requires "the presence of a proband diagnosed with sarcoma before age 45, a first-degree relative younger than age 45 with any cancer, and a first- or second-degree relative younger than age 45 with any cancer or a sarcoma at any age" [19]. In 2009, a version of the Chompret Criteria was proposed to help clinicians to recognize Li–Fraumeni syndrome, expanding on the old definition and based on three criteria (Table 1) [20]. The syndrome is linked to TP53 gene mutations, which typically prevent tumor growth and promote cell death and DNA repair. These mutations often result in the loss of p53 function or impede the cell death pathway, contributing to sarcoma and other cancers due to genomic instability. Germline mutations in the p53 gene are found in Li–Fraumeni syndrome (LFS), with 30% to 60% of soft tissue sarcomas showing somatic p53 mutations. Affected individuals are prone to a range of tumors, including breast, brain, adrenocortical tumors, and various leukemias. Children with rhabdomyosarcoma under age 3 may particularly carry TP53 germline mutations [21]. A study from 2004 to 2015 with 89 identified carriers showed that those who underwent a comprehensive surveillance protocol for Li–Fraumeni syndrome had better outcomes. Surveillance led to the detection of 40 asymptomatic tumors and was associated with higher 5-year survival rates compared to those who declined surveillance. The study suggests incorporating surveillance into clinical management for improved survival in TP53 variant carriers [22].

Table 1. 2009 Chompret Criteria for germline TP53 mutation screening.

Criterion	Description
I.	Proband with a tumor belonging to the LFS tumor spectrum (e.g., soft tissue sarcoma, osteosarcoma, brain tumor, premenopausal breast cancer, adrenocortical carcinoma, leukemia, and lung bronchoalveolar cancer) before the age of 46 years and at least one first- or second-degree relative with an LFS tumor (except breast cancer if the proband has breast cancer) before the age of 56 years or with multiple tumors.
II.	Proband with multiple tumors (except multiple breast tumors), two of which belong to the LFS tumor spectrum, and the first of which occurred before the age of 46 years.
III.	Patient with adrenocortical carcinoma or choroid plexus tumor, irrespective of family history.
Abbreviations	LFS, Li–Fraumeni syndrome

Neurofibromatosis (NF1)

Neurofibromatosis is a genetic disorder that progresses slowly, primarily affecting neuroectodermal tissues like skin, nerves, bones, and eyes. It increases the risk of developing malignant peripheral nerve sheath tumors (MPNST), with up to a 10% lifetime risk. NF1, a subtype of this condition, involves a heterozygous mutation in the neurofibromin gene on chromosome 17q11.2. Neurofibromin is a tumor suppressor that regulates the Ras/MAPK/AP-1 pathway, and its loss leads to heightened Ras activity, which can promote tumor growth. NF1 is associated with various tumors, including neuroblastoma, neurofibroma, thymoma, and breast cancer [23].

Retinoblastoma (Rb)

Retinoblastoma is a childhood eye cancer arising in the retina, with hereditary and non-hereditary forms. Survivors of hereditary retinoblastoma are at increased risk of secondary tumors like osteosarcoma, and radiation treatment heightens this risk [24].

Retinoblastoma arises from germline mutations that deactivate one RB1 gene allele. Tumors form when both RB1 alleles mutate, followed by other genetic changes. Hereditary retinoblastoma patients are more likely to develop osteosarcoma, with RB1 mutations detected in 30–75% of these tumors. Hereditary survivors face a greater risk of secondary osteosarcomas than the general population or those with non-hereditary retinoblastoma [25,26]. RB1 gene mutations not only predispose individuals to retinoblastoma but also increase the risk for secondary tumors like soft tissue sarcomas, melanoma, brain tumors, and some carcinomas, including those of the lung, breast, and bladder [27].

Bloom syndrome (BS)

Bloom syndrome (BS) is a rare autosomal recessive disorder more common among Ashkenazi Jews, characterized by genetic instability and increased sarcoma risk. It is caused by mutations in the BLM gene on chromosome 15, which encodes a RecQ family DNA helicase essential for genomic stability. Ashkenazi Jewish BS patients often have a specific frameshift mutation. Cells from BS individuals show numerous cytogenetic abnormalities, such as increased chromosome breaks, quadriradial chromatid exchanges, and, notably, a high rate of sister chromatid exchanges (SCEs), which is a key diagnostic marker for the syndrome [28].

Rothmund–Thompson syndrome

Rothmund–Thomson syndrome is an autosomal recessive disorder caused by mutations in the RECQL4 gene, which increases cancer risk, particularly osteosarcoma. Mutations leading to RECQL4 loss-of-function include nonsense mutations, frameshifts, splice site changes, and intron deletions. Located on chromosome 8q24.3, RECQL4, a RecQ DNA helicase gene, is often amplified in osteosarcoma, near the MYC gene. Those with Rothmund-Thomson syndrome have a higher chance of developing malignancies, especially osteosarcoma [29].

Werner syndrome

Werner syndrome (WS) is a rare autosomal recessive disorder that emerges in the late teens or early twenties, simulating symptoms of accelerated aging like heart disease, osteoporosis, hair loss, cataracts, diabetes, and hypogonadism. It is caused by mutations in the WRN gene, a RecQ helicase on chromosome 8p11.1. People with WS have a higher risk for several cancers, including osteosarcoma, soft tissue sarcoma, meningioma, myeloid disorders, melanoma, thyroid carcinoma, and various epithelial cancers. There are over 90 known mutations affecting the WRN gene, including base substitutions, insertions, deletions, and complex mutations that disrupt its function and reading frames [30].

1.3. Sex Differences in Sarcomas

In recent years, epidemiological studies have highlighted sexual dimorphism as a relevant factor in the incidence and survival of many cancers, including colorectal, skin, head, neck, esophageal, lung, and liver cancers. Indeed, incidence rates ranging from 1.26:1 to 4.86:1 have been reported to be significantly higher in men than in women, irrespective of the ethnicity of the population studied [31–34].

Overall, the variations in question have commonly been linked to occupational or behavioral aspects. However, researchers have also explored cellular and molecular influences, particularly emphasizing the impact of sex hormones. Notably, these hormones might potentially influence cancer cells, elements within the tumor's microenvironment, cellular metabolism, and the immune system [34].

Recently, Rong J et al. collected gastric GIST data from 2010 to 2016 through the SEER database, using, for the first time, propensity score matching (PSM) with a relatively large sample size. These authors aimed to investigate the relationship between sex and prognosis in patients with gastric GISTs using a method capable of reducing the influence of confounding factors.

The findings from this study emphasized the role of gender as a distinct factor influencing the prognosis of gastric GIST, revealing that males exhibit a heightened risk of mortality compared to females [35].

Indeed, compared to female gastric GIST, male patients were less likely to undergo surgical treatment (95.9% vs. 98.1%), more likely to have large tumors (>10.0 cm) (24.0% vs. 16.4%), and more likely to have a mitotic index greater than 10/50 HPF (14.1% vs. 9.7%). These data confirm the findings of previous studies reporting that the prognosis of gastric GIST is related to sex [36].

The link between variations in sex and prognosis might be attributed to the involvement of sex hormones. Evidence from other cancer studies indicates that the pathway associated with sex hormone signaling can impact susceptibility to cancer and the microenvironment of tumors, albeit operating through diverse mechanisms. These hormones play a role in controlling processes such as angiogenesis and inflammation, thereby affecting the progression of cancer differently among genders. For instance, there's been a noted reduction in ERß levels within cancer scenarios.

A recent study has highlighted the interconnectedness of estrogen with the mucosal barrier, gastrointestinal functionality, and the regulation of intestinal inflammation. This hormone appears to play a protective role, particularly in gastrointestinal tumor development, notably in the case of colorectal cancer.

Indeed, it has been reported that the use of anti-estrogen drugs such as tamoxifen may increase the risk of gastric adenocarcinoma [37].

A prospective study conducted by Freedman ND and colleagues, as part of the Shanghai Women's Health Study and based on population data, revealed an association between exposure to hormones like estrogen and a decreased likelihood of developing gastric cancer [38].

Additionally, exploring the notion that sex hormones might contribute to the development of esophageal or gastric adenocarcinoma, M Lindblad et al. conducted research on the protective effects of hormone replacement therapy (HRT) in women against these tumors. Their findings indicated that HRT usage correlates with a reduction of over 50% in the risk of gastric adenocarcinoma among users compared to non-users [39].

Newer studies have provided additional evidence supporting the notion that extended exposure to estrogen could potentially lower the chances of developing gastric cancer. Recent meta-analysis findings have revealed a noteworthy connection: the utilization of hormone replacement therapy (HRT), commonly employed to alleviate menopausal symptoms, was linked to a 28% decrease in gastric cancer risk compared to individuals not exposed to HRT. Moreover, a subgroup analysis focused on the type of HRT formulation demonstrated risk reductions in gastric cancer following the use of both estrogen-only therapy (pooled RR, 0.63; 95% CI: 0.51–0.77, $I^2 = 0\%$) and estrogen–progestin therapy (pooled RR, 0.70; 95% CI: 0.57–0.87; $I^2 = 0\%$) compared to non-users [39,40].

For soft tissue sarcomas (STS), the variability of incidence-based mortality by sex over the past decade has not been well studied, but many studies examining sex differences have highlighted that men have a higher incidence of STS compared with their female counterparts.

In England, an estimated 4295 cases of soft tissue sarcoma are diagnosed annually, representing a crude rate of 7.7 cases per 100,000 individuals (2017–2019). The distribution of soft tissue sarcoma cases in the UK appears to be relatively equal between females and males (1996–2010). Despite this equality, the incidence rates of soft tissue sarcoma, especially the European age-standardized (AS) rate, significantly differ between genders, being notably lower in females compared to males [33]. These results were confirmed by another analysis conducted by Hung and collaborators performed on the Taiwanese population according to the 2013 WHO classification.

The study highlighted a male predominance, particularly marked in Kaposi's sarcoma (SIRR, 5.4; 95% CI: 4.41–6.63, $p < 0.05$), as well as in other subtypes such as UPS, liposarcoma, angiosarcoma, fibrosarcoma, GIST, chondrosarcoma (CS), and NOS sarcoma with a SIRR of 1.2–2.1 [41].

Hsieh et al. previously documented similar outcomes, exploring discrepancies in the occurrence rates and patterns of STS across different racial and ethnic groups in adolescents and young adults aged 15–29 years, considering sex, age, and histological type. Their findings revealed a 34% higher incidence of all STSs combined in males compared to females (95% CI: 1.28, 1.39), as well as a 60% higher incidence in Black individuals compared to Caucasians (95% CI: 1.52, 1.68) [42].

The demographic associations linked to STS in recent years and the probability of developing comorbidities have recently been analyzed to understand the mechanisms underlying the disease and to focus on diagnostic and therapeutic strategies.

In their examination of patients with STS, Van Herk-Sukel and colleagues discovered a notable escalation in the risk of developing medical complications, specifically highlighting a heightened risk for cardiovascular disease [43].

The proportion of STS in male and female patients may vary significantly depending on the type of tumor. A study analyzing the incidence and survival of cutaneous soft tissue sarcomas (CSTS) in the US population highlighted that CSTS rates varied markedly over time and by race, sex, and histological type, supporting the notion that the histological variants of CSTS are etiologically distinct and appear to affect males more than females [44]. Patel SJ et al., in a recent investigation utilizing data sourced from the National Cancer Institute's Surveillance, Epidemiology, and End Results (SEER), analyzed mortality rates correlated with incidence between 2000 and 2016 across various categories such as tumor grades, gender, and racial demographics among individuals diagnosed with STS.

The authors confirmed that STS has a male sex predilection and that the male sex tends to have a higher incidence of mortality, regardless of tumor grade and race.

Indeed, this study highlights the higher incidence-based mortality rate in Caucasian males compared to African American males over the past 15 years, suggesting that soft tissue sarcomas in Caucasian males have worse outcomes [45].

As a result of these recent trends, the American Cancer Society predicts that 13,400 new soft tissue sarcomas will be diagnosed in the United States in 2023 (7400 in males and 6000 in females), with the majority of cases occurring in male patients. In addition, an estimated 5140 people (2720 males and 2420 females) will die from STS. Although the proportion of STS in male and female patients can vary significantly depending on the type of tumor, STS appears to affect males more than females [46].

As osteosarcoma is rarer than other sarcomas and is characterized by an incidence with a bimodal age distribution, little is known about its incidence in males and females.

Cole S et al. conducted a study utilizing data provided by the National Cancer Institute's Surveillance, Epidemiology, and End Results (SEER) program to delve into younger osteosarcoma cases and examine racial minorities. Despite evidence suggesting a higher incidence of osteosarcoma among individuals of African descent, the authors extensively compared osteosarcoma occurrence and survival rates across four distinct age groups (0–9, 10–24, 25–59, and >60 years old) based on race/ethnicity, gender, decade, pathological subtype, and tumor location.

Their investigation uncovered a total of 2312 osteosarcoma cases within the 10–24 age category, nearly accounting for half of all osteosarcoma cases in the SEER 18 database. Additionally, it was noted that osteosarcoma demonstrated a higher prevalence in males compared to females across all race/ethnicity categories, displaying an overall male-to-female ratio of 1.3:1 (males, IR, 8.1; 95% CI: 7.7–8.6 vs. females, IR, 6.2; 95% CI: 5.8–6.6). Furthermore, in the >60 age group, males exhibited a higher prevalence of osteosarcoma compared to females, both overall (1.3:1) and within specific race/ethnicity categories [47].

Ewing sarcoma ranks as the second most prevalent bone sarcoma, displaying an incidence rate of 0.3 cases per 100,000 individuals annually. Within childhood cancer cases, this particular subtype of bone sarcoma constitutes approximately 2%. Notably, it exhibits a higher prevalence among males compared to females, maintaining a male-to-female ratio of roughly 1.5, and typically peaks in incidence around the age of 15 [48].

Gender differences in the incidence of radiation-associated sarcomas are not widely documented and the evidence may vary by study. However, some literature suggests that there could be a higher incidence in females, especially following radiation treatment for breast cancer, which is one of the more common scenarios leading to RIS. The specific incidence rates and gender differences can depend on a multitude of factors, including the type and location of the primary cancer treated with radiation, genetic predispositions, and the individual radiation doses administered. A recent study on RIS provided insights into the distinct genomic landscapes of these sarcomas. The research included 82 patients, predominantly females (83%), with a median age of 64 years. It compared radiation-associated angiosarcomas (RT-AS) with other RIS histotypes and sporadic sarcomas. The study found notable differences in the mutation and copy number alteration profiles among various RIS histotypes. RT-AS, especially those derived from breast radiation, had a unique genomic landscape with frequent MYC, FLT4, CRKL, HRAS, and KMT2D alterations. In contrast, other RIS types had genomic features similar to their non-radiation counterparts. The findings suggest that potential molecular targets for treatment could be specific to each histotype [49]. However, in the Inoue YZ study which evaluates the analysis of the clinical–pathological characteristics and treatment of patients with post-irradiation sarcoma of the bones and soft tissues, there does not seem to be a clear gender bias in this condition. Indeed, the ratio of males to females was 8:5, but this difference vanished when tumors specific to one gender (such as those affecting the breast, cervix, testis, or ovary) were excluded from the analysis. Additionally, no racial predilection has been noted in the research literature [50].

Furthermore, the onset of sarcomas associated with genetic syndromes does not seem to have a clear gender disparity. However, the manifestation of certain types of sarcomas may vary with age and the specific genetic mutation involved. For example, individuals with Li–Fraumeni syndrome (LFS), which is characterized by germline mutations in the TP53 gene, tend to develop sarcomas and other cancers at a younger age compared to the general population. Specifically, sarcomas represent a significant proportion of tumors in TP53 mutation carriers, with most occurring before the age of 50. The type of sarcoma can also be correlated with the type of TP53 mutation present, with certain mutations being associated with early-onset sarcomas like rhabdomyosarcoma in individuals younger than 5 years and osteosarcoma at any age [51].

It is important to note that knowledge of these genetic predispositions is crucial for the clinical management of sarcoma patients. Identifying individuals with heritable cancer predisposition syndromes can help tailor treatment strategies to minimize toxicity and maximize efficacy. Furthermore, understanding a patient's genetic background can assist in the implementation of appropriate genetic counseling, as well as screening or surveillance strategies for both the patient and their relatives [52].

The research emphasizes that while treatment strategies for most sarcomas may not significantly differ between sporadic cases and those associated with predisposition syndromes, the recognition of genetic predisposition is vital for overall patient management [53].

1.4. Sex Differences in Sarcoma in Childhood

Significant sex differences have already been observed in childhood sarcomas. Among the tumors strongly associated with the male sex, childhood sarcomas have been identified as being particularly sensitive to sex biological differences.

Osteosarcoma (OS) ranks as the prevailing primary malignant bone tumor found among children and teenagers, with an estimated 4.8 new cases per million individuals under the age of 20 in the United States annually; this figure to roughly 450 cases per year in this age group, contributing to approximately 3% to 5% of childhood tumors. OS demonstrates a higher occurrence among males and African Americans. Following OS, the second most frequent primary malignant bone tumor among children and adolescents is Ewing sarcoma (ES), recording an estimated 2.9 new cases per million among individuals under the age of 20 in the United States annually.

Ewing sarcoma (ES) exhibits a slightly higher occurrence among males, and its frequency is nine-fold greater in Caucasians compared to African Americans [54].

The male predominance in childhood cancer incidence is well known, but few studies have focused on sex differences in incidence during childhood and adolescence.

Ognjanovic et al. conducted an analysis focusing on the incidence and survival patterns of rhabdomyosarcoma (RMS) among children and adolescents under the age of 20. This cancer type is typically categorized into two primary histological subtypes: embryonal RMS (ERMS) and alveolar RMS (ARMS), with ERMS accounting for 60–70% of cases and ARMS representing 20–30%. The investigation considered various demographic factors such as sex (male and female), age brackets (0–4 years, 5–9 years, 10–14 years, and 15–19 years), and racial backgrounds (white, black, American Indian/Alaskan Native, and Asian/Pacific Islander combined). In the nine SEER registers between 1975 and 2005

A total of 987 cases of RMS were diagnosed among children aged 0–19 years. Males displayed a higher incidence of RMS compared to females, with rates of 5.2 per 1,000,000 and 3.8 per 1,000,000, respectively, resulting in a rate ratio of 1.37 (95% CI: 1.21–1.56). Notably, the male predominance in RMS was predominantly seen in ERMS, with a male-to-female rate ratio of 1.51 (95% CI: 1.27–1.80). These findings support previous reports indicating high incidence rates of RMS, its early onset before the age of 10 in more than 50% of cases, and a distinct male predominance [55].

This disparity has been found in most pediatric cancers, acute lymphoblastic leukemia, non-Hodgkin's lymphoma, medulloblastoma, hepatic tumors, osteosarcoma, and germ cell tumors, showing that the direct effect of the male sex is significant for several tumor types. Furthermore the American Cancer Society reported a higher cancer incidence in males aged 0–14 years with a rate of 178.0 per 1,000,000, unlike females where a rate of 160.1 cases per 1,000,000 was observed, corresponding to a crude incidence rate ratio for childhood cancer for 1.11 for male versus female [56]. To shed light on potential mechanisms, such as hormonal fluctuations or periods of rapid growth, that may contribute to increased cancer incidence in males, Williams LA and collaborators used data from the Surveillance, Epidemiology, and End Results (SEER)'s 18 registries (2000–2015) to examine the association between male sex and childhood cancer by single year of age (0–19) and tumor type. The study found that male sex was positively associated with most cancers.

Particularly notable were the positive correlations found between the male gender and various types of cancers across different age groups. Neuroblastoma (IRR, 1.13; 95% CI: 1.07–1.19), retinoblastoma (IRR, 1.17; 95% CI: 1.08–1.26), and hepatoblastoma (IRR, 1.70; 95% CI: 1.53–1.86) exhibited significant associations with male sex across all ages. Notably, neuroblastoma showed a significant link with males at ages 1, 2 to 3, and 5, while retinoblastoma manifested this association at age 2. Hepatoblastoma, on the other hand, exhibited a notable correlation with males aged < 1 to 3. Furthermore, the male gender also displayed positive associations with nephroblastoma at ages < 1 to 1, and with various bone tumors (osteosarcoma IRR, 1.33; 95% CI: 1.25–1.41; chondrosarcoma IRR, 2.59; 95% CI: 2.08–3.10; Ewing sarcoma IRR, 1.69; 95% CI: 1.56–1.81), notably significant in osteosarcoma at ages 14 to 19. Ewing sarcoma consistently exhibited an association with male sex, significantly so from ages 9 to 19, except at age 1. "The Ewing sarcoma family" of tumors (Ewing tumors, Askin tumors, and pPNET) showed associations with male sex (IRR, 1.27; 1.06–1.48), similarly with rhabdomyosarcoma (IRR, 1.42; 95% CI: 1.33–1.51) and fibrosarcoma (IRR, 1.14; 95% CI: 1.00–1.28). Additionally, a strong association was observed between male sex and GCTs, particularly intracranial/intraspinal (IRR, 2.73; 95% CI: 2.51–2.95) and malignant gonadal GCTs (IRR, 2.35; 95% CI: 2.24–2.45) across all age groups. These higher incidence rates among males remained consistent from childhood through adolescence [57,58].

In a more recent analysis aiming to identify sex-based survival disparities in childhood cancers, the same researchers examined overall survival differences and estimated the risk of death in males versus females for 18 childhood cancers. This investigation utilized the Surveillance, Epidemiology, and End Results (SEER) program's 18 registries spanning from 2000 to 2014.

This study not only confirmed that males had worse overall survival than females, but also reported worse survival and an increased risk of death for males diagnosed with ependymoma, neuroblastoma, and osteosarcoma. Furthermore, disparities in five-year survival rates based on gender were evident, indicating lower rates among males (85% for males, 88% for females). This trend was notably pronounced in specific cancers: ependymoma (71% male, 78% female), neuroblastoma (74% male, 78% female), hepatoblastoma (77% male, 82% female), osteosarcoma (64% male, 71% female), and Ewing sarcoma (67% male, 71% female). Notably, males exhibited significantly poorer overall survival across a 15-year observational period for ependymoma (log-rank $p = 0.02$), neuroblastoma (log-rank $p = 0.003$), and osteosarcoma (log-rank $p = 0.004$) [59]. However the concept of gender disparities in the incidence of certain sarcomas among different age groups remain an intriguing aspect of cancer epidemiology]. During childhood, particularly in the prepubescent years, some sarcomas show a marked difference in incidence between genders. For instance, embryonal rhabdomyosarcoma (ERMS) is more commonly diagnosed in boys than in girls [60]. The reasons behind this gender inequality in incidence rates are not fully understood but may involve genetic, environmental, and hormonal factors, even though these hormonal influences are not as pronounced before puberty.

As individuals age, the incidence of sarcomas still shows gender variations, but the reasons for these differences may shift. In older adults, hormonal differences between genders diminish, particularly post-menopause in women, when levels of hormones such as estrogen and progesterone decrease significantly. Despite the reduced hormonal differences, gender disparities in sarcoma incidence persist in the elderly. This could suggest that factors other than hormones, such as genetics, lifestyle, or environmental exposures, may play a more significant role in the development of sarcomas in older age groups [61].

1.5. Sex Biological Differences in Sarcoma in Childhood

The intricate biological pathways contributing to the variations in childhood cancer occurrence between genders have largely eluded extensive investigation.

However, it is thought that the increased cancer diagnoses in boys are mainly due to hormonal, genetic, and immune factors.

Some sex differences in cancer may be due to differences in the hormonal milieu. While this facet might not hold as much significance during early childhood compared to adolescence and adulthood, certain cancers like osteosarcoma, fueled by swift bone development and/or hormonal changes during puberty, could potentially be affected by fluctuations in hormones [62].

Williams LA et al. observed that osteosarcomas and Ewing sarcoma have a male-to-female ratio that fluctuates according to pubertal timing, which coincides with peak bone growth dependent on estrogen [58].

Recent research emphasizes the crucial involvement of the growth hormone (GH)/insulin-like growth factor 1 (IGF-1) axis in bone growth during puberty. Studies suggest that estrogen boosts GH secretion in both boys and girls, while testosterone primarily impacts GH secretion by converting it into estrogen through aromatization [63].

Differences in the onset of puberty and estrogen levels in males and females may regulate the rate of bone growth, which may explain the higher incidence of osteosarcoma and Ewing sarcoma in males during adolescence [47].

Furthermore, the strikingly higher male predominance observed in chondrosarcoma compared to other bone tumors raises noteworthy considerations, suggesting a potential involvement of hormonal or growth-related factors in this observed pattern. Population-based studies indicate a protective effect for women in cancer survival, potentially linked to hormonal differences. In high-grade CS, the female gender, likely influenced by estrogen, significantly improves survival compared to males. This effect is age-dependent, diminishing post-menopause. Estrogen's role in bone and cartilage development and its presence in CS suggest a potential for anti-estrogen therapy. Despite inconsistent findings across studies, the age and gender impact on chondrosarcoma survival highlights the importance of further research into the interaction between sex hormones and high-grade CS [48,64].

The consistent prevalence of males in medulloblastoma is likely attributed to the asymmetric impact of sex on brain development. Animal studies have highlighted the contribution of prenatal hormones, sex chromosomes [65], and the immune system in influencing early neural sexual differentiation [66].

The elevated risk of childhood cancer among males may also be rooted in genetic factors. Although studies of sex-biased gene expression during human development are limited in adults, it has also been observed in children that genes on the X chromosome show higher variation in expression between the sexes than autosomal genes, and this may depend on sex differences in chromatin accessibility. Additional genetic mechanisms could involve somatic mutations occurring on the X or Y chromosome, affecting males and females differently. Over time, a preference toward one X chromosome may emerge, potentially confining mutations to the inactive copy due to selection pressure. In contrast, any mutation of the X chromosome in males is obligatory throughout their lifespan and could exert a more pronounced influence on gene expression and cancer development [67]. Given that the Y chromosome contains fewer than 100 genes, its role in increasing cancer risk in males cannot be ignored [68].

Finally, in both children and adults, disparities in cancer risk between the sexes may be linked to differences in the immune response to tumor development. Notably, males typically exhibit lower innate and adaptive immune responses than females, evidenced by higher rates of infectious diseases in males and elevated rates of autoimmune diseases in females.

This attribute likely arises from the abundance of immune-related genes located on the X chromosome and their distinct expression patterns, which potentially contribute to discernible differences in immune function between males and females [65].

2. Conclusions

Substantial evidence indicates that gender significantly contributes to the incidence, prognosis, and mortality of many cancers. Although sarcoma remains one of the least studied tumors in terms of sex differences, numerous studies using data collected from the SEER and USSEER databases on a large number of cohorts, and supported by clinical trials, have shown that women also have an advantage over men in sarcoma. Many studies suggest (Table 2) that males have a higher incidence of this type of cancer and are less likely to survive a sarcoma diagnosis.

Significant sex differences have already been observed in childhood malignancies sarcomas. Among the tumors strongly associated with the male sex, childhood sarcomas have been identified as being particularly sensitive to sex biological differences. Differences in incidence between males and females have also been found in different racial and age groups. Although the precise causes behind the disparities in cancer incidence and survival rates between sexes remain unclear, recent research has hinted at the potential involvement of intricate interactions between genetic and environmental factors. These complex interplays might contribute to the generally inferior prognosis observed among males in contrast to females, a trend notable even in soft tissue sarcoma.

Despite this evidence, further research is needed to gain a better understanding of sex differences in sarcoma etiology and prognosis in order to guarantee a therapeutic approach specifically designed for different sexes. In particular, a more systematic approach to examine the disparities in disease rates between male and female patients could be beneficial. This would involve dividing the study into different aspects and categories, such as bone versus soft tissue diseases and pediatric versus adult cases, to confirm the links between translocation and different phenotypes. Supplementing this analysis with fundamental scientific data could help clarify the reasons for these differences.

Table 2. Key landmark studies in sarcoma's gender disparity.

	Study Design, Data Source, Years Include	Relevant Study Population	Key Results
Part A: Evidence for adult			
Rong J et al., 2020 [35]	Retrospective registry-based cohort (CHINA-USA) 2010–2016	1050	Data from gastric GIST patients were collected from the SEER database. Propensity score matching (PSM) was performed to reduce confounding factors, and the clinicopathological features and prognosis of GIST patients were comprehensively evaluated. Gender could be a prognostic factor for gastric GIST survival, and male patients had a higher risk of death.
Mo Chen et al., 2018 [36]	Retrospective registry-based cohort (CHINA-USA) 1973–2013	6582	Data from gastric GIST patients were collected from the SEER database. The study investigated the impact of marital status on the overall survival (OS) and cancer-specific survival (CSS) of operable GIST cases. The marriage could be a protective prognostic factor for survival, and widowed patients had a higher risk of death.
Neal D Freedman et al., 2007 [38]	Retrospective registry-based cohort (USA)	154	The study investigated the association of menstrual and reproductive factors and gastric cancer risk. No associations were observed between gastric cancer risk and age of menarche, number of children, breast feeding, or oral contraceptive use. In contrast, associations were observed with age of menopause, years of fertility, years since menopause, and intrauterine device use.
M Lindblad et al., 2006 [39]	Retrospective (SWEDEN) 1994–2001	612	Esophageal and gastric adenocarcinoma share an unexplained male predominance, A nested case–control study of hormone replacement therapy (HRT) was conducted among 299 women with esophageal cancer, 313 with gastric cancer, and 3191 randomly selected control women. Among 1,619,563 person-years of follow-up, more than 50% reduced risk of gastric adenocarcinoa was found among users of HRT compared to non-users. This inverse association appeared to be stronger for gastric noncardia and weaker for gastric cardia tumors. There was no association between HRT and esophageal adenocarcinoma.
Giun-Yi Hung et al., 2019 [41]	Retrospective registry-based cohort (TAIWAN) 2007–2013	11,393	STS data were acquired from the population-based 2007–2013 Taiwan Cancer Registry of the Health and Welfare Data Science Center, Taiwan. In total, 11,393 patients with an age-standardized incidence rate of 5.62 per 100,000 person-years were identified. Overall, a male predominance and the rate increased with age, peaking at >75 years.
Mei-Chin Hsieh et al., 2013 [42]	Retrospective registry-based cohort (USA) 1995–2008	10,289	STS data were obtained from the North American Association of Central Cancer Registries (NAACCR). The incidence of all STSs combined was higher in males than females.
Rouhani P et al., 2008 [44]	Retrospective registry-based cohort (USA) 1992–2004	12,114	Data from cutaneous soft tissue sarcoma (CSTS) patients were collected from the SEER database confirmed that the incidence of all CSTSs combined was higher in males than females.
Part B: Evidence for childhood and adolescent			
Cole S et al., 2022 [47]	Retrospective registry-based cohort (USA) 1975–2017	5016	Data from osteosarcoma patients were collected from the SEER database. The findings confirm in cases 0 to 9 years old, incidence of primary osteosarcoma was similar between the sexes and increased significantly throughout the study period. Overall, survival rates for all cases have remained relatively unchanged over recent decades, with worse survival observed in males.
Ognjanovic S et al., 2009 [55]	Retrospective registry-based cohort (USA) 1975–2005	987	Data from childhood rhabdomyosarcoma (RMS) patients were collected from the SEER database. The findings revealed the incidence of an embryonal rhabdomyosarcoma (ERMS) was higher in male than females and, more specifically, a smaller peak of embryonal rhabdomyosarcoma (ERMS) incidence rates was observed during adolescence in males which may be related to only those sex-specific hormonal differences.

Table 2. *Cont.*

	Study Design, Data Source, Years Include	Relevant Study Population	Key Results
Ward E et al., 2014 [56]	Retrospective registry-based cohort (USA) 1975–2010	15,780	Data from children and adolescent patients diagnosed with cancer were collected from the SEER database and The North American Association of CentralCancer Registries (NAACCR). The findings confirm that gender disparity has been found in most pediatric cancers, acute lymphoblastic leukemia, non-Hodgkin's lymphoma, medulloblastoma, hepatic tumors, osteosarcoma, and germ cell tumors, showing that the direct effect of male sex is significant for several tumor types.
Williams LA et al., 2019 [69]	Retrospective registry-based cohort (USA) 2000–2015	71,906	Cancer cases aged 0–19 years were identified using the SEER Program. Male sex was positively associated with most cancers. The higher incidence rates observed in males remained consistent over the childhood and adolescent periods, suggesting that childhood and adolescent hormonal fluctuations may not be the primary driving factor for the sex disparities in childhood cancer. The observed incidence disparities may be due to sex differences in exposures, genetics, or immune responses.
Williams LA et al., 2019 [58]	Retrospective registry-based cohort (USA) 2000–2014	57,004	Cancer cases aged 0–19 years were identified using the the SEER program. Males had worse overall survival and a higher risk of death for acute lymphoblastic leukemia, ependymoma, neuroblastoma, osteosarcoma, thyroid carcinoma, and malignant melanoma.

Author Contributions: Conceptualization, I.C. and P.D.F.; methodology, I.C. and P.D.F.; formal analysis, I.C.; resources, S.M. and A.F.; writing—original draft preparation, I.C.; writing—review and editing, I.C. and P.D.F.; visualization, I.C. and P.D.F.; supervision, S.M. and A.F. All authors have read and agreed to the published version of the manuscript.

Funding: This research has received "Current Research" funds from the Italian Ministry of Health to cover publication costs.

Conflicts of Interest: The authors declare no conflicts of interest.

References

1. Burningham, Z.; Hashibe, M.; Spector, L.; Schiffman, J.D. The Epidemiology of Sarcoma. *Clin. Sarcoma Res.* **2012**, *2*, 14. [CrossRef] [PubMed]
2. Jo, V.Y.; Fletcher, C.D.M. WHO classification of soft tissue tumours: An update based on the 2013 (4th) edition. *Pathology* **2014**, *46*, 95–104. [CrossRef] [PubMed]
3. Mastrangelo, G.; Fadda, E.; Cegolon, L.; Montesco, M.C.; Ray-Coquard, I.; Buja, A.; Fedeli, U.; Frasson, A.; Spolaore, P.; Rossi, C.R. A European project on incidence, treatment, and outcome of sarcoma. *BMC Public Health* **2010**, *10*, 188. [CrossRef] [PubMed]
4. Ducimetière, F.; Lurkin, A.; Ranchère-Vince, D.; Decouvelaere, A.V.; Isaac, S.; Claret-Tournier, C.; Suignard, Y.; Salameire, D.; Cellier, D.; Alberti, L.; et al. Incidence rate, epidemiology of sarcoma and molecular biology. Preliminary results from EMS study in the Rhône-Alpes region. *Bull Cancer* **2010**, *97*, 629–641. [PubMed]
5. Laurino, S.; Omer, L.C.; Albano, F.; Marino, G.; Bianculli, A.; Solazzo, A.P.; Sgambato, A.; Falco, G.; Russi, S.; Bochicchio, A.M. Radiation-induced sarcomas: A single referral cancer center experience and literature review. *Front. Oncol.* **2022**, *12*, 986123. [CrossRef] [PubMed]
6. Xi, M.; Liu, M.Z.; Wang, H.X.; Cai, L.; Zhang, L.; Xie, C.F.; Li, Q.Q. Radiation-induced sarcoma in patients with nasopharyngeal carcinoma: A single-institution study. *Cancer* **2010**, *116*, 5479–5486. [CrossRef] [PubMed]
7. Williams, L.; Tmanova, L.; Mydlarz, W.K.; Page, B.; Richmon, J.D.; Quon, H.; Schmitt, N.C. Radiation-Associated Sarcoma of the Neck: Case Series and Systematic Review. *Ann. Otol. Rhinol. Laryngol.* **2018**, *127*, 735–740. [CrossRef]
8. Gladdy, R.A.; Qin, L.X.; Moraco, N.; Edgar, M.A.; Antonescu, C.R.; Alektiar, K.M.; Brennan, M.F.; Singer, S. Do radiation-associated soft tissue sarcomas have the same prognosis as sporadic soft tissue sarcomas? *J. Clin. Oncol.* **2010**, *28*, 2064–2069. [CrossRef]
9. Lazarides, A.L.; Burke, Z.D.C.; Gundavda, M.K.; Novak, R.; Ghert, M.; Wilson, D.A.; Rose, P.S.; Wong, P.; Griffin, A.M.; Ferguson, P.C.; et al. How Do the Outcomes of Radiation-Associated Pelvic and Sacral Bone Sarcomas Compare to Primary Osteosarcomas following Surgical Resection? *Cancers* **2022**, *14*, 2179. [CrossRef]
10. Snow, A.; Ring, A.; Struycken, L.; Mack, W.; Koç, M.; Lang, J.E. Incidence of radiation induced sarcoma attributable to radiotherapy in adults: A retrospective cohort study in the SEER cancer registries across 17 primary tumor sites. *Cancer Epidemiol.* **2021**, *70*, 101857. [CrossRef]
11. Hardell, L. Malignant mesenchymal tumours and exposure to phenoxy acids. A clinical observation. *Lakartidningen* **1977**, *74*, 7974021.

12. Hardell, L.; Eriksson, M.; Degerman, A. Meta-analysis of four Swedish case-control studies on exposure to pesticides as risk-factor for soft-tissue sarcoma including the relation to tumour localization and histopathological type. *Int. J. Oncol.* **1995**, *6*, 847–851. [PubMed]
13. Edwards, D.; Voronina, A.; Attwood, K.; Grand'Maison, A. Association between occupational exposures and sarcoma incidence and mortality: Systematic review and meta-analysis. *Syst. Rev.* **2021**, *10*, 231. [CrossRef] [PubMed]
14. Ward, E.; Boffetta, P.; Andersen, A.; Colin, D.; Comba, P.; Deddens, J.A.; De Santis, M.; Engholm, G.; Hagmar, L.; Langard, S.; et al. Update of the follow-up of mortality and cancer incidence among European workers employed in the vinyl chloride industry. *Epidemiology* **2001**, *12*, 710–718. [CrossRef] [PubMed]
15. Hossain, A.; McDuffie, H.H.; Bickis, M.G.; Pahwa, P. Case-control study on occupational risk factors for soft-tissue sarcoma. *J. Occup. Environ. Med.* **2007**, *49*, 1386–1393. [CrossRef] [PubMed]
16. Bhalla, A.D.; Landers, S.M.; Singh, A.K.; Landry, J.P.; Yeagley, M.G.; Myerson, G.S.B.; Delgado-Baez, C.B.; Dunnand, S.; Nguyen, T.; Ma, X.; et al. Experimental models of undifferentiated pleomorphic sarcoma and malignant peripheral nerve sheath tumor. *Lab. Investig.* **2022**, *102*, 658–666. [CrossRef] [PubMed]
17. Lahat, G.; Lazar, A.; Lev, D. Sarcoma Epidemiology and Etiology: Potential Environmental and Genetic Factors. *Surg. Clin. N. Am.* **2008**, *88*, 451–481. [CrossRef] [PubMed]
18. Patil, D.T.; Rubin, B.P. Genetics of Gastrointestinal Stromal Tumors a Heterogeneous Family of Tumors? *Surg. Pathol. Clin.* **2015**, *8*, 515–524. [CrossRef]
19. Li, F.P.; Fraumeni, J.F.; Mulvihill, J.J.; Blattner, W.A.; Dreyfus, M.G.; Tucker, M.A.; Miller, R.W. A Cancer Family Syndrome in Twenty-four Kindreds. *Cancer Res.* **1988**, *48*, 5358–5362.
20. Tinat, J.; Bougeard, G.; Baert-Desurmont, S.; Vasseur, S.; Martin, C.; Bouvignies, E.; Caron, O.; Bressac-de Paillerets, B.; Berthet, P.; Dugast, C.; et al. 2009 Version of the Chompret Criteria for Li Fraumeni Syndrome. *J. Clin. Oncol.* **2009**, *27*, e108–e109. [CrossRef]
21. Guha, T.; Malkin, D. Inherited TP53 mutations and the Li-fraumeni syndrome. *Cold Spring Harb. Perspect. Med.* **2017**, *7*, a026187. [CrossRef] [PubMed]
22. Villani, A.; Shore, A.; Wasserman, J.D.; Stephens, D.; Kim, R.H.; Druker, H.; Gallinger, B.; Naumer, A.; Kohlmann, W.; Novokmet, A.; et al. Biochemical and imaging surveillance in germline TP53 mutation carriers with Li-Fraumeni syndrome: 11 year follow-up of a prospective observational study. *Lancet Oncol.* **2016**, *17*, 1295–1305. [CrossRef] [PubMed]
23. Tamura, R. Current understanding of neurofibromatosis type 1, 2, and schwannomatosis. *Int. J. Mol. Sci.* **2021**, *22*, 5850. [CrossRef] [PubMed]
24. Wong, F.L. Cancer Incidence After Retinoblastoma. *JAMA* **1997**, *278*, 1262. [CrossRef]
25. Ottaviani, G.; Jaffe, N. The etiology of osteosarcoma. *Cancer Treat. Res.* **2009**, *152*, 15–32. [PubMed]
26. Hansen, M.F.; Koufos, A.; Gallie, B.L.; Phillips, R.A.; Fodstad, O.; Brøgger, A.; Gedde-Dahl, T.; Cavenee, W.K. Osteosarcoma and retinoblastoma: A shared chromosomal mechanism revealing recessive predisposition. *Proc. Natl. Acad. Sci. USA* **1985**, *82*, 6216–6220. [CrossRef] [PubMed]
27. Fletcher, O.; Easton, D.; Anderson, K.; Gilham, C.; Jay, M.; Peto, J. Lifetime risks of common cancers among retinoblastoma survivors. *J. Natl. Cancer Inst.* **2004**, *96*, 357–363. [CrossRef] [PubMed]
28. Cunniff, C.; Bassetti, J.A.; Ellis, N.A. Bloom's syndrome: Clinical spectrum, molecular pathogenesis, and cancer predisposition. *Mol. Syndromol.* **2017**, *8*, 4–23. [CrossRef]
29. Salih, A.; Inoue, S.; Onwuzurike, N. Rothmund-Thomson syndrome (RTS) with osteosarcoma due to RECQL4 mutation. *BMJ Case Rep.* **2018**, *2018*, bcr2017222384. [CrossRef]
30. Oshima, J.; Sidorova, J.M.; Monnat, R.J. Werner syndrome: Clinical features, pathogenesis and potential therapeutic interventions. *Ageing Res. Rev.* **2017**, *33*, 105–114. [CrossRef]
31. Whitmire, P.; Rickertsen, C.R.; Hawkins-Daarud, A.; Carrasco, E.; Lorence, J.; De Leon, G.; Curtin, L.; Bayless, S.; Clark-Swanson, K.; Peeri, N.C.; et al. Sex-specific impact of patterns of imageable tumor growth on survival of primary glioblastoma patients. *BMC Cancer* **2020**, *20*, 447. [CrossRef]
32. Miller, K.D.; Nogueira, L.; Devasia, T.; Mariotto, A.B.; Yabroff, K.R.; Jemal, A.; Kramer, J.; Siegel, R.L. Cancer treatment and survivorship statistics, 2022. *CA Cancer J. Clin.* **2022**, *72*, 409–436. [CrossRef] [PubMed]
33. Siegel, R.L.; Miller, K.D.; Fuchs, H.E.; Jemal, A. Cancer Research UK. Soft tissue sarcoma statistics. *CA Cancer J. Clin.* **2022**, *72*, 7–33. Available online: https://www.cancerresearchuk.org/health-professional/cancer-statistics/statistics-by-cancer-type/soft-tissue-sarcoma (accessed on 20 January 2022). [CrossRef] [PubMed]
34. Siegel, R.L.; Miller, K.D.; Wagle, N.S.; Jemal, A. Cancer statistics, 2023. *CA Cancer J. Clin.* **2023**, *73*, 17–48. [CrossRef] [PubMed]
35. Rong, J.; Chen, S.; Song, C.; Wang, H.; Zhao, Q.; Zhao, R.; He, Y.; Yan, L.; Song, Y.; Wang, F.; et al. The prognostic value of gender in gastric gastrointestinal stromal tumors: A propensity score matching analysis. *Biol. Sex. Differ.* **2020**, *11*, 1–7. [CrossRef] [PubMed]
36. Chen, M.; Wang, X.; Wei, R.; Wang, Z. The influence of marital status on the survival of patients with operable gastrointestinal stromal tumor: A SEER-based study. *Int. J. Health Plan. Manag.* **2019**, *34*, e447–e463. [CrossRef] [PubMed]
37. Nie, X.; Xie, R.; Tuo, B. Effects of Estrogen on the Gastrointestinal Tract. *Dig. Dis. Sci.* **2018**, *63*, 583–596. [CrossRef] [PubMed]
38. Freedman, N.D.; Chow, W.H.; Gao, Y.T.; Shu, X.O.; Ji, B.T.; Yang, G.; Lubin, J.H.; Li, H.L.; Rothman, N.; Zheng, W.; et al. Menstrual and reproductive factors and gastric cancer risk in a large prospective study of women. *Gut* **2007**, *56*, 1671–1677. [CrossRef]

39. Lindblad, M.; García Rodríguez, L.A.; Chandanos, E.; Lagergren, J. Hormone replacement therapy and risks of oesophageal and gastric adenocarcinomas. *Br. J. Cancer* **2006**, *94*, 136–141. [CrossRef]
40. Jang, Y.C.; Leung, C.Y.; Huang, H.L. Association of hormone replacement therapy with risk of gastric cancer: A systematic review and meta-analysis. *Sci. Rep.* **2022**, *12*, 1–7.
41. Hung, G.Y.; Horng, J.L.; Chen, P.C.H.; Lin, L.Y.; Chen, J.Y.; Chuang, P.H.; Chao, T.-C.; Yen, C.-C. Incidence of soft tissue sarcoma in Taiwan: A nationwide population-based study (2007–2013). *Cancer Epidemiol.* **2019**, *60*, 185–192. [CrossRef] [PubMed]
42. Hsieh, M.C.; Wu, X.C.; Andrews, P.A.; Chen, V.W. Racial and Ethnic Disparities in the Incidence and Trends of Soft Tissue Sarcoma among Adolescents and Young Adults in the United States, 1995–2008. *J. Adolesc. Young Adult Oncol.* **2013**, *2*, 89–94. [CrossRef] [PubMed]
43. Van Herk-Sukel, M.P.P.; Shantakumar, S.; Overbeek, L.I.H.; Van Boven, H.; Penning-Van Beest, F.J.A.; Herings, R.M.C. Occurrence of comorbidities before and after soft tissue sarcoma diagnosis. *Sarcoma* **2012**, *2012*, 402109. [CrossRef] [PubMed]
44. Rouhani, P.; Fletcher, C.D.M.; Devesa, S.S.; Toro, J.R. Cutaneous soft tissue sarcoma incidence patterns in the U.S. *Cancer* **2008**, *113*, 616–627. [CrossRef] [PubMed]
45. Patel, S.J.; Pappoppula, L.; Guddati, A.K. Analysis of trends in race and gender disparities in incidence-based mortality in patients diagnosed with soft tissue sarcomas from 2000 to 2016. *Int. J. Gen. Med.* **2021**, *14*, 3787–3791. [CrossRef] [PubMed]
46. American Cancer Society. Key Statistics for Soft Tissue Sarcomas. Available online: https://www.cancer.org/cancer/soft-tissue-sarcoma/about/key-statistics.html (accessed on 20 January 2022).
47. Cole, S.; Gianferante, D.M.; Zhu, B.; Mirabello, L. Osteosarcoma: A Surveillance, Epidemiology, and End Results program-based analysis from 1975 to 2017. *Cancer* **2022**, *128*, 2107–2118. [CrossRef] [PubMed]
48. Brown, H.K.; Schiavone, K.; Gouin, F.; Heymann, M.F.; Heymann, D. Biology of Bone Sarcomas and New Therapeutic Developments. *Calcif. Tissue Int.* **2018**, *102*, 174–195. [CrossRef] [PubMed]
49. Dermawan, J.K.; Chi, P.; Tap, W.D.; Rosenbaum, E.; D'Angelo, S.; Alektiar, K.M.; Antonescu, C.R. Distinct genomic landscapes in radiation-associated angiosarcoma compared with other radiation-associated sarcoma histologies. *J. Pathol.* **2023**, *260*, 465–477. [CrossRef]
50. Inoue, Y.Z.; Frassica, F.J.; Sim, F.H.; Unni, K.K.; Petersen, I.A.; McLeod, R.A. Clinicopathologic features and treatment of postirradiation sarcoma of bone and soft tissue. *J. Surg. Oncol.* **2000**, *75*, 42–50. [CrossRef]
51. Ognjanovic, S.; Olivier, M.; Bergemann, T.L.; Hainaut, P. Sarcomas in TP53 germline mutation carriers: A review of the IARC TP53 database. *Cancer* **2012**, *118*, 1387–1396. [CrossRef]
52. Farid, M.; Ngeow, J. Sarcomas Associated with Genetic Cancer Predisposition Syndromes: A Review. *Oncologist* **2016**, *21*, 1002–1013. [CrossRef] [PubMed]
53. Thomas, D.M.; Savage, S.A.; Bond, G.L. Hereditary and environmental epidemiology of sarcomas. *Clin. Sarcoma. Res.* **2012**, *2*, 1–3. [CrossRef] [PubMed]
54. Jawad, M.U.; Cheung, M.C.; Min, E.S.; Schneiderbauer, M.M.; Koniaris, L.G.; Scully, S.P. Ewing sarcoma demonstrates racial disparities in incidence-related and sex-related differences in outcome: An analysis of 1631 cases from the SEER database, 1973–2005. *Cancer* **2009**, *115*, 3526–3536. [CrossRef] [PubMed]
55. Ognjanovic, S.; Linabery, A.M.; Charbonneau, B.; Ross, J.A. Trends in childhood rhabdomyosarcoma incidence and survival in the United States, 1975–2005. *Cancer* **2009**, *115*, 4218–4226. [CrossRef] [PubMed]
56. American Cancer Society, Cancer in children & adolescents. *Spec. Sect. Cancer Child. Adolesc.* **2014**, *1*, 25–42.
57. Williams, L.A.; Richardson, M.; Kelum, R.D.; McLaughlin, C.C.; Mueller, B.A.; Chow, E.J.; Spector, L.G. The association between sex and most childhood cancers is not mediated by birthweight. *Cancer Epidemiol.* **2018**, *57*, 7–12. [CrossRef] [PubMed]
58. Williams, L.A.; Spector, L.G. Survival Differences between Males and Females Diagnosed with Childhood Cancer. *JNCI Cancer Spectr.* **2019**, *3*, pkz032. [CrossRef] [PubMed]
59. Smith, A.W.; Keegan, T.; Hamilton, A.; Lynch, C.; Wu, X.C.; Schwartz, S.M.; Kato, I.; Cress, R.; Harlan, L.; AYA HOPE Study Collaborative Group. Understanding care and outcomes in adolescents and young adult with Cancer: A review of the AYA HOPE study. *Pediatr. Blood Cancer* **2019**, *66*, e27486. [CrossRef]
60. Agaram, N.P. Evolving classification of rhabdomyosarcoma. *Histopathology* **2022**, *80*, 98–108. [CrossRef]
61. Locatelli, V.; Bianchi, V.E. Effect of GH/IGF-1 on Bone Metabolism and Osteoporsosis. *Int. J. Endocrinol.* **2014**, *2014*, 235060. [CrossRef]
62. Singh, D.; Sanyal, S.; Chattopadhyay, N. The role of estrogen in bone growth and formation: Changes at puberty. *Cell Health Cytoskelet.* **2011**, *3*, 2–12. [CrossRef]
63. Laitinen, M.K.; Albergo, J.I.; Stevenson, J.D.; Farfalli, G.L.; Aponte-Tinao, L.A.; Grimer, R.J.; Jeys, L.M.; Parry, M.C. Female gender in the hormonally active age group plays a major role in high-grade chondrosarcoma survival. *Acta Oncol.* **2020**, *59*, 242–246. [CrossRef] [PubMed]
64. Frank, G.R. Role of estrogen and androgen in pubertal skeletal physiology. *Med. Pediatr. Oncol.* **2003**, *41*, 217–221. [CrossRef] [PubMed]
65. Kukurba, K.R.; Parsana, P.; Balliu, B.; Smith, K.S.; Zappala, Z.; Knowles, D.A.; Favé, M.-J.; Davis, J.R.; Li, X.; Zhu, X.; et al. Impact of the X chromosome and sex on regulatory variation. *Genome Res.* **2016**, *26*, 768–777. [CrossRef] [PubMed]
66. Bianchi, I.; Lleo, A.; Gershwin, M.E.; Invernizzi, P. The X chromosome and immune associated genes. *J. Autoimmun.* **2012**, *38*, J187–J192. [CrossRef] [PubMed]

67. Libert, C.; Dejager, L.; Pinheiro, I. The X chromosome in immune functions: When a chromosome makes the difference. *Nat. Rev. Immunol.* **2010**, *10*, 594–604. [CrossRef] [PubMed]
68. Yin, Y.H.; Li, Y.Y.; Qiao, H.; Wang, H.C.; Yang, X.A.; Zhang, H.G.; Pang, X.-W.; Zhang, Y.; Chen, W.-F. TSPY is a cancer testis antigen expressed in human hepatocellular carcinoma. *Br. J. Cancer* **2005**, *93*, 458–463. [CrossRef] [PubMed]
69. Williams, L.A.; Richardson, M.; Marcotte, E.L.; Poynter, J.N.; Spector, L.G. Sex ratio among childhood cancers by single year of age. *Pediatr. Blood Cancer* **2019**, *66*, e27620. [CrossRef]

cancers

MDPI

Review

Genetic, Epigenetic and Transcriptome Alterations in Liposarcoma for Target Therapy Selection

Ekaterina A. Lesovaya [1,2,3], Timur I. Fetisov [1], Beniamin Yu. Bokhyan [1], Varvara P. Maksimova [1], Evgeny P. Kulikov [2], Gennady A. Belitsky [1], Kirill I. Kirsanov [1,3,†] and Marianna G. Yakubovskaya [1,3,*,†]

1 N.N. Blokhin Russian Cancer Research Center, Ministry of Health of Russia, 24 Kashirskoe Shosse, Moscow 115478, Russia; lesovenok@yandex.ru (E.A.L.); timkatryam@yandex.ru (T.I.F.); beniamin-bokhyan@mail.ru (B.Y.B.); lavarvar@gmail.com (V.P.M.); kkirsanov85@yandex.ru (K.I.K.)

2 Faculty of Oncology, I.P. Pavlov Ryazan State Medical University, Ministry of Health of Russia, 9 Vysokovol'tnaya St., Ryazan 390026, Russia; e.kulikov@rzgmu.ru

3 Laboratory of Single Cell Biology, Peoples' Friendship University of Russia, 6 Miklukho-Maklaya St., Moscow 117198, Russia

* Correspondence: mgyakubovskaya@mail.ru

† These authors have contributed equally to this work.

Simple Summary: Liposarcoma is the most widespread soft-tissue sarcoma in adults. This review summarizes the molecular genetics and epigenetics of the main liposarcoma subtypes and corresponding aberration in signaling forming the basis for targeted therapy selection. In recent years, specific inhibitors of *CDK4*/6 and *MDM2* and VEGFR/FGFR/PDGFR multi-kinase inhibitors have been proposed for the treatment of liposarcoma.

Abstract: Liposarcoma (LPS) is one of the most common adult soft-tissue sarcomas (STS), characterized by a high diversity of histopathological features as well as to a lesser extent by a spectrum of molecular abnormalities. Current targeted therapies for STS do not include a wide range of drugs and surgical resection is the mainstay of treatment for localized disease in all subtypes, while many LPS patients initially present with or ultimately progress to advanced disease that is either unresectable, metastatic or both. The understanding of the molecular characteristics of liposarcoma subtypes is becoming an important option for the detection of new potential targets and development novel, biology-driven therapies for this disease. Innovative therapies have been introduced and they are currently part of preclinical and clinical studies. In this review, we provide an analysis of the molecular genetics of liposarcoma followed by a discussion of the specific epigenetic changes in these malignancies. Then, we summarize the peculiarities of the key signaling cascades involved in the pathogenesis of the disease and possible novel therapeutic approaches based on a better understanding of subtype-specific disease biology. Although heterogeneity in liposarcoma genetics and phenotype as well as the associated development of resistance to therapy make difficult the introduction of novel therapeutic targets into the clinic, recently a number of targeted therapy drugs were proposed for LPS treatment. The most promising results were shown for *CDK4*/6 and *MDM2* inhibitors as well as for the multi-kinase inhibitors anlotinib and sunitinib.

Keywords: liposarcoma; well-differentiated liposarcoma; dedifferentiated liposarcoma; myxoid/round-cell liposarcoma; pleomorphic liposarcoma; myxoid pleomorphic liposarcoma; molecular genetic abnormalities; epigenetic changes; targeted therapy

Citation: Lesovaya, E.A.; Fetisov, T.I.; Bokhyan, B.Y.; Maksimova, V.P.; Kulikov, E.P.; Belitsky, G.A.; Kirsanov, K.I.; Yakubovskaya, M.G. Genetic, Epigenetic and Transcriptome Alterations in Liposarcoma for Target Therapy Selection. *Cancers* **2024**, *16*, 271. https://doi.org/10.3390/cancers16020271

Academic Editor: Catrin Sian Rutland

Received: 14 November 2023

Revised: 25 December 2023

Accepted: 25 December 2023

Published: 8 January 2024

1. Introduction

Liposarcoma (LPS) is a subtype of soft-tissue sarcoma (STS) further divided into five separate groups of malignancies characterized by distinct genetic and molecular aberrations, unique histologic appearance, therapy strategies and overall clinical outcome: well-differentiated liposarcoma (WDLPS), dedifferentiated liposarcoma (DDLSP), myxoid/round-cell liposarcoma (MLPS), pleomorphic liposarcoma (PLPS) and the recently isolated, separate sub-type myxoid pleomorphic liposarcoma (MPLPS), each harboring its own unique features [1]. Although surgical resection and radiotherapy remain the most frequent choices for treatment, chemotherapeutic options are also applied for the treatment of patients with advanced/metastatic clinically unresectable LPS. The specific patterns of disease pathogenesis and progression of each LPS subtype suggest different approaches to improve chemotherapy. An understanding of the genetic and epigenetic abnormalities and corresponding transcriptome changes is critical to the management of liposarcoma and further studies of the mechanisms of liposarcoma pathogenesis.

Well-differentiated (WDLPS) and dedifferentiated (DDLPS) liposarcoma are the most common subtypes of liposarcoma [2–4]. WDLPSs are slow-growing malignancies characterized by the presence of adipocytes [2–4]. DDLPS is characterized by a higher cellularity and elevated mitosis number [3,4]. DDLPS is a much more aggressive metastatic disease and associated with poor prognosis [5–7]. Both subtypes do not have specific age risk factors and usually develop in the retroperitoneum, extremities, paratesticular areas and trunk [3,4,8]. WDLPS and DDLPS are largely resistant to conventional cytotoxic chemotherapy and radiation therapy, and surgery remains the main option [2–4].

Myxoid/round-cell (MLPS) liposarcoma is a neoplasm with high cellularity and non-lipogenic, mesenchymal, round- to oval-shaped cells mixed with mature adipocytes [9,10] characterized by a more aggressive disease biology and worse clinical outcome [11]. MLPS is more common in younger patients and predominantly arises in the proximal lower extremities, as opposed to the retroperitoneum [12]. The tumor tends to recur locally and systemically, with a high risk of metastasis to the retroperitoneum, abdomen, chest and trunk [9]. Treatment for MLPS includes surgery and radio- and chemotherapy [13].

Pleomorphic liposarcoma (PLPS) is the most aggressive and histologically non-uniform subtype of liposarcoma. It is a high-grade, aggressive neoplasm consisting of pleomorphic lipoblasts and occasional multinucleated giant cells [2–4,9]. The median age of the patients is 55-65 years old and they most commonly present with disease in the lower extremities. These malignancies are highly resistant to all current treatment modalities [14,15].

Myxoid pleomorphic liposarcoma (MPLPS) is an exceedingly rare adipocytic malignancy developing in the mediastinum, followed by the limbs and the head and neck region. Morphologically, MPLPS shows features of both myxoid and pleomorphic liposarcoma with aggressive clinical behavior, including fast tumor growth and early metastasis to the lungs, bone and soft tissues [1]. Genetic and epigenetic results suggest a possible link with conventional pleomorphic liposarcoma [16].

In practice, distinguishing one liposarcoma subtype from another is rather challenging. Molecular studies should follow the histologic examination for more accuracy in diagnostics and optimal disease therapy course or enrollment into clinical trials.

2. Molecular Genetic Abnormalities and Corresponding Transcriptome Changes Specific to Liposarcomas and Their Possible Role as Therapeutic Targets

A number of the genetic abnormalities are specific to the whole set of LPSs: TOP2A, PTK7 and CHEK1 were overexpressed in 140 cases of liposarcoma [17]; point mutations in CTNNB1, CDH1, FBXW7 and EPHA1, C-MET and EGFR amplification and increased expression of C-KIT, EGFR, PD-L1 and PD-1 also represent potential oncogenic events in liposarcoma cells [18]. Loss of estrogen receptor expression may be involved in the pathogenesis of liposarcoma through an unknown mechanism [19]. The transcription factor TBX3, a critical developmental regulator, was shown to have a role as an oncogene/motogene in liposarcoma [20].

Additionally, the specific genetic alterations found were specific to several subtypes of liposarcoma.

2.1. Well-Differentiated Liposarcoma (WDLPS) and Dedifferentiated Liposarcoma (DDLSP): 12q13-15-Associated Chromosomal Aberrations as Major Driver of Pathogenesis

WDLPS and DDLPS usually share the same genetic aberration, represented by the distinctive ring and/or giant marker chromosomes from the 12q13-15 segment (Table 1 and [21]). This chromosome region bears more than 350 genes, including multiple proliferative genes [22]. In particular, the most common overamplified genes in WDLPS/DDLPS are the member of the High-Mobility Group A (HMGA) gene family *HMGA2*, encoding the transcriptional factor modulating the chromatin structure in the nucleus [21,23]; *CDK4*, gene of cyclin-dependent kinase 4 [24]; pro-proliferative genes from the *JUN* family [25]; and mouse double minute 2 (*MDM2*), encoding a well-studied inhibitor of the p53 tumor suppressor [24,26,27]. These genes are well-studied in the context of WDLPS/DDLSP and reveal several correlations with the type of malignancy, location, grade, node involvement, distant metastasis and recurrence-free survival [25]. Other genes frequently amplified within the 12q13-15 amplicon include tetraspanin 31 (TSPAN31), a gene with possible role in the proliferation, migration and inhibition of apoptosis [28,29]. YEATS4, a proliferative gene, and CPM, encoding carboxypeptidase M, a proteolytic enzyme inducing cleavage activation of growth factors, are genes commonly amplified within 12q13-15 that have been implicated in dedifferentiation [30]. FRS2, E2F1 and CDKN2A are also among the most upregulated genes in DDLPS and WDLPS [26,31]. Notable deletions were found in chromosome 1p (RUNX3, ARID1A), chromosome 11q (ATM, CHEK1) and chromosome 13q14.2 (MIR15A, MIR16-1) [30]. It was also demonstrated for WDLPS/DDLPS without the *CDK4* amplification that an alteration in the CDKN2A/CDKN2B/*CDK4*/CCND1 pathway is present in almost all cases without *CDK4* amplification and may play a pivotal role in oncogenesis [32].

Table 1. The most frequent genetic and epigenetic aberrations in LPS.

LPS Subtype	Cytogenetic Abnormality and Associated Genetic Aberration	Epigenetic-Related Change
WDLPS	Ring chromosome 12 12q13-15 region amplifications: *MDM2*, *CDK4*, *HMGA2*, *SAS*, *GL1*, *JUN* family genes [21–24,26,27]	Not described
DDLPS	Ring chromosome 12 12q13-15 region amplifications: *MDM2*, *CDK4*, *HMGA2*, *SAS*, *GL1*, *JUN* family genes [21–24,26,27]	Mutations in genes of epigenetic regulators (HDAC1) Aberrant methylation of tumor-promoting genes KLF4, CEBPA, CDKN2A Increased expression of miR-155 [33–36]
MLPS	t(12;16) (q13;p11), t(12;22) (q13;q12) *FUS-CHOP*, *EWS-CHOP* [18]	Specific methylation profile of 12q13-q14 region CpG-methylated APC locus and reduced APC expression Epigenetic regulation of increased expression of CDKN2A, MGMT, RASSF1A, MST1, MST2 Increased expression of microRNA-135b [37–40]
PLPS	13q14.2-5 deletion Rb/TP53 deletion Complex karyotype [21,26,41,42]	Not described
MPLPS	No specific changes Complex karyotype	Not described

In WDLPS/DDLPS, the molecular features of malignancies may vary between subtypes. In particular, insulinoma-associated protein 1 (INSM1) is a specific biomarker for neuroendocrine cancers, but its expression is also detected in liposarcomas. Moreover, INSM1 expression in WDLPS was significantly higher than in adipocytes and DDLPS cells. Significant differences in the expression of INSM1 in WDLPS and DDLPS may assist in the diagnosis, enriching the diagnostic index system of mesenchymal cancers [43]. Additional chromosomal abnormalities, more exclusive for DDLPS than for WDLPS, are recurrent amplifications of 1p32 and 6q23, in particular, overexpression of ASK1, DDR2, ERBB3, STAT6, FGFR1, MAP3K5, LGR5, MCL1, CALR, AQP7, ACACB, FZD4, GPD1, LEP and ROS1 [21,44–46]. Another set of core genes in DDLPS identified as significantly enriched in microarray profiling generated from DDLPS and normal fat controls include APP, *MDM2*, CDK1, PCNA, TKT, *CDK4*, CDC20, BUB1B, BARD1, ADRB2, LGALS3, CAV1, CCNA2 and CDKN2A. The pathways identified as enriched in DDLPS are the pyruvate pathway, cell cycle genes and molecular mechanisms associated with the DDLPS pathway and PPAR signaling pathway [47]. CTDSP1/2-DNM3OS fusion genes were identified in a subset of DDLPS tumors by integrating exome and RNA sequencing data [48].

Several genes located at 19p13.1-13.2 were highly expressed in DDLPS, including genes encoding CRT, the inhibitor of adipocyte differentiation, and CD47, tightly associated with malignant transformation [18]. The expression of the E3-ubiqutin ligase gene SIAH2 in DDLPS tumor-associated macrophages and other stromal cells indicates that SIAH2 expression may serve as a molecular marker distinguishing between DDLPS and WDLPS, but more complete evaluation of the role of SIAH2 in the DDLPS phenotype is limited by the availability of fresh tissues from these rare cancers [49]. In a study of the role of α-thalassemia/mental retardation syndrome X-linked (ATRX) or death domain-associated protein 6 (DAXX) gene expression in telomerase activation and alternative lengthening of telomeres, a 100% correlation was demonstrated between ATRX or DAXX and alternative telomere lengthening in DDLPS. It was also correlated with poor survival, suggesting the prognostic role of ATRX and DAXX in DDLPS [50]. Expression of the PD-1 gene, encoding the differentiation marker of the immune cells, was particularly high in DDLPS [51]. Another study reported a correlation between high expression of the centromere protein F (CENPF) gene and worse survival of DDLPS patients, therefore suggesting CENPF as a malignant indicator of tumor immune infiltration-related survival [52]. Rare DDLPS-specific alterations are mutations in the fibroblast growth factors FGFR1, FGFR2, FGFR3 and FGFR4, as well as in FGFR substrate 2 (FRS2), characterized by a poor prognosis [18,53–55].

For WDLPS pathogenesis, a second amplicon originating from 10p11-14 is described containing 62 genes, including oncogenes such as MLLT10, previously described in chimeric fusion with MLL in leukemias, NEBL and BMI1 [22]. SORBS1, KRT8 and MT1G are among the top downregulated genes in WDLPS and DDLPS [31]. MT1G was previously reported to be a tumor suppressor and was silenced in hepatocellular carcinoma [56]. Low SORBS1 expression is associated with promotion of invasion and metastasis as well as an overall poor prognosis in breast cancer [57]. CCAAT/enhancer binding protein (CEBPA) and PPAR-γ are reported to be downregulated in DD/WDLPS but more frequently in DDLPS [18].

2.2. *Myxoid and Round-Cell Liposarcoma (MLPS): DNA Damage-Associated Gene CHOP and Its Translocation Partners*

MLPS is characterized by unique chromosome rearrangements, namely, t(12;16) (q13;p11), that result in the *FUS-CHOP* (*FUS-DDIT3*) gene fusion in more than 95% of cases or the rarer translocation t(12;22) (q13;q12), leading to the formation of the *EWS*-CHOP oncogene in 5% of malignancies [18]. The gene CHOP encodes a growth arrest and DNA-damage inducible member of the C/EBP family of transcription factors, regulates adipogenesis and assists in growth arrest, but loses the function after the rearrangement and stimulates proliferation [58]. The CHOP translocation partners include a TLS gene of nuclear RNA-binding protein and an *EWS* gene with great similarity to TLS, whose protein product is involved in the development of a wide variety of cancers, including Ewing's sarcoma, melanoma and several neuroendocrine cancers [33]. Interestingly, the breakage in the introns of the CHOP gene with further formation of chimeric genes suggests the presence of a characteristic sequence in the breakpoint regions, including the mobile element Alu and palindromic oligomer sequences [34]. To date, eleven *FUS-CHOP* and five *EWS-CHOP* chimeric genes have been described [35]. The corresponding aberrant proteins interfere with normal adipocyte differentiation and are involved in the activation of several tyrosine kinase receptor pathways including MET, RET, IGFR, AXL, EGFR, PI3K/Akt and VEGFR2 specifically for round-cell liposarcoma [18]. Activating mutations or amplification of PIK3CA, P110α catalytic subunit mutations of PI3K are seen in approximately 15% of MLPS and are associated with a poor prognosis, whereas PTEN deletion has also been described [18,21]. MLPSs are also characterized by a high frequency of hotspot mutations (C228T or C250T) in the promoter region of telomerase reverse transcriptase (TERT), which encodes the TERT protein responsible for telomerase reactivation [36,59]. TERT mutation is associated with a poor prognosis in MLPS; however, it could not be depicted as a prognostic factor. Thus, in a retrospective study on 83 primary MLPS tumor samples, TERT hotspot mutations were observed in 77% of cases, but aberrant telomere lengthening was not detected. Furthermore, TERT promoter hotspot mutations did not correlate with patient survival [60], in contrast with ATRX/DAXX overexpression and alternative telomere lengthening in DDLPS [50]. Gene expression studies have reported the specific expression of CTAG1B, CTAG2, MAGEA9 and PRAME in myxoid and round-cell liposarcoma [61]. High expression of the CHSY1 gene encoding surface glycosaminoglycan could be an additional marker of malignant pathologic grade and poor clinical prognosis in soft-tissue sarcomas with myxoid substance [62]. STAT6 can also be overexpressed in myxoid liposarcoma [46].

2.3. Pleomorphic Liposarcoma (PLPS) and Myxoid Pleomorphic Liposarcoma (MPLPS): Complex Karyotype and Poor Prognosis

PLPS and MPLPS are usually characterized by complex karyotypic aberrations without specific genetic alterations. Comparative genomic hybridization analyses showed gains of 1p, 1q21-q32, 2q, 3p, 3q, 5p12-p15, 5q, 6p21, 7p, 7q22, 8q, 10q, 12q12-q24, 13q, 14q, 15q, 17p, 17q, 18p, 18q12, 19p12, 19q13, 20q, 22q and Xq21-q27 and losses of 1q, 2q, 3p, 4q, 10q, 11q, 12p13, 13q14, 13q21-qter, 14q23-24, 16q22, 17p13, 17q11.2 and 22q13 [41,42]. TP53 mutations are observed in 60% of PLPS patients [26], deletion of 13q14.2-5 (containing the tumor-suppressor gene RB1) in up to 50% [21] and loss of tumor the suppressor-gene NF1 in 5% of patients [30]. In a study of 155 patients diagnosed with PLPS, increased expression of PPARγ (adipogenic marker), BCL2 and survivin (survival factors), VEGF (angiogenic factor), MMP2 metalloprotease and other biomarkers was revealed [15,18]. Amplification of δ catenin on 5p and deregulation of genes involved in adipogenesis (CEBPA on 19q, EP300 on 22q13) associated with the promotion of metastasis and loss of adipocyte differentiation are also observed [41].

2.4. Conclusion on Liposarcoma Genetics

To sum up, some genetic alterations with oncogenic potential are described for all subtypes of liposarcoma. The most frequent WDLPS and DDLPS genetic aberration is represented by the 12q13-15 segment rearrangements, affecting the expression of more than 60 genes, including pro-proliferative ones. An additional frequent transcriptome abnormality for DDLPS is represented by the overexpression of several genes located at 19p13.1-13.2. Telomerase activation and alternative lengthening of telomeres were also demonstrated for DDLPS. Moreover, high expression of PD-1 was found in DDLPS tumor-associated macrophages. For WDLPS, a second amplicon originating from 10p11-14 is described. Several genes, including SORBS1, KRT8 and MT1G, are downregulated in WDLPS and DDLPS. MLPS is characterized by the translocation (12;16) (q13;p11), resulting in the *FUS-CHOP* gene fusion and affecting adipocyte differentiation and the activation of tyrosine kinases MET, RET, IGFR, AXL, EGFR, PI3K/Akt and VEGFR2. Additionally, overexpression of CTAG1B, CTAG2, MAGEA9, PRAME and CHSY1 was described in MLPS. As concerns tumor-suppressor genes, PTEN deletion is also not uncommon in this type of liposarcoma. PLPS and MPLPS are characterized by complex karyotype and simultaneous aberrations simultaneously with P53 mutations and the deletion of 13q14.2-5, including RB1.

3. Epigenetic Markers of Liposarcoma

Epigenetic regulation of gene expression occurs on multiple levels, including DNA methylation, histone mutations and modification, chromatin structure alterations and remodeling, the formation of alternative DNA structures as well as transcription regulation by specific subsets of long non-coding RNA (lncRNA) and miRNA [45,63]. Novel manners of cell communication and genetic exchange such as exosomes, macrovesicle, and apoptotic bodies containing miRNAs with LPS-relevant functions involve adjacent and distant recipient cells and add complexity to this situation [45]. It has to be noted that studies of LPS epigenetics have not been reported all LPS subtypes, and the use of epigenetic modulators in therapy for liposarcoma should develop a stronger basis. WDLPS and DDLPS are already characterized by a multi-component landscape of histone modifications and histone-modifying enzymes as well as by a specific miRNA profile. In contrast, there are no data describing the epigenetic changes in PLPS and MPLPS. Nevertheless, specific miRNAs, in particular, miR-215-5p, was shown to promote *MDM2* expression in liposarcoma without specificity to a certain subtype. In addition, it was found to promote cell proliferation, inhibit apoptosis, promote cell cycle progression and promote cell invasion and migration. Therefore, miR-215-5p could be considered a novel therapeutic target in liposarcoma [64]. Hypermethylation of H3K4me3 and H3K9me3 was found in a study of patient-derived xenografts from upper-abdominal soft-tissue liposarcoma. This epigenetic feature may be related to methionine addiction, a fundamental hallmark of cancer, termed the Hoffman effect [65]. The over-methylation of these histone marks requires excess methionine in the form of S-adenosylmethionine and may, at least in part, account for the excess methionine required by cancer cells [66]. Liposarcoma subtypes have their unique genetic and clinical characteristics, undoubtedly cross-talking with the epigenetic features of specific malignancies. Below, we review the current and potential future epigenetic prognostic markers and/or therapy targets.

3.1. Mutations in the Genes of Epigenetic Regulators and the Whole Set of Differentially Expressed miRNAs in WDLPS and DDLSP

In DDLPS, specific methylation profiles correlate with clinical outcomes [67,68]. In many cases, promoter elements are hypomethylated, while enhancers and coding sequences are hypermethylated, although the net consequences on transcription in vivo are not entirely predictable [67].

Mutations in genes of epigenetic regulators, specifically in histone deacetylase 1 HDAC1, were demonstrated for DDLPS, but the significance of HDAC1 mutations in DLPS remains to be fully defined at the biochemical level [69]. A comparative analysis of epigenetic modifications and the DNA methylation level in DDLPS identified 833 differentially methylated regions affecting the promoters of 677 genes [70]. Significant tumor-specific promoter methylation associated with downregulation was found in KLF4 and CEBPA, encoding two transcription factors associated with adipocyte differentiation. KLF4 regulates CEBPA, and loss of expression of these factors is considered to be tumorigenic [70]. A study of DNA methylation status and gene expression levels in a large and representative cohort of 80 untreated, primary high-grade sarcomas composed of eight subtypes revealed the prognostic value of DNA hypermethylation of CpG sites in the CDKN2A gene in PLPS and DDLPS [71]. p16INK4a gene promoter hypermethylation is considered to be a potential marker for DDLPS but not for WDLPS [72].

More than 40 miRNAs were found to be differentially expressed in DDLPS and WDLPS among themselves as well as in comparison to normal fat [18]. One of the most frequently upregulated miRNAs in DDLPS is miR-155, involved in malignization via the regulation of casein kinase 1α (CK1α), which results in the activation of the β-catenin pathway [73].

β-catenin and its downstream effector cyclin D1 were found to be overexpressed in all human DDLPS cell lines compared with preadipocytes and adipocytes and were also shown to induce DDLPS cell proliferation and cell cycle progression [73,74]. Knockdown of miR-155 inhibited DDLPS cell proliferation, decreased colony formation, induced cell cycle arrest in vitro and blocked tumor growth in xenografts [75]. MiR-193 family members were found to be downregulated in DDLPS compared with normal fat, and miR-193 expression is considered a favorable prognostic factor in WDLPS/DDLPS [76], as well as a therapeutic approach, as miR-193 targets PDGFRβ, SMAD4 and YAP1, belonging to strongly interacting pathways (focal adhesion, TGFβ and Hippo, respectively) [77]. Interestingly, the expression of miR-193b in liposarcoma cells was downregulated by promoter methylation, resulting at least in part from increased expression of the DNA methyltransferase DNMT1 in WDLS/DDLS, which leads researchers to also consider the immediate implication of demethylation agents for therapeutic exploration [76,78]. MiR-143, which is abundant in normal adipose tissue, was found to be underexpressed in WDLPS, and its expression decreased further as the tumor progressed to DDLPS. The signaling targets of miRNA-143 include BCL2, TOP2A and PLK1 [79]. The role of miR-145 and miR-451 in the suppression of tumor growth was demonstrated for DDLPS, as well as the tumor-promoting role of miR-26a in DDLPS/WDLPS [80]. Loss of miR-133a expression induces a metabolic shift due to a reduction in oxidative metabolism favoring a Warburg effect in DDLPS [81].

In a study of the expression of 1888 miRNAs in 25 human liposarcoma samples, a DDLPS-specific downregulated subset of miRNAs was described, including miR-144, miR-451, miR-29b-2, miR-365, miR29b miR-499-5b, miR-486-5p and miR-551 [82]. Further, the role of the miRNAs miR-133a, miR-199a-3p, miR25-3p and miR-92a-3p was investigated in DDLPS progression, but a correlation between the expression of miRNAs and tumor viability was shown only for miR-199a-3p [18]. MiR-133, miR-1 and miR-206 were significantly underexpressed in WDLPS and may function as tumor suppressors, as described in muscle-relevant rhabdomyosarcomas [83]. Tan et al. described other specific subsets of miRNAs in DDLPS and WDLPS: they confirmed the upregulation of miR-214-3p, miR-199a, miR-21-3p and miR-21-5p and downregulation of miR-10b, miR-126-3p, miR-126-5p, miR-143-3p, miR-143-5p, miR-145-5p and miR-193b-3p in WDLPS/DDLPS compared to benign lipoma [84]. MiR-3613-3p is upregulated in DDLPS patients and may serve as a potential specific biomarker for dedifferentiated liposarcoma [85]. The analysis of tissue and serum miRNA expression in DDLPS identified miR-1246, -4532, 4454, -619-5p and -6126 as biomarkers for DDLPS [86].

3.2. MLPS: FUS-CHOP-Associated Chromatin Remodeling and Changes in Specific miRNA Expression

The specific methylation profile of the 12q13-q14 region in MLPS with t(12;16) (q13;p11) translocation has been described [38]. Epigenetic analyses showed that 45% of myxoid/round-cell liposarcomas were CpG-methylated at the APC locus and had reduced APC expression [39]. Increases in expression of CDKN2A, MGMT, RASSF1A, MST1 and MST2 were also found to be epigenetically regulated by the DNA methylation level [40].

Chromatin remodeling plays a role in MLPS through interactions between *FUS-DDIT3* and components of the subfamily of ATP-dependent chromatin remodeling complexes SWI/SNF and polycomb repressive complex 2 PRC2 [87–89]. The histone code reader Spindlin1 (SPIN1) was shown to impair proliferation and increase apoptosis of liposarcoma cells in vitro and in xenograft mouse models. Using signaling pathway, genome-wide chromatin binding and transcriptome analyses, Franz et al. found that SPIN1 directly enhances the expression of GDNF, an activator of the RET signaling pathway, in cooperation with the transcription factor MAZ. Importantly, a mutation of SPIN1 within the reader domain interfering with chromatin binding reduces liposarcoma cell proliferation and survival. These data suggest SPIN1 as a novel target for chromatin-associated small-molecule inhibitors [90]. In a study of integral DNA methylation patterns in liposarcoma samples, it was demonstrated that ALDH1A3 was the most hypermethylated and downregulated gene for MLPS compared to normal fat [71]. ALDH1A3 is a member of the aldehyde dehydrogenase family with 19 isoenzymes that potentially plays a role in the detoxification of aldehydes in alcohol metabolism and lipid peroxidation [91]. High ALDH1 activity in sarcoma cell lines is associated with an increase in proliferation [92]. The EFEMP1 gene encoding fibulin-3, a member of the extracellular matrix glycoprotein family associated with lymph node metastasis, vascular invasion and poor prognosis, was also found to be hypermethylated and downregulated in MLPS compared to normal fat [71,93–95].

A lesser extent of specific miRNAs is described for MLPS. Thus, microRNA-135b (miR-135b) is described as a key regulator of the malignancy, promoting MLPS metastasis in vivo through the direct suppression of thrombospondin 2 (THBS2) and following an increase in the total amount of MMP2 [37]. Another study demonstrated the role of high expression of miR-9, miR-9* and miR-31 in the progression and metastasis of MLPS [96]. It was demonstrated that miR-486 expression was repressed in TLS-CHOP-expressing MLS tissues, so downregulation of miR-486 may be an important process for MLPS development [97].

3.3. Conclusions of Liposarcoma Epigenetics

In conclusion to the epigenetic section, it should be noted that almost all liposarcoma subtypes accumulate a number of epigenetic alterations, which could be considered possible therapy targets. In particular, the whole pool of target miRNAs in DDLPS, WDLPS and MLPS is described as drivers/markers of pathogenesis and are under extensive investigation. Hypermetylation of H3K4me3 and H3K9me3 in the abovementioned LPS subtypes may lead to the hyperexpression of cell cycle regulators and a decrease in the expression of tumor-suppressor genes such as APC. No data on epigenetic-specific features of PLPS and MPLPS have been described in the literature.

4. Changes in Signaling and Therapeutic Approaches

Treatment of liposarcoma typically involves surgery and radiation therapy, while the use of classic cytostatic treatment and targeted therapy frequently lead to the development of resistance at the advanced disease stage. However, multiple translational studies of novel therapies target various genetic and molecular aberrations in different subtypes of liposarcoma. In particular, WDLPS/DDLPS-specific aberrations in the 12q13-15 amplicon leading to the amplification of *MDM2* and *CDK4* and MLPS-specific *FUS-DDIT3/EWSR1-DDIT3* fusion represent potential therapeutic candidates. Moreover, several low-molecular-weight multi-kinase inhibitors targeting MET, AXL, IGF1R, EGFR, VEGFR2 and PDGFR-β could be effective in the types of liposarcoma characterized by abnormalities in PI3K/Akt/mTOR signaling and the associated deregulation of other cascades [98–102].

4.1. MDM2/p53 and CDK4 Signaling Aberrations as Well as Activation Mutations in Multiple Growth Factors in WDLPS and DDLPS

As described in the section "Molecular genetic abnormalities", *MDM2* and *CDK4* are frequently co-amplified in WDLPS and DDLPS [103]. Amplification of *MDM2* results in the inactivation of p53, and *CDK4* amplification leads to cell cycle progression [53,104]. Both alterations can be targeted by specific inhibitors (*MDM2* antagonists RG7388 and Nutlin 3A (RG7112); *CDK4/6* inhibitors palbociclib, ribociclib, abemaciclib and TQB3616) in experimental and clinical trials either used individually or in combination, especially in the therapy of DDLPS [98,100,103,105–112]. However, it has to be noted that combinations of *MDM2* and *CDK4* inhibition in DDLPS should be thoroughly investigated in clinical studies due to the possible combined toxicities of these drugs [106].

The orally bioavailable selective inhibitor of nuclear export selinexor has been demonstrated to have preclinical activity in various cancer types and is currently in phase I and II clinical trials for advanced cancers. It was shown in vitro that selinexor induces G1-arrest in liposarcoma cell lines with *MDM2* and *CDK4* amplification by increasing the protein level of p53 and p21, indicating a post-transcriptional effect. These results justify the exploration of selinexor in clinical trials targeting various sarcoma subtypes [113].

MDM2 inactivates p53 in a phosphorylated form. Dephosphorylation and depletion of *MDM2* by the inhibitor of HDAC resulted in increased apoptosis, anti-proliferative effects and cell cycle arrest in liposarcoma cell lines, warranting further evaluation of HDACi as a therapeutic option in *MDM2*-amplified LPS [114]. Another epigenetic approach to the treatment of DDLPS/WDLPS is the inhibition of specific miRNAs. Thus, promotion of *MDM2* expression, cell proliferation and invasion of the liposarcoma SW-872 cell line as well as inhibition of apoptosis by miR-215-5p is described in the literature. Targeting miR-215-5p may be a novel therapeutic strategy for the treatment of liposarcoma [64].

Following *MDM2*/P53 signaling, these molecules are linked to PTEN and the PI3K/Akt/mTOR pathway, regulating the pro-apoptotic and anti-apoptotic signals. Specifically, *MDM2* could be stabilized by Akt-mediated phosphorylation, and, in turn, inhibit PI3K/Akt activity via prevention of the nuclear localization of the tumor suppressor REST [104]. PTEN expression in patient samples correlates with poor survival [115]. The PTEN-controlled PI3K/Akt/mTOR pathway could be a therapeutic target for DDLPS, as PTEN protects p53 from *MDM2*-mediated degradation. Together with the inhibition of PI3K/Akt/mTOR signaling, it can augment P53-mediated apoptosis, as was demonstrated in multiple studies in vitro and in vivo [104]. PI3K/Akt/mTOR inhibitors, for example, BEZ235, could be an option for combined treatment of WDLPS/DDLPS [104,116]. Further, downstream Akt targets c-Jun N-terminal kinase (JNK) from the mitogen-activated protein kinase (MAPK) family and this cross-talk may be useful in the development of therapy approaches [117]. However, in phase II trials, the multi-kinase, dual-action inhibitor sorafenib demonstrated a lack of significant clinical efficacy in liposarcoma treatment [118].

Mutational events in the fibroblast growth factor receptors FGFR1, FGFR2, FGFR3 and FGFR4 and the FGFR substrate FRS2 suggest that FGFR signaling plays a role in the pathogenesis of liposarcoma, especially in the development of high-grade DDLPS [18,53]. Moreover, a combination of the FGFR inhibitors erdafitinib and NVP-BGJ398 together with the *MDM2* antagonist RG7388 was shown to be a promising strategy for the treatment of DDLPS and needs further investigation in clinical trials [55,119].

In addition to the TP53 and RB signaling pathways, other pathways may be involved in the dedifferentiation process from WDLPS to DDLPS, including mitogenic and motogenic Wnt and Hedgehog signaling cascades, as well as Notch signaling regulation the differentiation. Besides their overall tumorigenic properties, a specific association of Wnt, Hedgehog and Notch activation with malignant transformation was demonstrated in DDLPS and WDLPS. However, there is no clear evidence for a role of this pathway in regulating tumor progression and the dedifferentiation process [18]. Another Akt downstream target is insulin-like growth factor 1 receptor (IGF1R). IGF1R inhibitors are early-stage therapeutics, and their potential synergistic effect in combination with *CDK4*/6 inhibitors were predicted in silico and proved in vitro [120]. Several receptors, including MET, PDGFR, AXL, VEGFR and EGFR as well as Aurora kinase family proteins are overexpressed in WDLPS/DDLPS. All these receptors may act as targets and have already available small-molecule inhibitors, and some of them have already demonstrated anti-proliferative and proapoptotic effects in liposarcoma cells [18,45]. Noteworthily, the multi-kinase angiogenesis inhibitor anlotinib demonstrated in preclinical and clinical studies a higher efficacy compared to the multi-kinase inhibitors sorafenib, sunitinib and nintedatinib [102]. A phase II trial showed the promising efficacy and acceptable toxicity of anlotinib as maintenance treatment after first-line anthracycline-based chemotherapy [121,122].

Peroxisome proliferator-activated receptors (PPAR) regulate normal adipocyte differentiation. PPAR-gamma is regulated by c-*JUN* and induces the differentiation of normal preadipocytes. Hyperactivation of c-*JUN* blocks differentiation and may contribute to malignant transformation. PPAR-gamma agonists revealed antitumor activity in vitro in liposarcoma cell lines. In this sense, PPAR-gamma represents an attractive target, particularly for DDLPS, MLPS, and in some cases, PLS as a mechanism to revert these subtypes to a well-differentiated phenotype. However, clinical trials with the PPAR-gamma ligands demonstrated mixed results. The PPAR-gamma agonist troglitazone was used for the treatment of patients with advanced liposarcoma and demonstrated expression of several mRNA transcripts characteristic of adipocytic differentiation and a marked reduction in cancer cell proliferation [123]. Two other clinical trials with rosiglitazone and efatutazone demonstrated mixed results [124,125].

4.2. FUS-CHOP-Associated Abnormalities of PI3K/Akt/mTOR and Other Proliferative Signaling in MLPS

Fusion proteins from the chimeric oncogenes *FUS-DDIT3* and *EWS1R-DDIT3* act as aberrant transcription factors and may affect many signaling molecules. Thus, gene expression studies in MLPS have identified the recurrent upregulation of MET, RET, IGF-IR and PIK3CA, suggesting that these genes to be downstream targets of MLPS-specific fusion proteins [104]. Mutations in the PI3K catalytic subunit, IGF1R expression, amplification and mutations in PIK3Ca and loss of PTEN are reported in 12–18% of cases, therefore affecting multiple PTEN and PI3K/Akt/mTOR downstream genes [18,104]. The use of IGF-IR/PI3K/Akt/mTOR inhibitors in therapy for MLPS has a therapeutic potential and is currently under investigation. More specifically, treatment of myxoid liposarcoma cell lines in vitro and xenograft-bearing mice in vivo with several IGF-IR and PI3K/Akt/mTOR inhibitors resulted in significant growth inhibition [126,127]. One of the mechanisms of tumor heterogeneity and oncogenic potential maintenance is the phosphorylation of Interacts With SUPT6H (IWS1), a regulator of histone activity, by AKT. These findings support the use of the AKT/IWS1 axis as a novel prognostic factor and potential therapeutic target in liposarcoma therapy [128].

Another microarray analysis revealed overexpression of FGFR2 and other members of the FGF/FGFR family and the efficacy of the FGFR inhibitors PD173074, TKI258 (dovitinib) and BGJ398 in experiments in vitro [129]. In addition, the FUS-DDIT3 protein induces increased expression of the CAAT/enhancer-binding protein (C/EBP) and nuclear factor NFKBIZ, a member of the NF-kB family, colocalizing with FUS-DDIT3 [130]. A study of the kinome of cell lines and primary cell cultures from patients with metastatic myxoid liposarcoma revealed the activation of the kinase set associated with activation of the atypical nuclear factor-kappaB and the Src pathways. Moreover, in vitro NF-kB suppression by Casein kinase II inhibitor TBB and Src inhibition using dasatinib decreased cancer cell viability and offered potential therapeutic strategies for myxoid liposarcoma patients with advanced disease [131].

The Hippo pathway effector and transcriptional co-regulator YAP1 was shown to be a downstream target of FUS-DDIT3. In vitro studies demonstrated that FUS-DDIT3-driven IGF-IR/PI3K/AKT signaling promotes stability and nuclear accumulation of YAP1 via deregulation of the Hippo pathway. Gene expression profiling revealed gene signatures related to proliferation, cell cycle progression, apoptosis and adipogenesis. Therefore, FUS-DDIT3 involves IGF-IR/PI3K/AKT signals via Hippo/YAP1, and YAP1 may be an immunohistochemical marker for MLPS diagnostics. Moreover, these findings provide a rationale for the development of low-molecular-weight inhibitors of key components in Hippo/YAP1 signaling [132–134].

FUS-DDIT3-associated malignant transformation of adipocytes resulted in elevated levels of STAT3 and phosphorylated STAT3, suggesting the involvement of JAK/STAT signaling in the pathogenesis of MLPS [135]. Several inhibitors targeting JAK and GSK-3 caused downregulation of FUS-DDIT3 in vitro and reduced cell proliferation [136].

The components of the VEGF signaling pathway FLT1, PGF, VEGFA and VEGFB were shown to be indirect targets of FUS-DDIT3 in vitro. This could be a consequence of the ability of FUS-DDIT3 to reprogram primary adipocytes to a liposarcoma-like phenotype [137]. One case is reported in the literature of a 68-year-old Chinese woman initially diagnosed with advanced multiple intra-abdominal and pelvic round-cell liposarcomas who responded to therapy with the VEGFR2 inhibitor apatinib [138]. Further clinical trials are needed to confirm the efficacy and safety of VEFGR inhibitors in the treatment of MLPS.

MLPSs, like some other malignancies associated with chromosomal translocations resulting in expression of a fusion protein, are more responsive to trabectedin than other sarcoma types. Trabectedin does not act as an inhibitor of specific hyperactivated/overexpressed proteins; it binds covalently to the exocyclic amino group of guanines in the DNA minor groove, competing with the fusion protein and preventing its transcriptional activity. Several clinical studies have demonstrated the efficacy and favorable safety profile of trabectedin [18,112].

4.3. PLPS and MPLPS: No Specific Targets

No data on specific features in signaling and targeted treatments of pleomorphic liposarcoma are described in the literature. It is the rarest type of liposarcoma with poor prognosis, and its therapy involves mainly surgical management and the application of radiation. In addition, PLPS and MPLPS may respond to a doxorubicin and ifosfamide combination; trabectedin and eribulin are also options for advanced disease. A reduction in the primary tumor and the eradication of lung metastasis were reported in a clinical case of combined PLPS treatment with the multi-kinase inhibitor pazopanib, eribulin and dacarbazine [139]. Significant work remains to be done to develop novel therapies for this disease. To date, most studies have failed to identify targetable aberrations and have noted only consistent losses in p53 and Rb pathway proteins [18,140–142]

4.4. Perspectives of Targeted Therapy for Liposarcomas

Currently, CDK 4/6 and *MDM2* amplifications present the prospective targets for LPS therapy, and the efficacy of *CDK4/6* and *MDM2* inhibitors was proved in clinical trials on WDLPS and DDLPS (Table 2). Multi-kinase inhibitors including the well-studied sunitinib and the most recent, anlopanib, demonstrated mixed results, suggesting the necessity of further studies. To date, their combination with standard radiotherapy and conventional cytostatic approaches is still required until chemoresistance to the standard therapy appears. As nowadays chemoresistance prediction based on molecular genetics analysis seems insufficient, the development of experimental approaches for testing it ex vivo and in vitro may be useful for the exclusion of potentially ineffective targeted therapy courses and the choice of more promising treatment strategies.

Table 2. Targeted molecules proposed for LPS treatment.

Drug	Target/Mechanism of Action	LPS Subtype	References
Palbociclib	CDK 4/6 inhibitor	WDLPS, DDLPS	[100,112]
Abemaciclib	CDK 4/6 inhibitor		
Milademetan	*MDM2* inhibitor	WDLPS, DDLPS	[143–146]
BI 907828 (brigimadlin)	*MDM2* inhibitor		
Sunitinib	PDGFR/VEGFR inhibitor	Metastatic LPS	[98]
Lenvatinib	VEGFR/c-Kit/PDGFR/FGFR/RET inhibitor	LPS	[147]
Pazopanib	PDGFR/VEGFR/FGFR inhibitor	Metastatic LPS	[99]
Efatutazone	PPAR-α inhibitor	MLPS	[125]
Anlotinib	VEGFR/c-Kit/PDGFR/FGFR1 inhibitor	WDLPS/DDLPS	[102,121,122]
Selinexor	Inhibitor of nuclear transportation (inhibitor of exportin 1)	DDLPS	[148]

5. Conclusions/Future Direction in Therapy

Although multiple key genetic and epigenetic aberrations in liposarcoma have been explored, only a few of them have given rise to novel targeted therapy courses. The heterogeneity and very variable percentage of genetic and epigenetic abnormalities lead to an insufficient understanding of the complex signaling changes enabling tumor progression and high chance of development of tumor resistance. Notably, the reviewed data on specific genetic abnormalities taken together present a cluster of genetically characterized liposarcomas that may be considered for targeted therapies. The results of clinical trials of *CDK4* and *MDM2* inhibitors in the case of WDLPS/DDLPS and multi-kinase inhibitors targeting the FUS-CHOP downstream proteins seem promising. In many other cases, the complexity of sarcoma genetics could impede the diagnostics and may lead to tumor resistance and a poor prognosis. However, the combination with standard radiotherapy and conventional cytostatic approaches is still required until chemoresistance to the standard therapy appears. As nowadays chemoresistance prediction based on molecular genetics analysis seems insufficient, the development of experimental approaches for testing it ex vivo and in vitro may be useful for the exclusion of potentially ineffective targeted therapy courses and the choice of more promising treatment strategies. Overall, further data accumulation is required in the field of LPS molecular pathogenesis as well as in clinical trials of specific inhibitors, as the first target therapy applications gave rather promising results. A better understanding of the distinct genetic and molecular aberrations of liposarcoma subtypes may allow the development of several novel biology-driven therapies based on the specific molecular genetic profile of the disease.

Author Contributions: Conceptualization, E.A.L., K.I.K., G.A.B., B.Y.B. and M.G.Y.; analysis of the literature and table preparation, T.I.F., V.P.M. and E.P.K.; writing—original draft preparation, E.A.L., T.I.F. and V.P.M.; writing—review and editing, E.A.L., K.I.K., G.A.B., B.Y.B. and M.G.Y.; supervision, M.G.Y.; project administration, E.A.L. and K.I.K.; funding acquisition, M.G.Y. All authors have read and agreed to the published version of the manuscript.

Funding: This work was funded by the Russian Science Foundation, grant number 23-65-00003.

Conflicts of Interest: The authors declare no conflicts of interest.

Abbreviations

ACACB	Acetyl-Coa Carboxylase
ADRB2	Adrenoceptor Beta 2
Akt	AKT Serine/Threonine Kinase
ALDH1A3	Aldehyde Dehydrogenase 1 Family Member A3
APC	Adenomatous Polyposis Coli
APP	Amyloid Beta (A4) Precursor Protein
AQP7	Aquaporin-7
ARID1A	At-Rich Interaction Domain 1a
ASK1	Apoptosis Signal-Regulating Kinase 1
ATM	Ataxia Telangiectasia Mutated
ATRX	ATP-Dependent Helicase
AXL	Axl Receptor Tyrosine Kinase
BARD1	BRCA1 Associated RING Domain 1
BCL2	B-Cell Lymphoma 2
BMI1	BMI1 Proto-Oncogene, Polycomb Ring Finger
BUB1B	Mitotic Checkpoint Serine/Threonine-Protein Kinase BUB1 Beta
CALR	Calreticulin
CAV1	Caveolin 1
CCNA	Cyclin A
CCND1	Cyclin D1
CDC20	Cell Division Cycle Protein 20
CDH1	E-Cadherin
CDK	Cyclin-Dependent Kinase
CDKN	Cyclin Dependent Kinase Inhibitor 2a
CEBPA	CCAAT/Enhancer Binding Protein Alpha
CENPF	Centromere Protein F

CHEK1	Checkpoint Kinase 1
CHOP	C/EBP Homologous Protein Alpha
CpG	CG-Dinucleotides
CPM	Carboxypeptidase M
CTAG	Cancer/Testis Antigen
CTNNB1	Catenin Beta 1
DAXX	Death-Associated Protein 6
DDIT3	DNA Damage-Inducible Transcript 3,
DDLPS	Dedifferentiated Liposarcoma,
DDR2	Discoidin Domain Receptor Tyrosine Kinase 2
E2F1	E2f Transcription Factor 1
EGF	Epidermal Growth Factor
EGFR	Epidermal Growth Factor Receptor
EPHA1	Ephrin Type-A Receptor 1
ERBB3	Human Epidermal Growth Factor Receptor 3
EWS	Ewing Sarcoma Protein
FBXW7	F-Box and WD Repeat Domain Containing 7
FGF	Fibroblast Growth Factor
FGFR	Fibroblast Growth Factor Receptor
FLT1	FMS-Like Tyrosine Kinase 1
FRS2	Fibroblast Growth Factor Receptor Substrate 2
FUS	Fused In Sarcoma
FZD4	Frizzled Class Receptor 4
GPD1	Glycerol-3-Phosphate Dehydrogenase 1 (Soluble)
HDAC	Histone Deacetylase
HMGA	High-Mobility Group A
IGF	Insulin-Like Growth Factor
IGFR	Insulin-Like Growth Factor
INSM1	Insm Transcriptional Repressor 1
KLF4	Kruppel Like Factor 4
KRT8	Keratin 8
LEP	Leptin
LGALS3	Galectin-3
LGR5	Leucine-Rich Repeat Containing G Protein-Coupled Receptor 5
lncRNA	Long Non-Coding RNA
LPS	Liposarcoma
MDM2	Murin Double Minute 2
miRNA	Microrna
MAGEA9	Melanoma-Associated Antigen 9
MAP3K5	Mitogen-Activated Protein Kinase Kinase Kinase 5
MAPK	Mitogen-Activated Protein Kinase
MAZ	MYC-Associated Zinc Finger Protein
MCL1	Myeloid Leukemia and Chlamydia 1
MET	MET Proto-Oncogene, Receptor Tyrosine Kinase
MGMT	O6-Methylguanine DNA Methyltransferase
MLLT10	Myeloid/Lymphoid or Mixed-Lineage Leukemia (Trithorax Homolog, *Drosophila*); Translocated To 10
MLPS	Myxoid/Round-Cell Liposarcoma,
MMP2	Matrix Metalloproteinase 2
MPLPS	Myxoid Pleomorphic Liposarcoma
MST1	Mammalian Sterile 20-Like 1 Kinase
MT1G	Metallothionein 1G
mTOR	Mammalian Target of Rapamycin
NEBL	Nebulette
PCNA	Proliferating Cell Nuclear Antigen
PD-1	Programmed Cell Death Protein 1
PD-L1	Programmed Death Ligand 1
PDGF	Platelet-Derived Growth Factor
PDGFR	Platelet-Derived Growth Factor Receptor

PGF	Placental Growth Factor
PGFR	Placental Growth Factor Receptor
PI3K	Phosphatidylinositol 3-Kinase
PLPS	Pleomorphic Liposarcoma
PPAR	Peroxisome Proliferator-Activated Receptor
PRAME	Preferentially Expressed Antigen in Melanoma
PRC2	Polycomb Repressive Complex 2
PTEN	Phosphatase and Tensin Homolog
PTK7	Protein Tyrosine Kinase 7
RASSF1A	Ras-Association Domain Family 1 Isoform A
REST	Re1 Silencing Transcription Factor
RET	Rearranged During Transfection
ROS1	ROS Proto-Oncogene 1, Receptor Tyrosine Kinase
RUNX3	Runt-Related Transcription Factor 3
SAS	Stranded At Second
SIAH2	Seven In Absentia Homolog (SIAH) 2
SMAD4	Similar To The Gene Products Of The *Drosophila* Gene Mothers Against Decapentaplegic 4
SORBS1	Sorbin and SH3 Domain-Containing Protein 1
SPIN1	Spindlin 1
STAT6	Signal Transducer and Activator Of Transcription 6
STS	Soft-Tissue Sarcoma
SWI/SNF	Switch/Sucrose Non-Fermentable
TBX3	T-Box Transcription Factor 3
TERT	Telomerase Reverse Transcriptase
TGF	Transforming Growth Factor
THBS2	Thrombospondin 2
TKT	Transketolase
TOP2A	DNA Topoisomerase II Alpha
TSPAN31	Tetraspanin 31
VEGF	Vascular Endothelial Growth Factor
VEGFR	Vascular Endothelial Growth Factor Receptor
WDLPS	Well-Differentiated Liposarcoma
YAP1	Yes-Associated Protein 1
YEATS4	YEATS Domain Containing 4

References

1. Sbaraglia, M.; Bellan, E.; Dei Tos, A.P. The 2020 WHO Classification of Soft Tissue Tumours: News and Perspectives. *Pathologica* **2021**, *113*, 70–84. [CrossRef] [PubMed]
2. Dodd, L.G. Update on Liposarcoma: A Review for Cytopathologists. *Diagn. Cytopathol.* **2012**, *40*, 1122–1131. [CrossRef] [PubMed]
3. Lam, S.W.; Silva, T.M.; Bovée, J.V.M.G. New Molecular Entities of Soft Tissue and Bone Tumors. *Curr. Opin. Oncol.* **2022**, *34*, 354–361. [CrossRef] [PubMed]
4. Schaefer, I.-M.; Gronchi, A. WHO Pathology: Highlights of the 2020 Sarcoma Update. *Surg. Oncol. Clin. N. Am.* **2022**, *31*, 321–340. [CrossRef] [PubMed]
5. Singer, S.; Antonescu, C.R.; Riedel, E.; Brennan, M.F. Histologic Subtype and Margin of Resection Predict Pattern of Recurrence and Survival for Retroperitoneal Liposarcoma. *Ann. Surg.* **2003**, *238*, 358–370, discussion 370–371. [CrossRef]
6. Lahat, G.; Anaya, D.A.; Wang, X.; Tuvin, D.; Lev, D.; Pollock, R.E. Resectable Well-Differentiated versus Dedifferentiated Liposarcomas: Two Different Diseases Possibly Requiring Different Treatment Approaches. *Ann. Surg. Oncol.* **2008**, *15*, 1585–1593. [CrossRef]
7. Zhou, M.Y.; Bui, N.Q.; Charville, G.W.; Ganjoo, K.N.; Pan, M. Treatment of De-Differentiated Liposarcoma in the Era of Immunotherapy. *Int. J. Mol. Sci.* **2023**, *24*, 9571. [CrossRef]
8. Ghadimi, M.P.; Al-Zaid, T.; Madewell, J.; Peng, T.; Colombo, C.; Hoffman, A.; Creighton, C.J.; Zhang, Y.; Zhang, A.; Lazar, A.J.; et al. Diagnosis, Management, and Outcome of Patients with Dedifferentiated Liposarcoma Systemic Metastasis. *Ann. Surg. Oncol.* **2011**, *18*, 3762–3770. [CrossRef]
9. Hoffman, A.; Lazar, A.J.; Pollock, R.E.; Lev, D. New Frontiers in the Treatment of Liposarcoma, a Therapeutically Resistant Malignant Cohort. *Drug Resist. Updates* **2011**, *14*, 52–66. [CrossRef]
10. Ho, T.P. Myxoid Liposarcoma: How to Stage and Follow. *Curr. Treat. Options Oncol.* **2023**, *24*, 292–299. [CrossRef]

11. Fiore, M.; Grosso, F.; Lo Vullo, S.; Pennacchioli, E.; Stacchiotti, S.; Ferrari, A.; Collini, P.; Lozza, L.; Mariani, L.; Casali, P.G.; et al. Myxoid/Round Cell and Pleomorphic Liposarcomas: Prognostic Factors and Survival in a Series of Patients Treated at a Single Institution. *Cancer* **2007**, *109*, 2522–2531. [CrossRef] [PubMed]
12. De Vreeze, R.S.; De Jong, D.; Tielen, I.H.; Ruijter, H.J.; Nederlof, P.M.; Haas, R.L.; Van Coevorden, F. Primary Retroperitoneal Myxoid/Round Cell Liposarcoma Is a Nonexisting Disease: An Immunohistochemical and Molecular Biological Analysis. *Mod. Pathol.* **2009**, *22*, 223–231. [CrossRef] [PubMed]
13. Jones, R.L.; Fisher, C.; Al-Muderis, O.; Judson, I.R. Differential Sensitivity of Liposarcoma Subtypes to Chemotherapy. *Eur. J. Cancer* **2005**, *41*, 2853–2860. [CrossRef] [PubMed]
14. Hornick, J.L.; Bosenberg, M.W.; Mentzel, T.; McMenamin, M.E.; Oliveira, A.M.; Fletcher, C.D.M. Pleomorphic Liposarcoma: Clinicopathologic Analysis of 57 Cases. *Am. J. Surg. Pathol.* **2004**, *28*, 1257–1267. [CrossRef] [PubMed]
15. Ghadimi, M.P.; Liu, P.; Peng, T.; Bolshakov, S.; Young, E.D.; Torres, K.E.; Colombo, C.; Hoffman, A.; Broccoli, D.; Hornick, J.L.; et al. Pleomorphic Liposarcoma: Clinical Observations and Molecular Variables. *Cancer* **2011**, *117*, 5359–5369. [CrossRef]
16. Creytens, D.; Folpe, A.L.; Koelsche, C.; Mentzel, T.; Ferdinande, L.; van Gorp, J.M.; Van der Linden, M.; Raman, L.; Menten, B.; Fritchie, K.; et al. Myxoid Pleomorphic Liposarcoma-a Clinicopathologic, Immunohistochemical, Molecular Genetic and Epigenetic Study of 12 Cases, Suggesting a Possible Relationship with Conventional Pleomorphic Liposarcoma. *Mod. Pathol.* **2021**, *34*, 2043–2049. [CrossRef] [PubMed]
17. Gobble, R.M.; Qin, L.-X.; Brill, E.R.; Angeles, C.V.; Ugras, S.; O'Connor, R.B.; Moraco, N.H.; DeCarolis, P.L.; Antonescu, C.; Singer, S. Expression Profiling of Liposarcoma Yields a Multigene Predictor of Patient Outcome and Identifies Genes That Contribute to Liposarcomagenesis. *Cancer Res.* **2011**, *71*, 2697–2705. [CrossRef] [PubMed]
18. Saponara, M.; Stacchiotti, S.; Gronchi, A. Pharmacological Therapies for Liposarcoma. *Expert Rev. Clin. Pharmacol.* **2017**, *10*, 361–377. [CrossRef]
19. Li, X.-Q.; Hisaoka, M.; Hashimoto, H. Expression of Estrogen Receptors α and β in Soft Tissue Sarcomas: Immunohistochemical and Molecular Analysis: ERα and ERβ in Soft Tissue Sarcomas. *Pathol. Int.* **2003**, *53*, 671–679. [CrossRef]
20. Willmer, T.; Cooper, A.; Sims, D.; Govender, D.; Prince, S. The T-Box Transcription Factor 3 Is a Promising Biomarker and a Key Regulator of the Oncogenic Phenotype of a Diverse Range of Sarcoma Subtypes. *Oncogenesis* **2016**, *5*, e199. [CrossRef]
21. Lee, A.T.J.; Thway, K.; Huang, P.H.; Jones, R.L. Clinical and Molecular Spectrum of Liposarcoma. *J. Clin. Oncol.* **2018**, *36*, 151–159. [CrossRef] [PubMed]
22. Pedeutour, F.; Maire, G.; Pierron, A.; Thomas, D.M.; Garsed, D.W.; Bianchini, L.; Duranton-Tanneur, V.; Cortes-Maurel, A.; Italiano, A.; Squire, J.A.; et al. A Newly Characterized Human Well-Differentiated Liposarcoma Cell Line Contains Amplifications of the 12q12-21 and 10p11-14 Regions. *Virchows Arch.* **2012**, *461*, 67–78. [CrossRef] [PubMed]
23. Campos Gudiño, R.; McManus, K.J.; Hombach-Klonisch, S. Aberrant *HMGA2* Expression Sustains Genome Instability That Promotes Metastasis and Therapeutic Resistance in Colorectal Cancer. *Cancers* **2023**, *15*, 1735. [CrossRef]
24. Carmagnani Pestana, R.; Moyers, J.T.; Roszik, J.; Sen, S.; Hong, D.S.; Naing, A.; Herzog, C.E.; Fu, S.; Piha-Paul, S.A.; Rodon, J.; et al. Impact of Biomarker-Matched Therapies on Outcomes in Patients with Sarcoma Enrolled in Early-Phase Clinical Trials (SAMBA 101). *Clin. Cancer Res.* **2023**, *29*, 1708–1718. [CrossRef] [PubMed]
25. Saâda-Bouzid, E.; Burel-Vandenbos, F.; Ranchère-Vince, D.; Birtwisle-Peyrottes, I.; Chetaille, B.; Bouvier, C.; Château, M.-C.; Peoc'h, M.; Battistella, M.; Bazin, A.; et al. Prognostic Value of *HMGA2*, *CDK4*, and *JUN* Amplification in Well-Differentiated and Dedifferentiated Liposarcomas. *Mod. Pathol.* **2015**, *28*, 1404–1411. [CrossRef] [PubMed]
26. Dei Tos, A.P. Liposarcomas: Diagnostic Pitfalls and New Insights. *Histopathology* **2014**, *64*, 38–52. [CrossRef] [PubMed]
27. Oda, Y.; Yamamoto, H.; Kohashi, K.; Yamada, Y.; Iura, K.; Ishii, T.; Maekawa, A.; Bekki, H. Soft Tissue Sarcomas: From a Morphological to a Molecular Biological Approach. *Pathol. Int.* **2017**, *67*, 435–446. [CrossRef] [PubMed]
28. Takashima, Y.; Komatsu, S.; Ohashi, T.; Kiuchi, J.; Kamiya, H.; Shimizu, H.; Arita, T.; Konishi, H.; Shiozaki, A.; Kubota, T.; et al. Overexpression of Tetraspanin31 Contributes to Malignant Potential and Poor Outcomes in Gastric Cancer. *Cancer Sci.* **2022**, *113*, 1984–1998. [CrossRef]
29. Wang, J.; Zhou, Y.; Li, D.; Sun, X.; Deng, Y.; Zhao, Q. TSPAN 31 Is a Critical Regulator on Transduction of Survival and Apoptotic Signals in Hepatocellular Carcinoma Cells. *FEBS Lett.* **2017**, *591*, 2905–2918. [CrossRef]
30. Kanojia, D.; Nagata, Y.; Garg, M.; Lee, D.H.; Sato, A.; Yoshida, K.; Sato, Y.; Sanada, M.; Mayakonda, A.; Bartenhagen, C.; et al. Genomic Landscape of Liposarcoma. *Oncotarget* **2015**, *6*, 42429–42444. [CrossRef]
31. Liu, T.; Wang, J.; Yang, H.; Jin, Q.; Wang, X.; Fu, Y.; Luan, Y.; Wang, Q.; Youngblood, M.W.; Lu, X.; et al. Enhancer Coamplification and Hijacking Promote Oncogene Expression in Liposarcoma. *Cancer Res.* **2023**, *83*, 1517–1530. [CrossRef] [PubMed]
32. Louis-Brennetot, C.; Coindre, J.-M.; Ferreira, C.; Pérot, G.; Terrier, P.; Aurias, A. The CDKN2A/CDKN2B/*CDK4*/CCND1 Pathway Is Pivotal in Well-Differentiated and Dedifferentiated Liposarcoma Oncogenesis. An Analysis of 104 Tumors. *Genes. Chromosomes Cancer* **2011**, *50*, 896–907. [CrossRef] [PubMed]
33. Ohno, T.; Ouchida, M.; Lee, L.; Gatalica, Z.; Rao, V.N.; Reddy, E.S. The *EWS* Gene, Involved in Ewing Family of Tumors, Malignant Melanoma of Soft Parts and Desmoplastic Small Round Cell Tumors, Codes for an RNA Binding Protein with Novel Regulatory Domains. *Oncogene* **1994**, *9*, 3087–3097. [PubMed]
34. Xiang, H.; Wang, J.; Hisaoka, M.; Zhu, X. Characteristic Sequence Motifs Located at the Genomic Breakpoints of the Translocation t(12;16) and t(12;22) in Myxoid Liposarcoma. *Pathology* **2008**, *40*, 547–552. [CrossRef] [PubMed]

35. Powers, M.P.; Wang, W.-L.; Hernandez, V.S.; Patel, K.S.; Lev, D.C.; Lazar, A.J.; López-Terrada, D.H. Detection of Myxoid Liposarcoma-Associated FUS-DDIT3 Rearrangement Variants Including a Newly Identified Breakpoint Using an Optimized RT-PCR Assay. *Mod. Pathol.* **2010**, *23*, 1307–1315. [CrossRef] [PubMed]
36. Kunieda, J.; Yamashita, K.; Togashi, Y.; Baba, S.; Sakata, S.; Inamura, K.; Ae, K.; Matsumoto, S.; Machinami, R.; Kitagawa, M.; et al. High Prevalence of TERT Aberrations in Myxoid Liposarcoma: TERT Reactivation May Play a Crucial Role in Tumorigenesis. *Cancer Sci.* **2022**, *113*, 1078–1089. [CrossRef]
37. Nezu, Y.; Hagiwara, K.; Yamamoto, Y.; Fujiwara, T.; Matsuo, K.; Yoshida, A.; Kawai, A.; Saito, T.; Ochiya, T. miR-135b, a Key Regulator of Malignancy, Is Linked to Poor Prognosis in Human Myxoid Liposarcoma. *Oncogene* **2016**, *35*, 6177–6188. [CrossRef]
38. Paulien, S.; Turc-Carel, C.; Dal Cin, P.; Jani-Sait, S.; Sreekantaiah, C.; Leong, S.P.; Vogelstein, B.; Kinzler, K.W.; Sandberg, A.A.; Gemmill, R.M. Myxoid Liposarcoma with t(12;16) (Q13;P11) Contains Site-Specific Differences in Methylation Patterns Surrounding a Zinc-Finger Gene Mapped to the Breakpoint Region on Chromosome 12. *Cancer Res.* **1990**, *50*, 7902–7907.
39. Sievers, S.; Fritzsch, C.; Lehnhardt, M.; Zahn, S.; Kutzner, N.; Kuhnen, C.; Müller, O. Hypermethylation of the APC Promoter but Lack of APC Mutations in Myxoid/Round-Cell Liposarcoma. *Int. J. Cancer* **2006**, *119*, 2347–2352. [CrossRef]
40. Bennani-Baiti, I.M. Epigenetic and Epigenomic Mechanisms Shape Sarcoma and Other Mesenchymal Tumor Pathogenesis. *Epigenomics* **2011**, *3*, 715–732. [CrossRef]
41. Taylor, B.S.; Barretina, J.; Socci, N.D.; DeCarolis, P.; Ladanyi, M.; Meyerson, M.; Singer, S.; Sander, C. Functional Copy-Number Alterations in Cancer. *PLoS ONE* **2008**, *3*, e3179. [CrossRef] [PubMed]
42. Guillou, L.; Aurias, A. Soft Tissue Sarcomas with Complex Genomic Profiles. *Virchows Arch.* **2010**, *456*, 201–217. [CrossRef] [PubMed]
43. Zhang, Q.; Dong, Y.; Zhou, M.; Guo, Y.; Lou, L.; Qu, Z.; Zheng, Y.; Duan, Y. INSM1 Expression in Mesenchymal Tumors and Its Clinicopathological Significance. *BioMed Res. Int.* **2022**, *2022*, 1580410. [CrossRef]
44. Shannon, N.B.; Tan, Q.X.; Tan, J.W.-S.; Hendrikson, J.; Ng, W.H.; Ng, G.; Liu, Y.; Tan, G.H.C.; Wong, J.S.M.; Soo, K.C.; et al. Gene Expression Changes Associated with Dedifferentiation in Liposarcoma Predict Overall Survival. *Cancers* **2021**, *13*, 3049. [CrossRef] [PubMed]
45. Bill, K.L.J.; Casadei, L.; Prudner, B.C.; Iwenofu, H.; Strohecker, A.M.; Pollock, R.E. Liposarcoma: Molecular Targets and Therapeutic Implications. *Cell. Mol. Life Sci.* **2016**, *73*, 3711–3718. [CrossRef] [PubMed]
46. Tariq, M.U.; Din, N.U.; Abdul-Ghafar, J.; Park, Y.-K. The Many Faces of Solitary Fibrous Tumor; Diversity of Histological Features, Differential Diagnosis and Role of Molecular Studies and Surrogate Markers in Avoiding Misdiagnosis and Predicting the Behavior. *Diagn. Pathol.* **2021**, *16*, 32. [CrossRef] [PubMed]
47. Yu, H.; Pei, D.; Chen, L.; Zhou, X.; Zhu, H. Identification of Key Genes and Molecular Mechanisms Associated with Dedifferentiated Liposarcoma Based on Bioinformatic Methods. *OncoTargets Ther.* **2017**, *10*, 3017–3027. [CrossRef] [PubMed]
48. Hirata, M.; Asano, N.; Katayama, K.; Yoshida, A.; Tsuda, Y.; Sekimizu, M.; Mitani, S.; Kobayashi, E.; Komiyama, M.; Fujimoto, H.; et al. Integrated Exome and RNA Sequencing of Dedifferentiated Liposarcoma. *Nat. Commun.* **2019**, *10*, 5683. [CrossRef]
49. Dang, T.N.; Tiongco, R.P.; Brown, L.M.; Taylor, J.L.; Lyons, J.M.; Lau, F.H.; Floyd, Z.E. Expression of the Preadipocyte Marker ZFP423 Is Dysregulated between Well-Differentiated and Dedifferentiated Liposarcoma. *BMC Cancer* **2022**, *22*, 300. [CrossRef]
50. Lee, J.-C.; Jeng, Y.-M.; Liau, J.-Y.; Tsai, J.-H.; Hsu, H.-H.; Yang, C.-Y. Alternative Lengthening of Telomeres and Loss of ATRX Are Frequent Events in Pleomorphic and Dedifferentiated Liposarcomas. *Mod. Pathol.* **2015**, *28*, 1064–1073. [CrossRef]
51. Gatalica, Z.; Snyder, C.; Maney, T.; Ghazalpour, A.; Holterman, D.A.; Xiao, N.; Overberg, P.; Rose, I.; Basu, G.D.; Vranic, S.; et al. Programmed Cell Death 1 (PD-1) and Its Ligand (PD-L1) in Common Cancers and Their Correlation with Molecular Cancer Type. *Cancer Epidemiol. Biomark. Prev.* **2014**, *23*, 2965–2970. [CrossRef] [PubMed]
52. Chen, J.; Lian, Y.; Zhao, B.; Han, J.; Li, X.; Wu, J.; Hou, M.; Yue, M.; Zhang, K.; Liu, G.; et al. Deciphering the Prognostic and Therapeutic Significance of Cell Cycle Regulator CENPF: A Potential Biomarker of Prognosis and Immune Microenvironment for Patients with Liposarcoma. *Int. J. Mol. Sci.* **2023**, *24*, 7010. [CrossRef] [PubMed]
53. Zhang, K.; Chu, K.; Wu, X.; Gao, H.; Wang, J.; Yuan, Y.-C.; Loera, S.; Ho, K.; Wang, Y.; Chow, W.; et al. Amplification of FRS2 and Activation of FGFR/FRS2 Signaling Pathway in High-Grade Liposarcoma. *Cancer Res.* **2013**, *73*, 1298–1307. [CrossRef] [PubMed]
54. Li, C.; Shen, Y.; Ren, Y.; Liu, W.; Li, M.; Liang, W.; Liu, C.; Li, F. Oncogene Mutation Profiling Reveals Poor Prognosis Associated with FGFR1/3 Mutation in Liposarcoma. *Hum. Pathol.* **2016**, *55*, 143–150. [CrossRef] [PubMed]
55. Dadone-Montaudié, B.; Laroche-Clary, A.; Mongis, A.; Chamorey, E.; Mauro, I.D.; Chaire, V.; Finetti, P.; Schiappa, R.; Le Loarer, F.; Birtwisle-Peyrottes, I.; et al. Novel Therapeutic Insights in Dedifferentiated Liposarcoma: A Role for FGFR and *MDM2* Dual Targeting. *Cancers* **2020**, *12*, 3058. [CrossRef] [PubMed]
56. Zeng, J.; Zhang, N.; Zhao, G.; Xu, L.; Yang, Y.; Xu, X.; Chen, M.; Wang, H.; Zheng, S.X.; Li, X. MT1G Is Silenced by DNA Methylation and Contributes to the Pathogenesis of Hepatocellular Carcinoma. *J. Cancer* **2018**, *9*, 2807–2816. [CrossRef] [PubMed]
57. Song, L.; Chang, R.; Dai, C.; Wu, Y.; Guo, J.; Qi, M.; Zhou, W.; Zhan, L. SORBS1 Suppresses Tumor Metastasis and Improves the Sensitivity of Cancer to Chemotherapy Drug. *Oncotarget* **2017**, *8*, 9108–9122. [CrossRef] [PubMed]
58. Crozat, A.; Åman, P.; Mandahl, N.; Ron, D. Fusion of CHOP to a Novel RNA-Binding Protein in Human Myxoid Liposarcoma. *Nature* **1993**, *363*, 640–644. [CrossRef]
59. Ferreira, M.S.V.; Crysandt, M.; Braunschweig, T.; Jost, E.; Voss, B.; Bouillon, A.-S.; Knuechel, R.; Brümmendorf, T.H.; Beier, F. Presence of TERT Promoter Mutations Is a Secondary Event and Associates with Elongated Telomere Length in Myxoid Liposarcomas. *Int. J. Mol. Sci.* **2018**, *19*, 608. [CrossRef]

60. Bettegowda, C.; Yip, S.; Jiang, B.; Wang, W.-L.; Clarke, M.J.; Lazary, A.; Gambarotti, M.; Zhang, M.; Sciubba, D.M.; Wolinsky, J.-P.; et al. Prognostic Significance of Human Telomerase Reverse Transcriptase Promoter Region Mutations C228T and C250T for Overall Survival in Spinal Chordomas. *Neuro-Oncol.* **2019**, *21*, 1005–1015. [CrossRef]
61. Hemminger, J.A.; Toland, A.E.; Scharschmidt, T.J.; Mayerson, J.L.; Guttridge, D.C.; Iwenofu, O.H. Expression of Cancer-Testis Antigens MAGEA1, MAGEA3, ACRBP, PRAME, SSX2, and CTAG2 in Myxoid and Round Cell Liposarcoma. *Mod. Pathol.* **2014**, *27*, 1238–1245. [CrossRef] [PubMed]
62. Momose, T.; Yoshimura, Y.; Harumiya, S.; Isobe, K.; Kito, M.; Fukushima, M.; Kato, H.; Nakayama, J. Chondroitin Sulfate Synthase 1 Expression Is Associated with Malignant Potential of Soft Tissue Sarcomas with Myxoid Substance. *Hum. Pathol.* **2016**, *50*, 15–23. [CrossRef] [PubMed]
63. Kanojia, D.; Kirtonia, A.; Srujana, N.S.V.; Jeevanandan, S.P.; Shyamsunder, P.; Sampath, S.S.; Dakle, P.; Mayakonda, A.; Kaur, H.; Yanyi, J.; et al. Transcriptome Analysis Identifies TODL as a Novel lncRNA Associated with Proliferation, Differentiation, and Tumorigenesis in Liposarcoma through FOXM1. *Pharmacol. Res.* **2022**, *185*, 106462. [CrossRef] [PubMed]
64. Song, Z.; Bai, J.; Jiang, R.; Wu, J.; Yang, W. MicroRNA-215-5p Promotes Proliferation, Invasion, and Inhibits Apoptosis in Liposarcoma Cells by TargetingMDM2. *Cancer Med.* **2023**, *12*, 13455–13470. [CrossRef] [PubMed]
65. Lauinger, L.; Kaiser, P. Sensing and Signaling of Methionine Metabolism. *Metabolites* **2021**, *11*, 83. [CrossRef] [PubMed]
66. Aoki, Y.; Yamamoto, J.; Tome, Y.; Hamada, K.; Masaki, N.; Inubushi, S.; Tashiro, Y.; Bouvet, M.; Endo, I.; Nishida, K.; et al. Over-Methylation of Histone H3 Lysines Is a Common Molecular Change Among the Three Major Types of Soft-Tissue Sarcoma in Patient-Derived Xenograft (PDX) Mouse Models. *Cancer Genom.-Proteom.* **2021**, *18*, 715–721. [CrossRef] [PubMed]
67. Abeshouse, A.; Adebamowo, C.; Adebamowo, S.N.; Akbani, R.; Akeredolu, T.; Ally, A.; Anderson, M.L.; Anur, P.; Appelbaum, E.L.; Armenia, J.; et al. Comprehensive and Integrated Genomic Characterization of Adult Soft Tissue Sarcomas. *Cell* **2017**, *171*, 950–965.e28. [CrossRef] [PubMed]
68. Nacev, B.A.; Jones, K.B.; Intlekofer, A.M.; Yu, J.S.E.; Allis, C.D.; Tap, W.D.; Ladanyi, M.; Nielsen, T.O. The Epigenomics of Sarcoma. *Nat. Rev. Cancer* **2020**, *20*, 608–623. [CrossRef]
69. Meltzer, P.S.; Helman, L.J. Genomic Investigation of Dedifferentiated Liposarcoma Suggests a Role for Therapeutic Targeting of the Tumor Epigenome. *Cancer Discov.* **2011**, *1*, 555–556. [CrossRef]
70. Taylor, B.S.; DeCarolis, P.L.; Angeles, C.V.; Brenet, F.; Schultz, N.; Antonescu, C.R.; Scandura, J.M.; Sander, C.; Viale, A.J.; Socci, N.D.; et al. Frequent Alterations and Epigenetic Silencing of Differentiation Pathway Genes in Structurally Rearranged Liposarcomas. *Cancer Discov.* **2011**, *1*, 587–597. [CrossRef]
71. Renner, M.; Wolf, T.; Meyer, H.; Hartmann, W.; Penzel, R.; Ulrich, A.; Lehner, B.; Hovestadt, V.; Czwan, E.; Egerer, G.; et al. Integrative DNA Methylation and Gene Expression Analysis in High-Grade Soft Tissue Sarcomas. *Genome Biol.* **2013**, *14*, r137. [CrossRef] [PubMed]
72. He, M.; Aisner, S.; Benevenia, J.; Patterson, F.; Harrison, L.E.; Hameed, M. Epigenetic Alteration of p16INK4a Gene in Dedifferentiation of Liposarcoma. *Pathol.-Res. Pract.* **2009**, *205*, 386–394. [CrossRef] [PubMed]
73. Zhang, P.; Bill, K.; Liu, J.; Young, E.; Peng, T.; Bolshakov, S.; Hoffman, A.; Song, Y.; Demicco, E.G.; Terrada, D.L.; et al. MiR-155 Is a Liposarcoma Oncogene That Targets Casein Kinase-1α and Enhances β-Catenin Signaling. *Cancer Res.* **2012**, *72*, 1751–1762. [CrossRef]
74. Boro, A.; Bauer, D.; Born, W.; Fuchs, B. Plasma Levels of miRNA-155 as a Powerful Diagnostic Marker for Dedifferentiated Liposarcoma. *Am. J. Cancer Res.* **2016**, *6*, 544–552. [PubMed]
75. Vincenzi, B.; Iuliani, M., Zoccoli, A.; Pantano, F.; Fioramonti, M.; De Lisi, D.; Frezza, A.M.; Rabitti, C.; Perrone, G.; Muda, A.O.; et al. Deregulation of Dicer and Mir-155 Expression in Liposarcoma. *Oncotarget* **2015**, *6*, 10586–10591. [CrossRef] [PubMed]
76. Mazzu, Y.Z.; Hu, Y.; Soni, R.K.; Mojica, K.M.; Qin, L.-X.; Agius, P.; Waxman, Z.M.; Mihailovic, A.; Socci, N.D.; Hendrickson, R.C.; et al. miR-193b–Regulated Signaling Networks Serve as Tumor Suppressors in Liposarcoma and Promote Adipogenesis in Adipose-Derived Stem Cells. *Cancer Res.* **2017**, *77*, 5728–5740. [CrossRef] [PubMed]
77. Mazzu, Y.Z.; Hu, Y.; Shen, Y.; Tuschl, T.; Singer, S. miR-193b Regulates Tumorigenesis in Liposarcoma Cells via PDGFR, TGFβ, and Wnt Signaling. *Sci. Rep.* **2019**, *9*, 3197. [CrossRef]
78. Vos, M.; Boers, R.; Vriends, A.L.M.; Boers, J.; van Kuijk, P.F.; van Houdt, W.J.; van Leenders, G.J.L.H.; Wagrodzki, M.; van IJcken, W.F.J.; Gribnau, J.; et al. MicroRNA Expression and DNA Methylation Profiles Do Not Distinguish between Primary and Recurrent Well-Differentiated Liposarcoma. *PLoS ONE* **2020**, *15*, e0228014. [CrossRef]
79. Ugras, S.; Brill, E.; Jacobsen, A.; Hafner, M.; Socci, N.D.; Decarolis, P.L.; Khanin, R.; O'Connor, R.; Mihailovic, A.; Taylor, B.S.; et al. Small RNA Sequencing and Functional Characterization Reveals MicroRNA-143 Tumor Suppressor Activity in Liposarcoma. *Cancer Res.* **2011**, *71*, 5659–5669. [CrossRef]
80. Gits, C.M.M.; Van Kuijk, P.F.; Jonkers, M.B.E.; Boersma, A.W.M.; Smid, M.; Van Ijcken, W.F.; Coindre, J.-M.; Chibon, F.; Verhoef, C.; Mathijssen, R.H.J.; et al. MicroRNA Expression Profiles Distinguish Liposarcoma Subtypes and Implicate miR-145 and miR-451 as Tumor Suppressors: MiRNA Expression Profiling of Liposarcoma Subtypes. *Int. J. Cancer* **2014**, *135*, 348–361. [CrossRef]
81. Yu, P.Y.; Lopez, G.; Braggio, D.; Koller, D.; Bill, K.L.J.; Prudner, B.C.; Zewdu, A.; Chen, J.L.; Iwenofu, O.H.; Lev, D.; et al. miR-133a Function in the Pathogenesis of Dedifferentiated Liposarcoma. *Cancer Cell Int.* **2018**, *18*, 89. [CrossRef] [PubMed]
82. Zhou, Y.; Zhang, Y.; Huang, Y.; Tan, R.; Liu, T.; Zhuang, R.; Zhu, M.; Han, W.; Hou, Y.; Liu, J.; et al. Liposarcoma miRNA Signatures Identified from Genome-Wide miRNA Expression Profiling. *Future Oncol.* **2014**, *10*, 1373–1386. [CrossRef] [PubMed]

83. Yu, P.Y.; Balkhi, M.Y.; Ladner, K.J.; Alder, H.; Yu, L.; Mo, X.; Kraybill, W.G.; Guttridge, D.C.; Hans Iwenofu, O. A Selective Screening Platform Reveals Unique Global Expression Patterns of microRNAs in a Cohort of Human Soft-Tissue Sarcomas. *Lab. Investig.* **2016**, *96*, 481–491. [CrossRef] [PubMed]
84. Tan, H.M.; Cheng, H.; Tang, Y.C.; Leong, S.M.; Teo, P.Y.; Lee, C.K.; Lee, V.K.M.; Hue, S.S.-S. MicroRNAs as Potential Biomarkers in the Differential Diagnosis of Lipomatous Tumors and Their Mimics. *Int. J. Mol. Sci.* **2022**, *23*, 7804. [CrossRef] [PubMed]
85. Fricke, A.; Cimniak, A.F.V.; Ullrich, P.V.; Becherer, C.; Bickert, C.; Pfeifer, D.; Heinz, J.; Stark, G.B.; Bannasch, H.; Braig, D.; et al. Whole Blood miRNA Expression Analysis Reveals miR-3613-3p as a Potential Biomarker for Dedifferentiated Liposarcoma. *Cancer Biomark.* **2018**, *22*, 199–207. [CrossRef] [PubMed]
86. Kohama, I.; Asano, N.; Matsuzaki, J.; Yamamoto, Y.; Yamamoto, T.; Takahashi, R.-U.; Kobayashi, E.; Takizawa, S.; Sakamoto, H.; Kato, K.; et al. Comprehensive Serum and Tissue microRNA Profiling in Dedifferentiated Liposarcoma. *Oncol. Lett.* **2021**, *22*, 623. [CrossRef] [PubMed]
87. Yu, J.S.E.; Colborne, S.; Hughes, C.S.; Morin, G.B.; Nielsen, T.O. The FUS-DDIT3 Interactome in Myxoid Liposarcoma. *Neoplasia* **2019**, *21*, 740–751. [CrossRef]
88. Lindén, M.; Thomsen, C.; Grundevik, P.; Jonasson, E.; Andersson, D.; Runnberg, R.; Dolatabadi, S.; Vannas, C.; Luna Santamaría, M.; Fagman, H.; et al. FET Family Fusion Oncoproteins Target the SWI / SNF Chromatin Remodeling Complex. *EMBO Rep.* **2019**, *20*, e45766. [CrossRef]
89. Zullow, H.J.; Sankar, A.; Ingram, D.R.; Samé Guerra, D.D.; D'Avino, A.R.; Collings, C.K.; Lazcano, R.; Wang, W.-L.; Liang, Y.; Qi, J.; et al. The FUS::DDIT3 Fusion Oncoprotein Inhibits BAF Complex Targeting and Activity in Myxoid Liposarcoma. *Mol. Cell* **2022**, *82*, 1737–1750.e8. [CrossRef]
90. Franz, H.; Greschik, H.; Willmann, D.; Ozretić, L.; Jilg, C.A.; Wardelmann, E.; Jung, M.; Buettner, R.; Schüle, R. The Histone Code Reader SPIN1 Controls RET Signaling in Liposarcoma. *Oncotarget* **2015**, *6*, 4773–4789. [CrossRef]
91. Marcato, P.; Dean, C.A.; Pan, D.; Araslanova, R.; Gillis, M.; Joshi, M.; Helyer, L.; Pan, L.; Leidal, A.; Gujar, S.; et al. Aldehyde Dehydrogenase Activity of Breast Cancer Stem Cells Is Primarily Due to Isoform ALDH1A3 and Its Expression Is Predictive of Metastasis. *Stem Cells* **2011**, *29*, 32–45. [CrossRef] [PubMed]
92. Lohberger, B.; Rinner, B.; Stuendl, N.; Absenger, M.; Liegl-Atzwanger, B.; Walzer, S.M.; Windhager, R.; Leithner, A. Aldehyde Dehydrogenase 1, a Potential Marker for Cancer Stem Cells in Human Sarcoma. *PLoS ONE* **2012**, *7*, e43664. [CrossRef] [PubMed]
93. En-lin, S.; Sheng-guo, C.; Hua-qiao, W. The Expression of EFEMP1 in Cervical Carcinoma and Its Relationship with Prognosis. *Gynecol. Oncol.* **2010**, *117*, 417–422. [CrossRef] [PubMed]
94. Timpl, R.; Sasaki, T.; Kostka, G.; Chu, M.-L. Fibulins: A Versatile Family of Extracellular Matrix Proteins. *Nat. Rev. Mol. Cell Biol.* **2003**, *4*, 479–489. [CrossRef] [PubMed]
95. Song, E.; Hou, Y.; Yu, S.; Chen, S.; Huang, J.; Luo, T.; Kong, L.; Xu, J.; Wang, H. EFEMP1 Expression Promotes Angiogenesis and Accelerates the Growth of Cervical Cancer In Vivo. *Gynecol. Oncol.* **2011**, *121*, 174–180. [CrossRef] [PubMed]
96. Renner, M.; Czwan, E.; Hartmann, W.; Penzel, R.; Brors, B.; Eils, R.; Wardelmann, E.; Büttner, R.; Lichter, P.; Schirmacher, P.; et al. MicroRNA Profiling of Primary High-Grade Soft Tissue Sarcomas. *Genes. Chromosomes Cancer* **2012**, *51*, 982–996. [CrossRef]
97. Borjigin, N.; Ohno, S.; Wu, W.; Tanaka, M.; Suzuki, R.; Fujita, K.; Takanashi, M.; Oikawa, K.; Goto, T.; Motoi, T.; et al. TLS-CHOP Represses miR-486 Expression, Inducing Upregulation of a Metastasis Regulator PAI-1 in Human Myxoid Liposarcoma. *Biochem. Biophys. Res. Commun.* **2012**, *427*, 355–360. [CrossRef] [PubMed]
98. Mahmood, S.T.; Agresta, S.; Vigil, C.E.; Zhao, X.; Han, G.; D'Amato, G.; Calitri, C.E.; Dean, M.; Garrett, C.; Schell, M.J.; et al. Phase II Study of Sunitinib Malate, a Multitargeted Tyrosine Kinase Inhibitor in Patients with Relapsed or Refractory Soft Tissue Sarcomas. Focus on Three Prevalent Histologies: Leiomyosarcoma, Liposarcoma and Malignant Fibrous Histiocytoma. *Int. J. Cancer* **2011**, *129*, 1963–1969. [CrossRef]
99. Samuels, B.L.; Chawla, S.P.; Somaiah, N.; Staddon, A.P.; Skubitz, K.M.; Milhem, M.M.; Kaiser, P.E.; Portnoy, D.C.; Priebat, D.A.; Walker, M.S.; et al. Results of a Prospective Phase 2 Study of Pazopanib in Patients with Advanced Intermediate-grade or High-grade Liposarcoma. *Cancer* **2017**, *123*, 4640–4647. [CrossRef]
100. Dickson, M.A.; Koff, A.; D'Angelo, S.P.; Gounder, M.M.; Keohan, M.L.; Kelly, C.M.; Chi, P.; Antonescu, C.R.; Landa, J.; Qin, L.-X.; et al. Phase 2 Study of the *CDK4* Inhibitor Abemaciclib in Dedifferentiated Liposarcoma. *J. Clin. Oncol.* **2019**, *37*, 11004. [CrossRef]
101. Riedel, R.F.; Ballman, K.V.; Lu, Y.; Attia, S.; Loggers, E.T.; Ganjoo, K.N.; Livingston, M.B.; Chow, W.; Wright, J.; Ward, J.H.; et al. A Randomized, Double-Blind, Placebo-Controlled, Phase II Study of Regorafenib Versus Placebo in Advanced/Metastatic, Treatment-Refractory Liposarcoma: Results from the SARC024 Study. *Oncologist* **2020**, *25*, e1655–e1662. [CrossRef]
102. Chi, Y.; Fang, Z.; Hong, X.; Yao, Y.; Sun, P.; Wang, G.; Du, F.; Sun, Y.; Wu, Q.; Qu, G.; et al. Safety and Efficacy of Anlotinib, a Multikinase Angiogenesis Inhibitor, in Patients with Refractory Metastatic Soft-Tissue Sarcoma. *Clin. Cancer Res.* **2018**, *24*, 5233–5238. [CrossRef] [PubMed]
103. Laroche-Clary, A.; Chaire, V.; Algeo, M.-P.; Derieppe, M.-A.; Loarer, F.L.; Italiano, A. Combined Targeting of *MDM2* and *CDK4* Is Synergistic in Dedifferentiated Liposarcomas. *J. Hematol. Oncol. J. Hematol. Oncol.* **2017**, *10*, 123. [CrossRef] [PubMed]
104. Stefano, S.; Giovanni, S. The PTEN Tumor Suppressor Gene in Soft Tissue Sarcoma. *Cancers* **2019**, *11*, 1169. [CrossRef]
105. Fuchs, J.W.; Schulte, B.C.; Fuchs, J.R.; Agulnik, M. Targeted Therapies for the Treatment of Soft Tissue Sarcoma. *Front. Oncol.* **2023**, *13*, 1122508. [CrossRef] [PubMed]
106. Mariño-Enríquez, A.; Bovée, J.V.M.G. Molecular Pathogenesis and Diagnostic, Prognostic and Predictive Molecular Markers in Sarcoma. *Surg. Pathol. Clin.* **2016**, *9*, 457–473. [CrossRef] [PubMed]

107. Klein, M.E.; Dickson, M.A.; Antonescu, C.; Qin, L.-X.; Dooley, S.J.; Barlas, A.; Manova, K.; Schwartz, G.K.; Crago, A.M.; Singer, S.; et al. PDLIM7 and CDH18 Regulate the Turnover of *MDM2* during *CDK4*/6 Inhibitor Therapy-Induced Senescence. *Oncogene* **2018**, *37*, 5066–5078. [CrossRef]
108. Assi, T.; Kattan, J.; Rassy, E.; Nassereddine, H.; Farhat, F.; Honore, C.; Le Cesne, A.; Adam, J.; Mir, O. Targeting *CDK4* (Cyclin-Dependent Kinase) Amplification in Liposarcoma: A Comprehensive Review. *Crit. Rev. Oncol. Hematol.* **2020**, *153*, 103029. [CrossRef]
109. Schettini, F.; De Santo, I.; Rea, C.G.; De Placido, P.; Formisano, L.; Giuliano, M.; Arpino, G.; De Laurentiis, M.; Puglisi, F.; De Placido, S.; et al. CDK 4/6 Inhibitors as Single Agent in Advanced Solid Tumors. *Front. Oncol.* **2018**, *8*, 608. [CrossRef]
110. Clark, A.S.; Karasic, T.B.; DeMichele, A.; Vaughn, D.J.; O'Hara, M.; Perini, R.; Zhang, P.; Lal, P.; Feldman, M.; Gallagher, M.; et al. Palbociclib (PD0332991)—A Selective and Potent Cyclin-Dependent Kinase Inhibitor: A Review of Pharmacodynamics and Clinical Development. *JAMA Oncol.* **2016**, *2*, 253. [CrossRef]
111. Konecny, G.E. Cyclin-Dependent Kinase Pathways as Targets for Women's Cancer Treatment. *Curr. Opin. Obstet. Gynecol.* **2016**, *28*, 42–48. [CrossRef] [PubMed]
112. Dickson, M.A.; Schwartz, G.K.; Keohan, M.L.; D'Angelo, S.P.; Gounder, M.M.; Chi, P.; Antonescu, C.R.; Landa, J.; Qin, L.-X.; Crago, A.M.; et al. Progression-Free Survival Among Patients with Well-Differentiated or Dedifferentiated Liposarcoma Treated with *CDK4* Inhibitor Palbociclib: A Phase 2 Clinical Trial. *JAMA Oncol.* **2016**, *2*, 937. [CrossRef] [PubMed]
113. Nakayama, R.; Zhang, Y.-X.; Czaplinski, J.T.; Anatone, A.J.; Sicinska, E.T.; Fletcher, J.A.; Demetri, G.D.; Wagner, A.J. Preclinical Activity of Selinexor, an Inhibitor of XPO1, in Sarcoma. *Oncotarget* **2016**, *7*, 16581–16592. [CrossRef] [PubMed]
114. Ou, W.-B.; Zhu, J.; Eilers, G.; Li, X.; Kuang, Y.; Liu, L.; Mariño-Enríquez, A.; Yan, Z.; Li, H.; Meng, F.; et al. HDACi Inhibits Liposarcoma via Targeting of the *MDM2*-P53 Signaling Axis and PTEN, Irrespective of P53 Mutational Status. *Oncotarget* **2015**, *6*, 10510–10520. [CrossRef]
115. Smith, K.B.; Tran, L.M.; Tam, B.M.; Shurell, E.M.; Li, Y.; Braas, D.; Tap, W.D.; Christofk, H.R.; Dry, S.M.; Eilber, F.C.; et al. Novel Dedifferentiated Liposarcoma Xenograft Models Reveal PTEN Down-Regulation as a Malignant Signature and Response to PI3K Pathway Inhibition. *Am. J. Pathol.* **2013**, *182*, 1400–1411. [CrossRef] [PubMed]
116. Laroche, A.; Chaire, V.; Algeo, M.-P.; Karanian, M.; Fourneaux, B.; Italiano, A. *MDM2* Antagonists Synergize with PI3K/mTOR Inhibition in Well-Differentiated/Dedifferentiated Liposarcomas. *Oncotarget* **2017**, *8*, 53968–53977. [CrossRef]
117. Zhao, H.-F.; Wang, J.; To, S.-S.T. The Phosphatidylinositol 3-Kinase/Akt and c-Jun N-Terminal Kinase Signaling in Cancer: Alliance or Contradiction? (Review). *Int. J. Oncol.* **2015**, *47*, 429–436. [CrossRef]
118. von Mehren, M.; Rankin, C.; Goldblum, J.R.; Demetri, G.D.; Bramwell, V.; Ryan, C.W.; Borden, E. Phase 2 Southwest Oncology Group-Directed Intergroup Trial (S0505) of Sorafenib in Advanced Soft Tissue Sarcomas. *Cancer* **2012**, *118*, 770–776. [CrossRef]
119. Hanes, R.; Grad, I.; Lorenz, S.; Stratford, E.W.; Munthe, E.; Reddy, C.C.S.; Meza-Zepeda, L.A.; Myklebost, O. Preclinical Evaluation of Potential Therapeutic Targets in Dedifferentiated Liposarcoma. *Oncotarget* **2016**, *7*, 54583–54595. [CrossRef]
120. Miller, M.L.; Molinelli, E.J.; Nair, J.S.; Sheikh, T.; Samy, R.; Jing, X.; He, Q.; Korkut, A.; Crago, A.M.; Singer, S.; et al. Drug Synergy Screen and Network Modeling in Dedifferentiated Liposarcoma Identifies *CDK4* and IGF1R as Synergistic Drug Targets. *Sci. Signal.* **2013**, *6*, ra85. [CrossRef]
121. Xu, B.; Pan, Q.; Pan, H.; Li, H.; Li, X.; Chen, J.; Pang, D.; Zhang, B.; Weng, D.; Peng, R.; et al. Anlotinib as a Maintenance Treatment for Advanced Soft Tissue Sarcoma after First-Line Chemotherapy (ALTER-S006): A Multicentre, Open-Label, Single-Arm, Phase 2 Trial. *eClinicalMedicine* **2023**, *64*, 102240. [CrossRef] [PubMed]
122. Available online: https://Clinicaltrials.Gov/Study/NCT03890068 (accessed on 14 November 2023).
123. Demetri, G.D.; Fletcher, C.D.M.; Mueller, E.; Sarraf, P.; Naujoks, R.; Campbell, N.; Spiegelman, B.M.; Singer, S. Induction of Solid Tumor Differentiation by the Peroxisome Proliferator-Activated Receptor-γ Ligand Troglitazone in Patients with Liposarcoma. *Proc. Natl. Acad. Sci. USA* **1999**, *96*, 3951–3956. [CrossRef] [PubMed]
124. Debrock, G.; Vanhentenrijk, V.; Sciot, R.; Debiec-Rychter, M.; Oyen, R.; Van Oosterom, A. A Phase II Trial with Rosiglitazone in Liposarcoma Patients. *Br. J. Cancer* **2003**, *89*, 1409–1412. [CrossRef] [PubMed]
125. Pishvaian, M.J.; Marshall, J.L.; Wagner, A.J.; Hwang, J.J.; Malik, S.; Cotarla, I.; Deeken, J.F.; He, A.R.; Daniel, H.; Halim, A.-B.; et al. A Phase 1 Study of Efatutazone, an Oral Peroxisome Proliferator-Activated Receptor Gamma Agonist, Administered to Patients with Advanced Malignancies: PPARγ Agonist Efatutazone. *Cancer* **2012**, *118*, 5403–5413. [CrossRef]
126. Trautmann, M.; Menzel, J.; Bertling, C.; Cyra, M.; Isfort, I.; Steinestel, K.; Elges, S.; Grünewald, I.; Altvater, B.; Rossig, C.; et al. FUS–DDIT3 Fusion Protein-Driven IGF-IR Signaling Is a Therapeutic Target in Myxoid Liposarcoma. *Clin. Cancer Res.* **2017**, *23*, 6227–6238. [CrossRef] [PubMed]
127. Trautmann, M.; Cyra, M.; Isfort, I.; Jeiler, B.; Krüger, A.; Grünewald, I.; Steinestel, K.; Altvater, B.; Rossig, C.; Hafner, S.; et al. Phosphatidylinositol-3-Kinase (PI3K)/Akt Signaling Is Functionally Essential in Myxoid Liposarcoma. *Mol. Cancer Ther.* **2019**, *18*, 834–844. [CrossRef] [PubMed]
128. Wang, Y.; Zhang, H.; La Ferlita, A.; Sp, N.; Goryunova, M.; Sarchet, P.; Hu, Z.; Sorkin, M.; Kim, A.; Huang, H.; et al. Phosphorylation of IWS1 by AKT Maintains Liposarcoma Tumor Heterogeneity through Preservation of Cancer Stem Cell Phenotypes and Mesenchymal-Epithelial Plasticity. *Oncogenesis* **2023**, *12*, 30. [CrossRef]
129. Künstlinger, H.; Fassunke, J.; Schildhaus, H.-U.; Brors, B.; Heydt, C.; Ihle, M.A.; Mechtersheimer, G.; Wardelmann, E.; Büttner, R.; Merkelbach-Bruse, S. FGFR2 Is Overexpressed in Myxoid Liposarcoma and Inhibition of FGFR Signaling Impairs Tumor Growth In Vitro. *Oncotarget* **2015**, *6*, 20215–20230. [CrossRef]

130. Göransson, M.; Andersson, M.K.; Forni, C.; Ståhlberg, A.; Andersson, C.; Olofsson, A.; Mantovani, R.; Åman, P. The Myxoid Liposarcoma FUS-DDIT3 Fusion Oncoprotein Deregulates NF-κB Target Genes by Interaction with NFKBIZ. *Oncogene* **2009**, *28*, 270–278. [CrossRef]
131. Bidère, N.; Ngo, V.N.; Lee, J.; Collins, C.; Zheng, L.; Wan, F.; Davis, R.E.; Lenz, G.; Anderson, D.E.; Arnoult, D.; et al. Casein Kinase 1α Governs Antigen-Receptor-Induced NF-κB Activation and Human Lymphoma Cell Survival. *Nature* **2009**, *458*, 92–96. [CrossRef]
132. Berthold, R.; Isfort, I.; Erkut, C.; Heinst, L.; Grünewald, I.; Wardelmann, E.; Kindler, T.; Åman, P.; Grünewald, T.G.P.; Cidre-Aranaz, F.; et al. Fusion Protein-Driven IGF-IR/PI3K/AKT Signals Deregulate Hippo Pathway Promoting Oncogenic Cooperation of YAP1 and FUS-DDIT3 in Myxoid Liposarcoma. *Oncogenesis* **2022**, *11*, 20. [CrossRef] [PubMed]
133. Trautmann, M.; Cheng, Y.; Jensen, P.; Azoitei, N.; Brunner, I.; Hüllein, J.; Slabicki, M.; Isfort, I.; Cyra, M.; Berthold, R.; et al. Requirement for YAP1 Signaling in Myxoid Liposarcoma. *EMBO Mol. Med.* **2019**, *11*, e9889. [CrossRef] [PubMed]
134. Regina, C.; Hettmer, S. Myxoid Liposarcoma: It's a Hippo's World. *EMBO Mol. Med.* **2019**, *11*, e10470. [CrossRef] [PubMed]
135. Dolatabadi, S.; Jonasson, E.; Andersson, L.; Luna Santamaría, M.; Lindén, M.; Österlund, T.; Åman, P.; Ståhlberg, A. FUS-DDIT3 Fusion Oncoprotein Expression Affects JAK-STAT Signaling in Myxoid Liposarcoma. *Front. Oncol.* **2022**, *12*, 816894. [CrossRef]
136. Svec, D.; Dolatabadi, S.; Thomsen, C.; Cordes, N.; Shannon, M.; Fitzpatrick, P.; Landberg, G.; Åman, P.; Ståhlberg, A. Identification of Inhibitors Regulating Cell Proliferation and FUS-DDIT3 Expression in Myxoid Liposarcoma Using Combined DNA, mRNA, and Protein Analyses. *Lab. Invest.* **2018**, *98*, 957–967. [CrossRef]
137. Andersson, M.K.; Göransson, M.; Olofsson, A.; Andersson, C.; Åman, P. Nuclear Expression of FLT1 and Its Ligand PGF in FUS-DDIT3 Carrying Myxoid Liposarcomas Suggests the Existence of an Intracrine Signaling Loop. *BMC Cancer* **2010**, *10*, 249. [CrossRef] [PubMed]
138. Dong, M.; Bi, J.; Liu, X.; Wang, B.; Wang, J. Significant Partial Response of Metastatic Intra-Abdominal and Pelvic Round Cell Liposarcoma to a Small-Molecule VEGFR-2 Tyrosine Kinase Inhibitor Apatinib: A Case Report. *Medicine* **2016**, *95*, e4368. [CrossRef] [PubMed]
139. Payandeh, M.; Karami, A.; Rahimi, S.; Karami, N. A Revolution in the Treatment of Aggressive Liposarcoma by Pazopanib Combined with Chemotherapy—A Case Report. *Biomed. Res. Ther.* **2019**, *6*, 3180–3183. [CrossRef]
140. Eilber, F.C.; Eilber, F.R.; Eckardt, J.; Rosen, G.; Riedel, E.; Maki, R.G.; Brennan, M.F.; Singer, S. The Impact of Chemotherapy on the Survival of Patients with High-Grade Primary Extremity Liposarcoma. *Ann. Surg.* **2004**, *240*, 686–695, discussion 695–697. [CrossRef]
141. Dalal, K.M.; Kattan, M.W.; Antonescu, C.R.; Brennan, M.F.; Singer, S. Subtype Specific Prognostic Nomogram for Patients with Primary Liposarcoma of the Retroperitoneum, Extremity, or Trunk. *Ann. Surg.* **2006**, *244*, 381–391. [CrossRef]
142. Crago, A.M.; Dickson, M.A. Liposarcoma. *Surg. Oncol. Clin. N. Am.* **2016**, *25*, 761–773. [CrossRef]
143. Gounder, M.M.; Bauer, T.M.; Schwartz, G.K.; LoRusso, P.; Kumar, P.; Kato, K.; Tao, B.; Hong, Y.; Patel, P.; Hong, D. Milademetan, an Oral *MDM2* Inhibitor, in Well-Differentiated/ Dedifferentiated Liposarcoma: Results from a Phase 1 Study in Patients with Solid Tumors or Lymphomas. *Eur. J. Cancer* **2020**, *138*, S3–S4. [CrossRef]
144. Available online: https://Clinicaltrials.Gov/Study/NCT04979442 (accessed on 14 November 2023).
145. Available online: https://Clinicaltrials.Gov/Study/NCT03449381 (accessed on 14 November 2023).
146. LoRusso, P.; Yamamoto, N.; Patel, M.R.; Laurie, S.A.; Bauer, T.M.; Geng, J.; Davenport, T.; Teufel, M.; Li, J.; Lahmar, M.; et al. The *MDM2*-P53 Antagonist Brigimadlin (BI 907828) in Patients with Advanced or Metastatic Solid Tumors: Results of a Phase Ia, First-in-Human, Dose-Escalation Study. *Cancer Discov.* **2023**, *13*, 1802–1813. [CrossRef]
147. Chen, T.W.-W.; Hsu, C.-L.; Hong, R.-L.; Lee, J.-C.; Chang, K.; Yu, C.-W.; Chen, S.-C.; Guo, J.-C.; Chen, M.-L.; Hsu, M.-C.; et al. A Single-Arm Phase Ib/II Study of Lenvatinib plus Eribulin in Advanced Liposarcoma and Leiomyosarcoma. *Clin. Cancer Res.* **2022**, *28*, 5058–5065. [CrossRef]
148. Gounder, M.M.; Razak, A.A.; Somaiah, N.; Chawla, S.; Martin-Broto, J.; Grignani, G.; Schuetze, S.M.; Vincenzi, B.; Wagner, A.J.; Chmielowski, B.; et al. Selinexor in Advanced, Metastatic Dedifferentiated Liposarcoma: A Multinational, Randomized, Double-Blind, Placebo-Controlled Trial. *J. Clin. Oncol.* **2022**, *40*, 2479–2490. [CrossRef]

cancers

MDPI

Article

Treatment Pathways and Prognosis in Advanced Sarcoma with Peritoneal Sarcomatosis

Fabian Klingler [1,†], Hany Ashmawy [1,†], Lena Häberle [2], Irene Esposito [2], Lars Schimmöller [3], Wolfram Trudo Knoefel [1,*] and Andreas Krieg [1,*]

1 Department of Surgery (A), Heinrich-Heine-University and University Hospital Duesseldorf, D-40225 Duesseldorf, Germany
2 Institute of Pathology, Heinrich-Heine-University and University Hospital Duesseldorf, D-40225 Duesseldorf, Germany
3 Department of Diagnostic and Interventional Radiology, Medical Faculty, University Duesseldorf, D-40225 Duesseldorf, Germany
* Correspondence: knoefel@med.uni-duesseldorf.de (W.T.K.); andreas.krieg@med.uni-duesseldorf.de (A.K.); Tel.: +49-0211-81-17351 (W.T.K.); +49-0211-81-19251 (A.K.)
† These authors contributed equally to this work.

Simple Summary: Presentation of sarcoma inside the peritoneal cavity is a rare finding to begin with. In such a rare incidence, there are a multitude of sarcoma subtypes that can be identified, with each of these subtypes presenting with different characteristics in terms of prognosis and treatment options. Considering these factors and the resulting lack of strong data to guide treatment plans, this study aims to share our experiences with cases of peritoneal sarcomatosis to increase the knowledge about possible options and outcomes. We report on 19 cases of surgery in patients with peritoneal sarcomatosis, ranging from palliative procedures to major multivisceral resections, and highlight their course of disease, treatment, and outcome. Hereby, we aspire to increase the cumulative experience with challenging cases like these and support a more informed tailoring of treatment plans for future cases to come.

Abstract: Sarcomas represent a heterogeneous group of mesenchymal malignancies that most commonly occur in the extremities, retroperitoneum, and head and neck. Intra-abdominal manifestations are rare and prove particularly difficult to treat when peritoneal sarcomatosis is present. Because of the overall poor prognosis of the disease, a tailored approach to surgical management is essential to achieve satisfactory outcomes with limited morbidity. We present the perioperative and long-term outcomes of 19 cases of sarcoma with peritoneal sarcomatosis treated surgically at our hospital. Treatment pathways were reviewed and clinical follow-up was performed. Patient characteristics, medical history, tumor subtype, surgical approach, hospital stay, complications, follow-up, and overall survival (OS) were assessed. Our patients were 9 women and 10 men with a median age of 45.9 years (18–88) and a median survival of 30 months (0–200). In most cases, peritoneal sarcomatosis was either discovered during surgery or the procedure was performed with palliative intent from the beginning. The surgical approach in these cases is very heterogeneous and should consider a variety of factors to tailor an approach for each patient. Sharing our experiences will help to increase knowledge about this rare disease and provide insight into the management of future cases.

Keywords: sarcomatosis; sarcoma; peritoneal

Citation: Klingler, F.; Ashmawy, H.; Häberle, L.; Esposito, I.; Schimmöller, L.; Knoefel, W.T.; Krieg, A. Treatment Pathways and Prognosis in Advanced Sarcoma with Peritoneal Sarcomatosis. *Cancers* **2023**, *15*, 1340. https://doi.org/10.3390/cancers15041340

Academic Editor: Catrin Sian Rutland

Received: 16 January 2023
Revised: 15 February 2023
Accepted: 17 February 2023
Published: 20 February 2023

1. Introduction

Soft tissue sarcomas (STSs) account for less than 1% of malignancies and are thus a rare entity [1]. In industrialized nations, STSs occur with an incidence of 4–5/100,000 residents per year. Owing to their wide distribution in localization and with more than 80 further specified histological subtypes, they comprise a very heterogeneous group of malignancies [1,2]. However, continuous efforts to characterize and subclassify STSs, taking molecular pathology into account, proves to be the most promising approach for individualized treatment efforts [3,4]. The growing understanding of the molecular differences between tumors that used to be viewed as indiscriminative offers great chances for future efforts to characterize these subtypes clinically and develop a more customized treatment approach. Previous attempts to develop standardized treatment protocols have faced difficulties in addressing all of the individual requirements and it is challenging to obtain reliable data for a specific subtype. Especially in advanced cases with recurrent or metastasized disease, finding the fine line between undertreatment and overtreatment can be a huge challenge. STS with the involvement of the peritoneum is even more uncommon than other sites of metastasis, hence there is very limited experience from which to draw [5]. Therefore, evidence-based treatment for these cases has not yet been established.

Treatment options for patients with peritoneal sarcomatosis (PS) range from palliative treatment to aggressive systemic therapy and include extensive cytoreductive surgery (CRS) and hyperthermic intraperitoneal chemotherapy (HIPEC) in selected patients. Given the high morbidity and differential survival benefit of cytoreduction and HIPEC, the approach to PS remains controversial [6,7]. As a result of the lack of sufficient data, current recommendations for treatment pathways mostly consist of consensus statements [8]. While most studies focus on a single histologic subtype or attempt to bundle sarcoma patients as a whole, there is a distinct lack of data focusing on advanced stages such as PS. In addition, to the best of our knowledge, there is no study that has focused on the role of surgery as a palliative modality in PS. Thus, the aim of our study was to present cases with PS that were treated at our department in order to illustrate the wide variety of therapeutic options available in the development of a tailored approach for each individual patient and to provide examples for future treatment plans. We further aim to showcase the highly heterogeneous outcome for patients with different STSs and PS to encourage continuous efforts to develop evidence-based treatment protocols.

2. Materials and Methods

Of the 291 sarcoma cases we have treated since September 2003, we identified 19 patients who underwent surgical procedures for PS. In all cases, an individualized approach was chosen depending on the disease course, prognosis, and patient requirements. We conducted a preoperative interdisciplinary discussion in our tumor board specialized in sarcomas, in which recommendations for the therapeutic regimen were made, considering histology, clinical findings, physical performance, and patient preferences.

Patients were selected from a retrospectively collected and prospectively maintained database of sarcoma patients treated at our university hospital. For the study, we included only surgically treated patients with histologically confirmed PS either before or at the time of their surgical procedure. We obtained approval for this study from the ethics committees of the Medical Faculty of Heinrich Heine University, Düsseldorf (study number: 2022–2010).

The following information was obtained from the patients' medical records: age at surgical intervention, sex, time between initial diagnosis and diagnosis of PS, disease status at presentation (primary or recurrent), prior treatment, change of treatment centre, tumor subtype, details of treatment (palliative or curative intent, resected structures, surgical margins, reconstructive technique, Clavien–Dindo surgical complications, length of hospital stay), and follow-up.

Follow-up and survival times after discovery of PS and after primary diagnosis of sarcoma were calculated and corresponding Kaplan–Meier survival curves were generated. The statistical analyses were conducted with the software R (version 1.4.1106) utilizing the packages readxl, survminer, and survival [9–11].

3. Results

3.1. Patient Characteristics

Detailed patient characteristics are summarized in Table S1. The median age at the time of PS diagnosis was 45.9 years, with large differences between tumor types. Nine of the patients were female and ten were male. Among the 19 patients we identified, there were 8 different tumor entities (Table 1).

Table 1. Tumor subtypes and patient characteristics.

Tumor Subtype	Number of Patients	Median Age (Range)	Sex Male/Female
NOS	5	67 (46–88)	3/2
DSRCT	4	18 (18–24)	2/2
Dedifferentiated liposarcoma	3	65 (36–84)	1/2
Myxoid liposarcoma	2	61 (44–64)	2/0
Leiomyosarcoma	2	51 (44–58)	0/2
MPNST	1	24	0/1
SFT	1	68	1/0
Osteosarcoma	1	26	1/0
Total	**19**	**45.9 (18–88)**	**10/9**

Desmoplastic small round cell tumor (DSRCT), pleomorphic sarcomas—not otherwise specified (NOS), malignant peripheral nerve sheath tumor (MPNST), solitary fibrous tumor (SFT).

3.2. State of Disease

Twelve of the patients were found to have a primary tumor in the abdominal cavity. Four patients had PS at the time of their primary diagnosis, with ten patients treated for more than 12 months before sarcomatosis occurred. The median time between initial diagnosis and PS was 16.2 months. We noted that 11 of the patients had been treated for their disease at another institution before being treated at our hospital. Thirteen of the patients had already undergone surgical resection prior to PS diagnosis and, in eight of these patients, the tumors were not completely removed from healthy tissue microscopically or macroscopically. In two cases, the primary tumor originated from the retroperitoneal space, and in both cases, R1/R2 resection was documented before the discovery of PS. In ten cases, distant metastases were already present at the time of PS discovery (Table 2, Figure 1).

Table 2. State of disease at discovery of peritoneal sarcomatosis.

Tumor Subtype	Number of Patients	Mean Time Since Primary Diagnosis (SD)	Prior External Treatment	Mean Prior Resections (Range)	Prior R1/R2	Distant Metastasis
NOS	5	18 (26.1)	40%	3.4 (0–10)	60%	20%
DSRCT	4	6 (9.6)	75%	0.3 (0–1)	25%	50%
Dedifferentiated liposarcoma	3	7 (10)	33%	1.7 (0–4)	66%	100%
Myxoid liposarcoma	2	40 (7.4)	100%	1.5 (1–2)	0%	100%
Leiomyosarcoma	2	9 (11.8)	100%	1.5 (1–2)	ND	50%
MPNST	1	13	100%	3	100%	100%
SFT	1	27	0%	1	0%	0%
Osteosarcoma	1	34	0%	2	0%	100%
Total	**19**	**16.2 (17.94)**	**57.9%**	**1.8**	**42.1%**	**52.6%**

Peritoneal sarcomatosis (PS), standard deviation (SD), desmoplastic small round cell tumor (DSRCT), pleomorphic sarcomas—not otherwise specified (NOS), malignant peripheral nerve sheath tumor (MPNST), solitary fibrous tumor (SFT), not defined (ND).

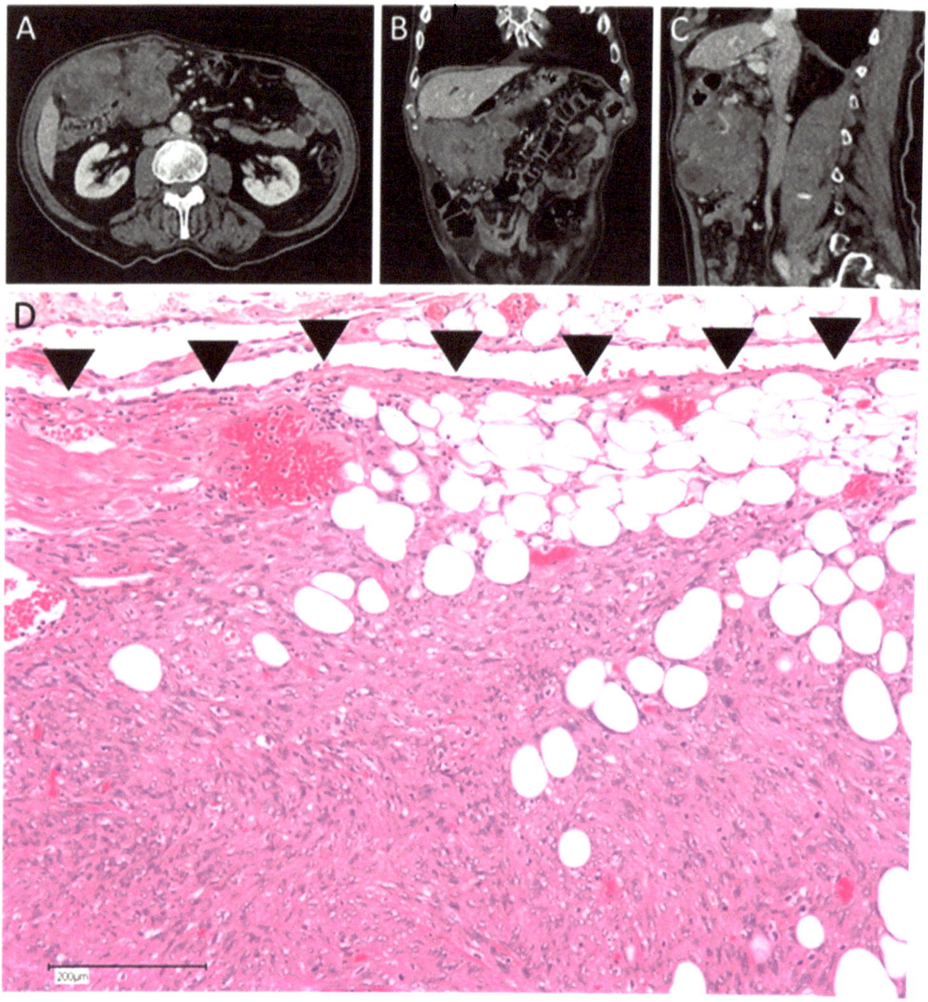

Figure 1. Example for diagnosis of a peritoneal metastasized soft tissue sarcoma. The patient was an 87-year-old male who presented with a palpable growing mass in his abdomen. (**A–C**) Abdominal CT-scan showed a large tumor in his right upper quadrant and suggested infiltration of the tumor into the distal stomach and ascending colon. (**D**) Histophotograph of the tumor shows a neoplasm consisting of atypical spindle cells infiltrating the peritoneum (arrowheads: mesothelial lining) (100×, H&E). Histology report showed a high-grade pleomorphic sarcoma—not otherwise specified (NOS).

3.3. Treatment

Each patient underwent surgical resection of the tumor mass at some point, with an average of 3.8 resections during the course of the disease (Table 3). The list of operations ranged from diagnostic procedures and bypass surgery to metastasectomy and major multiorgan resections (Figure 2).

At the time of diagnosis of PS, we treated nine patients with curative intent, sometimes including resection of the symptomatic tumor mass, and at other times to prevent future complications. In two cases, palliative treatment was chosen at another institution and resection with curative intent was performed after referral to our hospital.

Table 3. Treatment.

Tumor Subtype	Number of Patients	Total Number of Resections per Patient	Surgical Treatment Spectrum			
			Exploration and Biopsy	Palliative Procedure	Limited Resection	Multivisceral Resection
NOS	5	5			15	11
DSRCT	4	2.75	2		1	7
Dedifferentiated liposarcoma	3	2	2	1	2	3
Myxoid liposarcoma	2	5		1	3	6
Leiomyosarcoma	2	6	2		5	6
MPNST	1	4			2	2
SFT	1	2	1		1	1
Osteosarcoma	1	3			2	1
Total	**19**	**3.8**	**7**	**2**	**31**	**37**

Desmoplastic small round cell tumor (DSRCT), pleomorphic sarcomas—not otherwise specified (NOS), malignant peripheral nerve sheath tumor (MPNST), solitary fibrous tumor (SFT).

Four patients received CRS and cisplatin-based HIPEC for PS, all of whom had desmoplastic round cell tumors (DSRCTs).

3.4. Follow-Up

The mean follow-up was 22 months ranging from 0 to 172 months (Table 4). Six patients achieved a survival of greater than 5 years after primary diagnosis of sarcoma, while five patients are still alive at the time of publication of this study. The median follow-up of these surviving patients was 13 months, ranging from 2 to 172 months. Complications were common in these patients, with only five patients without any surgical complications. Seven patients developed wound healing issues and two patients died from terminal respiratory insufficiency after a complicative postoperative course.

Table 4. Follow-up.

Tumor Subtype	Number of Patients	Mean Survival in Months Since Primary Diagnosis (SD)	Mean Survival in Months after PS	Major Complications (Clavien–Dindo 3 or 4)	Mean Follow-Up in Months (SD)
NOS	5	38 (31.2)	20 (27.8)	60%	17 (25.3)
DSRCT	4	24 (1.0)	17 (9.2)	25%	13 (7.4)
Dedifferentiated liposarcoma	3	10 (10.3)	3 (2.5)	33%	3 (2.3)
Myxoid liposarcoma	2	63 (18.3)	22 (10.9)	0%	20 (9.6)
Leiomyosarcoma	2	151 (76.5)	143 (88.3)	0%	115 (80.4)
MPNST	1	15	2	100%	2
SFT	1	62	36	100%	3
Osteosarcoma	1	43	9	0%	8
Total	**19**	**46 (47.4)**	**30 (47.9)**	**31.6%**	**22 (40.2)**

Peritoneal sarcomatosis (PS), standard deviation (SD), desmoplastic small round cell tumor (DSRCT), pleomorphic sarcomas—not otherwise specified (NOS), malignant peripheral nerve sheath tumor (MPNST), solitary fibrous tumor (SFT), standard deviation (SD).

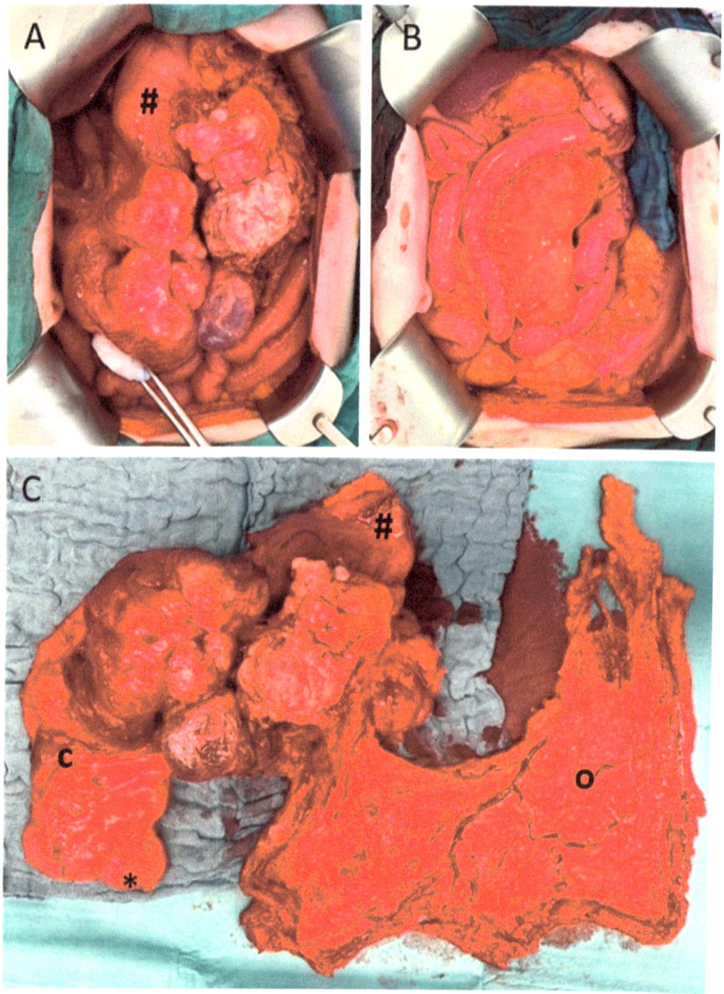

Figure 2. Example of palliative resection. Intraoperative findings of the previously mentioned 87-year-old male with pleomorphic sarcoma (Figure 1). (**A**) Surgical exploration showed diffuse peritoneal sarcomatosis and a massive tumor formation. (**B**) Palliative resection and reconstruction with gastrojejunostomy, Roux–en–Y reconstruction, and terminal ileostomy were performed, and the patient was discharged three weeks later. (**C**) The resected specimen contained the tumor with the infiltrated distal stomach (#) and ascending colon (**c**) and terminal ileum (*), as well as the greater omentum (**o**).

The patients we treated with curative intent were on average 40 years old, in contrast to an average age of 55 years for patients who received palliative surgery. Survival rates differed significantly between the groups and showed a longer overall survival after primary diagnosis as well as after diagnosis of PS for patients treated with a curative intent (Figure 3). Interestingly, patients receiving curative therapy tended to have fewer severe complications. In contrast to the group of patients who underwent curative surgery, all of whom survived the hospital stay, we observed an in-hospital mortality rate of 20% in patients who underwent palliative surgery. However, hospital stay did not differ between the two groups (Table 5).

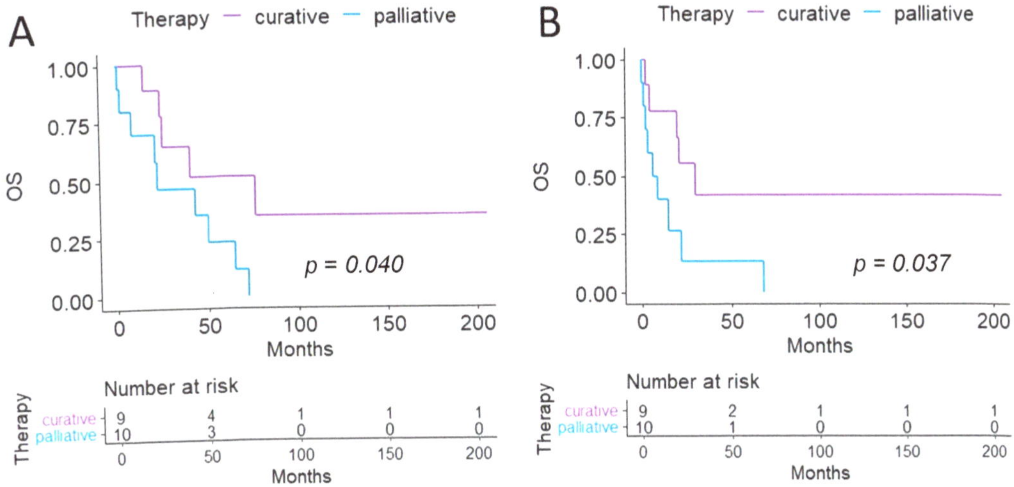

Figure 3. Overall survival. (**A**) Kaplan–Meier survival curve after primary diagnosis of sarcoma is shown. (**B**) These Kaplan–Meier curves show survival after the diagnosis of peritoneal sarcomatosis has been made.

Table 5. Treatment intention and outcome.

	Number of Patients	Age in Years	Major Complications (Clavien–Dindo 3 or 4)	Hospital Stay in Days (SD)
Curative Intention	9	40	22%	29 (16)
Palliative Intention	10	55	40%	27 (21)

Peritoneal sarcomatosis (PS), standard deviation (SD).

The median overall survival after primary diagnosis of sarcoma was 76 months in patients treated with a curative intent versus 22 months in patients with a palliative treatment plan, and the 5-year survival rates were 52% versus 23%. After diagnosis of PS, the median survival was 30 months and 7.5 months when comparing curative and palliative treated patients, respectively, with a 5-year survival rate of 40.2% versus 13%, respectively.

4. Discussion

Surgery with complete en bloc resection of all adjacent tissues and organs, combined with or without radiation therapy, remains the primary and only truly curative treatment option for localized and clinically resectable STSs. It is also recommended for advanced or metastatic STSs. In this context, factors such as isolated oligometastatic disease, long disease-free interval, favorable histology, response to chemotherapy, and high likelihood of complete resection make the argument for surgery even in advanced and recurrent disease [12]. Interestingly, a recently published meta-analysis also suggests that CRS with HIPEC may improve prognosis in a selected group of PS patients [13]. In addition, patients with advanced, primarily inoperable STS may also be offered palliative surgery for symptom control of tumor-related complications such as pain, bleeding, or bowel obstruction [14]. While the adjuvant and neoadjuvant treatment of patients with STS is still controversial and no standardized regimen exists, anthracycline-based chemotherapy in combination with ifosfamide is used as a first-line therapy in the treatment of advanced STS [15,16]. A study by Gough et al. demonstrated that palliative chemotherapy with doxorubicin alone or in combination with ifosfamide, as well as combination chemotherapy with gemcitabine and docetaxel, significantly reduced pain and sleep disturbance while worsening fatigue [17]. However, not all patients with advanced or metastatic STS benefit from conventional chemotherapy, and targeted therapy may play the most important role in the treatment of patients who are resistant to conventional chemotherapy or in whom conventional chemotherapy has failed. Accordingly, in recent years, increasing numbers of preclinical studies have been conducted to explore the pathogenesis and potential therapeutic targets of STS [4,18,19], and clinical trials have been initiated to target different molecules in distinct histologic subtypes [20], hopefully opening new doors in the clinical management of patients with advanced or metastatic disease in the future. Although the proportion of patients receiving outpatient palliative care for STS is likely to be very small [21], specialized palliative care interventions have recently been shown to result in significant symptom relief in patients with advanced STS, and early integration of palliative care in these patients is thus recommended [22]. Nevertheless, the treatment decision should be made in the context of a multidisciplinary discussion.

However, treatment recommendations for sarcoma patients increasingly depend on their individual tumor subtype [23], while scientific advances have provided increasing rationale for differentiating STSs with a focus on genetics and potential molecular targets for individualized treatment options. The increasing distinction of subtypes with different biological and clinical aspects makes it difficult to draw conclusions for individualized treatment pathways for each sarcoma subtype, especially for smaller studies [24]. As we were confronted with the same problem in our study, we aimed to provide a largely descriptive overview of our experience with PS. Because one-third of our patients are the only patients with their respective histologic subtype, we cannot claim to present conclusive evidence of standardized treatment. Nevertheless, we believe that any contribution to the collective knowledge of these difficult cases is valuable.

The highest number of patients with a tumor subtype in our study was generated by pleomorphic sarcomas—not otherwise specified (NOS), formerly known as malignant fibrous histiocytomas. This group accounts for 10–15% of STSs and is typical of low-differentiated tumors prone to aggressiveness and early metastasis. It is diagnosed less frequently today than twenty years ago, as technological improvements often make it possible to determine a line of differentiation [25]. Nevertheless, the group remains inhomogeneous in terms of age, tumor location, disease progression, and overall survival. Surgical intent was mostly palliative and complication rates were high. One case stood out with significantly higher overall survival—in this case, the primary diagnosis was sarcoma from a morcellated uterus during hysterectomy. We can only speculate whether the mere location of the primary tumor played a role in the favourable outcome or whether there was an affiliation with uterine leiomyosarcomas that could not be determined. Nevertheless, further progress in determining a histologic subtype of these low-differentiated tumors should be helpful in identifying the ideal treatment for these patients.

One of the more uniform patient groups in our study was those with DSRCTs. As DSRCTs typically occur in adolescent patients, we found our youngest patients in this group [26]. These often physically fit and motivated patients are dealing with a very aggressive disease with a 5-year overall survival rate of 15% to 30% [26,27]. In light of this, aggressive interdisciplinary treatment with perioperative chemotherapy following Ewing protocol and radical surgery with peritonectomy offers the best chance for patients [28]. As most cases of DSRCTs have peritoneal seeding, there is more information about PS in this type of sarcoma. Additional HIPEC after CRS is being investigated in DSRCTs and seems to provide additional benefit, which is why these patients were the only ones to receive this treatment in this study [29]. Other treatment options such as radiotherapy, targeted therapies, or the use of regional deep hyperthermia protocols should be discussed in serial tumor board meetings [26,30]. The patients with DSRCTs that we treated received aggressive treatment with multivisceral resection and HIPEC, often with repeated surgeries during the course of their disease, but incredible recovery and only minor complications were demonstrated in three of four patients.

Differentiating the extent of peritoneal involvement of a malignancy is an established concept for peritoneal carcinomatosis. The peritoneal carcinomatosis index (PCI) has been widely used for prognostic reasons or to evaluate aggressive treatment options such as cytoreductive surgery (CRS) and HIPEC [31,32]. The concept has also been used for PS, but its prognostic value for a benefit of CRS and HIPEC in PS remains controversial [33–35]. If future studies succeed in defining indications for CRS and HIPEC outside of DSRCTs, PCI is likely to become more important in the treatment of peritoneal metastatic sarcomas.

With more than 20% of all STSs in Germany, liposarcoma occupies a significant role in this group of rare diseases [1]. Even though there are more data available in relation to other STS types, scientific approaches to understand PS in liposarcoma are just beginning to take shape [5]. Well-differentiated liposarcomas have low metastatic rates, whereas dedifferentiated liposarcomas are more aggressive and are associated with a worse prognosis. Myxoid liposarcomas account for approximately 20–30% of all liposarcomas and typically carry a genetic translocation resulting in a higher rate of primary multifocal appearance and metastasis, but still have a significantly better overall survival rate than dedifferentiated liposarcomas [1,4,36]. Our results support these data, as we only studied PS in dedifferentiated and myxoid liposarcomas. The two patients we treated for myxoid liposarcoma both underwent multiple surgical resections, received various chemotherapies, and had an overall survival of more than 50 months. In contrast, the overall survival of the patients we treated for dedifferentiated liposarcoma was much lower, the rate of distant metastases was high, and recurrence was rapid. Looking at one of the patients we treated with palliative small bowel bypass anastomosis for tumor-associated ileus, we managed to discharge the patient without surgical complications, but unfortunately, he died the following month.

Leiomyosarcoma accounts for approximately 10–25% of STSs in Germany [1]. Both patients had a history of hysterectomy, making uterine primary disease likely, although histologic evidence of uterine tumor was obtained in only one case. Five-year survival rates for uterine leiomyosarcomas are 20–60%, which compares favourably with many of the previously mentioned tumor types, and our data support these findings with overall survival since primary diagnosis of 205 and 97 months, respectively, in our patients, both of whom are alive to date. It is worth noting that both patients received multiple surgical resections and systemic therapies during the course of their disease, but achieved long periods of tumor control. In one case, the patient was considered palliative owing to recurrent locally advanced disease, but opted for surgical resection with local peritonectomy. The patient received perioperative chemotherapy with a regional deep hyperthermia protocol and, to date, has been disease-free for 14 years without further treatment. Uterine leiomyosarcoma appears to reward persistent efforts to control the disease and offers opportunities for successful systemic and surgical treatment even with advanced disease.

The patient we treated for MPNST was in the expected age group for patients with neurofibromatosis type 1, and because MPNST is a highly malignant sarcoma, tumor recurrence and distant metastases ensured a rapid disease progression [1].

SFT is a subtype with a low metastatic burden, but also low sensitivity to chemotherapy, so the focus is on local control of the tumor [37]. We treated the patient with total parietal peritonectomy, right hemicolectomy with resection of the terminal ileum, radical lymphadenectomy, and partial resection of the urinary bladder. Although the postoperative course was not without complications, the patient is still alive and well 41 months later.

In one case, we treated a patient with advanced osteosarcoma with peritoneal involvement. Intra-abdominal metastases in osteosarcoma are very rare and often occur late in the disease course. A recent study of abdominally metastatic osteosarcomas found that 75% of patients died within 6 months of diagnosis of peritoneal involvement [38]. Nevertheless, resectable disease can be treated surgically and may be beneficial to patients. In our case, the patient could be discharged after 17 days of hospitalization and lived for 9 months after extensive abdominal surgery.

The tumor site from which sarcoma originated varied widely between cases and did not appear to play a role once PS was detected. Two of the documented cases started in the retroperitoneal space. In both cases, there was a history of previous incomplete resection prior to the development of PS. This supports the existing view that a clean surgical technique with clear resection margins is one of the most important factors in avoiding PS in surgical treatment of retroperitoneal STS [39].

Because most of our patients received surgical treatment for their malignancy before a diagnosis of PS, it is not surprising that a high number of R1/R2 resections lead to PS. Our data underscore the principle of negative surgical margins, supporting one of the well-established principles in sarcoma treatment [40,41].

During data collection, we found that most patients were referred to more than one centre during the course of their disease and that treatment plans changed frequently when the physician changed. In fact, a study from France demonstrated that more than 40% of histologic diagnoses were changed after obtaining a second opinion [42]. Numerous data suggest that patients benefit from treatment in a larger institution where multidisciplinary tumor boards discuss therapeutic options [43,44].

By focusing on advanced-stage sarcoma with PS, we hope to provide valuable experience for surgeons trying to find the right treatment path for their patients, taking into account the disease course and histologic subtype of the sarcoma. In contrast to other studies that focus on curative therapy for PS, we also investigated the importance of palliative surgery in this patient population. Importantly, to the best of our knowledge, there are no studies specifically evaluating the value of palliative surgery in patients with PS. However, there are a limited number of studies that focus exclusively on the quality of life and oncologic benefit of palliative resection for retroperitoneal or intra-abdominal STSs. In the study by Yeh and co-workers, in a collective of 1084 patients with intra-abdominal STS, palliative procedures were performed in 112 patients and were surgically performed in 82% of cases [45]. Of these, palliative procedures were most commonly undertaken in the gastrointestinal tract (44%). Symptom burden improved in 71% of patients 30 days after surgery, while only 54% of patients were symptom-free at 100 days. Moreover, 54% of obstructive gastrointestinal symptoms were successfully resolved at 30 days and 23% of patients were also symptom-free at 100 days. While the overall morbidity in this study was 29%, the postoperative mortality was 12%. A study by the U.S. Sarcoma Collaborative analyzed the results of palliative resection in 70 patients with retroperitoneal STS [46]. The predominant indication for palliative surgery was pain or bowel obstruction. However, the authors also observed a relatively high morbidity rate of 38% in their study. Unfortunately, these relatively high morbidity and mortality rates are consistent with our results, as we also observed a mortality rate of 20% and a major complication rate of 40% in the palliative surgery group in our own cohort of patients with PS. While it may be tempting to treat an obvious surgical emergency, it can be devastating for the patient to spend valuable time recovering from surgery or its complications while the disease inexorably progresses. Therefore, although palliative surgery for PS may improve symptom burden in these patients, given the limited oncologic benefit and increased postoperative morbidity and mortality, surgical therapy for symptom control should be considered only after careful selection and risk–benefit analysis in specialized centers.

5. Conclusions

Overall, patients with PS have a high rate of surgical complications and poor survival rates, and indications for further surgical intervention should take this into account. It should be noted that, in many cases, PS could be detected only during surgery. Nevertheless, in selected patients, considering tumor subtype, physical performance, and patient preference, surgical resection may be beneficial even if PS is detected.

The individualization of a treatment pathway always has the potential to be superior to a standardized approach by providing additional information. However, if the information available is based on unrepresentative personal experience or is not tailored to the specific case, there is also a risk of over- or undertreatment of the patient. Sharing experiences with colleagues, making treatment recommendations based on scientific evidence and consensus, and discussing all options with the patient are essential in the management of PS.

In the future, the use of large STS databases to generate reliable data on individual STS subtypes and stages of progression would be necessary to allow a scientific-based tailored treatment plan for difficult cases such as these. In our view, however, surgical control remains crucial in patients with PS and warrants a greater level of effort.

Supplementary Materials: The following are available online at https://www.mdpi.com/article/10.3390/cancers15041340/s1, Table S1. Detailed patient characteristics.

Author Contributions: Conceptualization, F.K. and A.K.; methodology, F.K. and A.K.; software, F.K. and A.K.; validation, H.A. and A.K.; formal analysis, F.K. and A.K.; investigation, F.K.; resources, F.K., H.A., L.H., I.E., L.S., and W.T.K.; data curation, F.K.; writing—original draft preparation, F.K.; writing—review and editing, H.A. and A.K.; visualization, F.K.; supervision, A.K.; project administration, A.K. All authors have read and agreed to the published version of the manuscript.

Funding: The article processing charge was funded by the Open Access Fond of the Heinrich-Heine-University Duesseldorf.

Institutional Review Board Statement: The study was conducted in accordance with the Declaration of Helsinki and approved by the Institutional Ethics Committee of the Medical Faculty of Heinrich Heine University, Düsseldorf (study number: 2022–2010, date of approval: 25 October 2022).

Informed Consent Statement: Patient consent was waived owing to the unidentifiable data. The waiver did not affect the rights and welfare of the subjects, and the research topic was not associated with any risk to the subjects.

Data Availability Statement: The data presented in this study are available upon request from the corresponding author.

Conflicts of Interest: The authors declare no conflict of interest.

References

1. Ressing, M.; Wardelmann, E.; Hohenberger, P.; Jakob, J.; Kasper, B.; Emrich, K.; Eberle, A.; Blettner, M.; Zeissig, S.R. Strengthening health data on a rare and heterogeneous disease: Sarcoma incidence and histological subtypes in Germany. *BMC Public Health* **2018**, *18*, 235. [CrossRef]
2. Gatta, G.; van der Zwan, J.M.; Casali, P.G.; Siesling, S.; Dei Tos, A.P.; Kunkler, I.; Otter, R.; Licitra, L.; Mallone, S.; Tavilla, A.; et al. Rare cancers are not so rare: The rare cancer burden in Europe. *Eur. J. Cancer* **2011**, *47*, 2493–2511. [CrossRef]
3. Lam, S.W.; Silva, T.M.; Bovée, J.V.M.G. New molecular entities of soft tissue and bone tumors. *Curr. Opin. Oncol.* **2022**, *34*, 354–361. [CrossRef]
4. Vay, C.; Schlünder, P.M.; Dizdar, L.; Esposito, I.; Ghadimi, M.P.H.; Knoefel, W.T.; Krieg, A. Targeting abundant survivin expression in liposarcoma: Subtype dependent therapy responses to YM155 treatment. *J. Cancer Res. Clin. Oncol.* **2022**, *148*, 633–645. [CrossRef]
5. Mersch, S.; Riemer, J.C.; Schlünder, P.M.; Ghadimi, M.P.; Ashmawy, H.; Möhlendick, B.; Topp, S.A.; Arent, T.; Kröpil, P.; Stoecklein, N.H.; et al. Peritoneal sarcomatosis: Site of origin for the establishment of an in vitro and in vivo cell line model to study therapeutic resistance in dedifferentiated liposarcoma. *Tumour Biol.* **2016**, *37*, 2341–2351. [CrossRef]
6. Pilati, P.; Rossi, C.R.; Mocellin, S.; Foletto, M.; Scagnet, B.; Pasetto, L.; Lise, M. Multimodal treatment of peritoneal carcinomatosis and sarcomatosis. *Eur. J. Surg. Oncol.* **2001**, *27*, 125–134. [CrossRef]
7. Sugarbaker, P.H. Intraperitoneal chemotherapy and cytoreductive surgery for the prevention and treatment of peritoneal carcinomatosis and sarcomatosis. *Semin. Surg. Oncol.* **1998**, *14*, 254–261. [CrossRef]
8. Rossi, C.R.; Casali, P.; Kusamura, S.; Baratti, D.; Deraco, M. The consensus statement on the locoregional treatment of abdominal sarcomatosis. *J. Surg. Oncol.* **2008**, *98*, 291–294. [CrossRef]
9. Wickham, H.; Bryan, J. Readxl: Read Excel Files R Package Version 1.3.1. Available online: https://cran.r-project.org/web/packages/readxl/index.html (accessed on 13 March 2019).
10. Kassambara, A.; Kosinski, M.; Biecek, P.; Scheipl, F. Survminer: Drawing Survival Curves using 'ggplot2'. Available online: https://cran.r-project.org/web/packages/survminer/index.html (accessed on 9 March 2021).
11. Therneau, T.M. A Package for Survival Analysis in R. R package version 3.4-0. Available online: https://cran.r-project.org/web/packages/survival/index.html (accessed on 9 August 2022).
12. Gronchi, A. Surgery in soft tissue sarcoma: The thin line between a surgical or more conservative approach. *Future Oncol.* **2021**, *17*, 3–6. [CrossRef]
13. Wong, L.C.K.; Li, Z.; Fan, Q.; Tan, J.W.; Tan, Q.X.; Wong, J.S.M.; Ong, C.J.; Chia, C.S. Cytoreductive surgery (CRS) with hyperthermic intraperitoneal chemotherapy (HIPEC) in peritoneal sarcomatosis—A systematic review and meta-analysis. *Eur. J. Surg. Oncol.* **2022**, *48*, 640–648. [CrossRef]

14. Zerhouni, S.; Van Coevorden, F.; Swallow, C.J. The role and outcomes of palliative surgery for retroperitoneal sarcoma. *J. Surg. Oncol.* **2018**, *117*, 105–110. [CrossRef] [PubMed]
15. Ratan, R.; Patel, S.R. Chemotherapy for soft tissue sarcoma. *Cancer* **2016**, *122*, 2952–2960. [CrossRef] [PubMed]
16. Schmitz, E.; Nessim, C. Retroperitoneal Sarcoma Care in 2021. *Cancers* **2022**, *14*, 1293. [CrossRef] [PubMed]
17. Gough, N.; Koffman, J.; Ross, J.R.; Riley, J.; Judson, I. Does palliative chemotherapy really palliate and are we measuring it correctly? A mixed methods longitudinal study of health related quality of life in advanced soft tissue sarcoma. *PLoS ONE* **2019**, *14*, e0210731. [CrossRef]
18. Grad, I.; Hanes, R.; Ayuda-Durán, P.; Kuijjer, M.L.; Enserink, J.M.; Meza-Zepeda, L.A.; Myklebost, O. Discovery of novel candidates for anti-liposarcoma therapies by medium-scale high-throughput drug screening. *PLoS ONE* **2021**, *16*, e0248140. [CrossRef]
19. Ghadimi, M.P.; Young, E.D.; Belousov, R.; Zhang, Y.; Lopez, G.; Lusby, K.; Kivlin, C.; Demicco, E.G.; Creighton, C.J.; Lazar, A.J.; et al. Survivin is a viable target for the treatment of malignant peripheral nerve sheath tumors. *Clin. Cancer Res.* **2012**, *18*, 2545–2557. [CrossRef]
20. Yuan, J.; Li, X.; Yu, S. Molecular targeted therapy for advanced or metastatic soft tissue sarcoma. *Cancer Control* **2021**, *28*, 10732748211038424. [CrossRef]
21. Loosen, S.H.; Krieg, S.; Eschrich, J.; Luedde, M.; Krieg, A.; Schallenburger, M.; Schwartz, J.; Neukirchen, M.; Luedde, T.; Kostev, K.; et al. The Landscape of Outpatient Palliative Care in Germany: Results from a Retrospective Analysis of 14,792 Patients. *Int. J. Environ. Res. Public Health* **2022**, *19*, 14885. [CrossRef]
22. Brandes, F.; Striefler, J.K.; Dörr, A.; Schmiester, M.; Märdian, S.; Koulaxouzidis, G.; Kaul, D.; Behzadi, A.; Thuss-Patience, P.; Ahn, J.; et al. Impact of a specialised palliative care intervention in patients with advanced soft tissue sarcoma—A single-centre retrospective analysis. *BMC Palliat. Care* **2021**, *20*, 16. [CrossRef]
23. Ray-Coquard, I.; Serre, D.; Reichardt, P.; Martín-Broto, J.; Bauer, S. Options for treating different soft tissue sarcoma subtypes. *Future Oncol.* **2018**, *14*, 25–49. [CrossRef]
24. Katz, D.; Palmerini, E.; Pollack, S.M. More Than 50 Subtypes of Soft Tissue Sarcoma: Paving the Path for Histology-Driven Treatments. *Am. Soc. Clin. Oncol. Educ. Book* **2018**, *38*, 925–938. [CrossRef] [PubMed]
25. Kelleher, F.C.; Viterbo, A. Histologic and genetic advances in refining the diagnosis of "undifferentiated pleomorphic sarcoma". *Cancers* **2013**, *5*, 218–233. [CrossRef] [PubMed]
26. Hayes-Jordan, A.; LaQuaglia, M.P.; Modak, S. Management of desmoplastic small round cell tumor. *Semin. Pediatr. Surg.* **2016**, *25*, 299–304. [CrossRef] [PubMed]
27. Gerald, W.L.; Ladanyi, M.; de Alava, E.; Cuatrecasas, M.; Kushner, B.H.; LaQuaglia, M.P.; Rosai, J. Clinical, pathologic, and molecular spectrum of tumors associated with t(11;22)(p13;q12): Desmoplastic small round-cell tumor and its variants. *J. Clin. Oncol.* **1998**, *16*, 3028–3036. [CrossRef]
28. Lal, D.R.; Su, W.T.; Wolden, S.L.; Loh, K.C.; Modak, S.; La Quaglia, M.P. Results of multimodal treatment for desmoplastic small round cell tumors. *J. Pediatr. Surg.* **2005**, *40*, 251–255. [CrossRef]
29. Hayes-Jordan, A.A.; Coakley, B.A.; Green, H.L.; Xiao, L.; Fournier, K.F.; Herzog, C.E.; Ludwig, J.A.; McAleer, M.F.; Anderson, P.M.; Huh, W.W. Desmoplastic Small Round Cell Tumor Treated with Cytoreductive Surgery and Hyperthermic Intraperitoneal Chemotherapy: Results of a Phase 2 Trial. *Ann. Surg. Oncol.* **2018**, *25*, 872–877. [CrossRef]
30. Wessalowski, R.; Schneider, D.T.; Mils, O.; Hannen, M.; Calaminus, G.; Engelbrecht, V.; Pape, H.; Willers, R.; Engert, J.; Harms, D.; et al. An approach for cure: PEI-chemotherapy and regional deep hyperthermia in children and adolescents with unresectable malignant tumors. *Klin. Padiatr.* **2003**, *215*, 303–309. [CrossRef]
31. Harmon, R.L.; Sugarbaker, P.H. Prognostic indicators in peritoneal carcinomatosis from gastrointestinal cancer. *Int. Semin. Surg. Oncol.* **2005**, *2*, 3. [CrossRef]
32. Jacquet, P.; Sugarbaker, P.H. Clinical research methodologies in diagnosis and staging of patients with peritoneal carcinomatosis. *Cancer Treat. Res.* **1996**, *82*, 359–374. [CrossRef]
33. Salti, G.I.; Ailabouni, L.; Undevia, S. Cytoreductive surgery and hyperthermic intraperitoneal chemotherapy for the treatment of peritoneal sarcomatosis. *Ann. Surg. Oncol.* **2012**, *19*, 1410–1415. [CrossRef]
34. Berthet, B.; Sugarbaker, T.A.; Chang, D.; Sugarbaker, P.H. Quantitative methodologies for selection of patients with recurrent abdominopelvic sarcoma for treatment. *Eur. J. Cancer* **1999**, *35*, 413–419. [CrossRef] [PubMed]
35. Baratti, D.; Pennacchioli, E.; Kusamura, S.; Fiore, M.; Balestra, M.R.; Colombo, C.; Mingrone, E.; Gronchi, A.; Deraco, M. Peritoneal sarcomatosis: Is there a subset of patients who may benefit from cytoreductive surgery and hyperthermic intraperitoneal chemotherapy? *Ann. Surg. Oncol.* **2010**, *17*, 3220–3228. [CrossRef]
36. Haddox, C.L.; Riedel, R.F. Recent advances in the understanding and management of liposarcoma. *Fac. Rev.* **2021**, *10*, 1. [CrossRef] [PubMed]
37. Martin-Broto, J.; Mondaza-Hernandez, J.L.; Moura, D.S.; Hindi, N. A Comprehensive Review on Solitary Fibrous Tumor: New Insights for New Horizons. *Cancers* **2021**, *13*, 2913. [CrossRef] [PubMed]
38. Serpico, R.; Brown, J.; Blank, A.; Jones, K.; Randall, R.L.; Groundland, J. Metastasis of Osteosarcoma to the Abdomen: A Report of Two Cases and a Review of the Literature. *Case Rep. Oncol.* **2021**, *14*, 647–658. [CrossRef] [PubMed]

39. Toulmonde, M.; Bonvalot, S.; Méeus, P.; Stoeckle, E.; Riou, O.; Isambert, N.; Bompas, E.; Jafari, M.; Delcambre-Lair, C.; Saada, E.; et al. Retroperitoneal sarcomas: Patterns of care at diagnosis, prognostic factors and focus on main histological subtypes: A multicenter analysis of the French Sarcoma Group. *Ann. Oncol.* **2014**, *25*, 735–742. [CrossRef] [PubMed]
40. Fairweather, M.; Gonzalez, R.J.; Strauss, D.; Raut, C.P. Current principles of surgery for retroperitoneal sarcomas. *J. Surg. Oncol.* **2018**, *117*, 33–41. [CrossRef]
41. Bredbeck, B.C.; Delaney, L.D.; Kathawate, V.G.; Harter, C.A.; Wilkowski, J.; Chugh, R.; Cuneo, K.C.; Dossett, L.A.; Sabel, M.S.; Angeles, C.V. Factors associated with disease-free and abdominal recurrence-free survival in abdominopelvic and retroperitoneal sarcomas. *J. Surg. Oncol.* **2022**, *125*, 1292–1300. [CrossRef]
42. Ray-Coquard, I.; Montesco, M.C.; Coindre, J.M.; Dei Tos, A.P.; Lurkin, A.; Ranchère-Vince, D.; Vecchiato, A.; Decouvelaere, A.V.; Mathoulin-Pélissier, S.; Albert, S.; et al. Sarcoma: Concordance between initial diagnosis and centralized expert review in a population-based study within three European regions. *Ann. Oncol.* **2012**, *23*, 2442–2449. [CrossRef]
43. Meyer, M.; Seetharam, M. First-Line Therapy for Metastatic Soft Tissue Sarcoma. *Curr. Treat. Options Oncol.* **2019**, *20*, 6. [CrossRef]
44. Blay, J.Y.; Soibinet, P.; Penel, N.; Bompas, E.; Duffaud, F.; Stoeckle, E.; Mir, O.; Adam, J.; Chevreau, C.; Bonvalot, S.; et al. Improved survival using specialized multidisciplinary board in sarcoma patients. *Ann. Oncol.* **2017**, *28*, 2852–2859. [CrossRef] [PubMed]
45. Yeh, J.J.; Singer, S.; Brennan, M.F.; Jaques, D.P. Effectiveness of palliative procedures for intra-abdominal sarcomas. *Ann. Surg. Oncol.* **2005**, *12*, 1084–1089. [CrossRef] [PubMed]
46. Thalji, S.Z.; Tsai, S.; Gamblin, T.C.; Clarke, C.; Christians, K.; Charlson, J.; Ethun, C.G.; Poultsides, G.; Grignol, V.P.; Roggin, K.K.; et al. Outcomes of palliative-intent surgery in retroperitoneal sarcoma-Results from the US Sarcoma Collaborative. *J. Surg. Oncol.* **2020**, *121*, 1140–1147. [CrossRef] [PubMed]

cancers

MDPI

Article

The First-In-Class Anti-AXL × CD3ε Pronectin™-Based Bispecific T-Cell Engager Is Active in Preclinical Models of Human Soft Tissue and Bone Sarcomas

Nicoletta Polerà [1,†], Antonia Mancuso [1,†], Caterina Riillo [1], Daniele Caracciolo [1], Stefania Signorelli [1], Katia Grillone [1], Serena Ascrizzi [1], Craig A. Hokanson [2], Francesco Conforti [3], Nicoletta Staropoli [1], Luigia Gervasi [1], Maria Teresa Di Martino [1], Mariamena Arbitrio [4], Giuseppe Nisticò [5], Roberto Crea [2,5], Pierosandro Tagliaferri [1], Giada Juli [1,*,‡] and Pierfrancesco Tassone [1,6,*,‡]

1 Department of Experimental and Clinical Medicine, Magna Græcia University, 88100 Catanzaro, Italy
2 Protelica, Inc., Hayward, CA 94545, USA
3 Pathology Unit, Annunziata Hospital, 87100 Cosenza, Italy
4 Institute of Research and Biomedical Innovation (IRIB), Italian National Council (CNR), 88100 Catanzaro, Italy
5 Renato Dulbecco Institute, 88046 Lamezia Terme, Italy
6 College of Science and Technology, Temple University, Philadelphia, PA 19122, USA
* Correspondence: giadajuli@libero.it (G.J.); tassone@unicz.it (P.T.)
† These authors contributed equally to this work.
‡ These authors contributed equally to this work.

Simple Summary: Sarcomas are a group of heterogeneous diseases with a poor prognosis and scarce therapeutic options. Innovative approaches based on novel therapeutic targets are eagerly awaited. AXL, a TAM family tyrosine kinase receptor, recently emerged as an interesting target for several type of sarcomas. Here, we propose an innovative immunotherapeutic strategy based on the targeting of AXL, using a first-in-class Pronectin™-based Bispecific T-Cell Engager (pAXL×CD3ε) for the treatment of sarcomas. Our results demonstrate that pAXL×CD3ε redirects T cells toward AXL-expressing sarcoma cell lines, leading a dose-dependent and T cell-mediated cytotoxicity in vitro. Moreover, pAXL×CD3ε inhibits the in vivo growth of human sarcoma xenografts and improves survival in immunocompromised mice, thus representing a new-generation strategy for the treatment of a still-incurable disease.

Abstract: Sarcomas are heterogeneous malignancies with limited therapeutic options and a poor prognosis. We developed an innovative immunotherapeutic agent, a first-in-class Pronectin™-based Bispecific T-Cell Engager (pAXL×CD3ε), for the targeting of AXL, a TAM family tyrosine kinase receptor highly expressed in sarcomas. AXL expression was first analyzed by flow cytometry, qRT-PCR, and Western blot on a panel of sarcoma cell lines. The T-cell-mediated pAXL×CD3ε cytotoxicity against sarcoma cells was investigated by flow cytometry, luminescence assay, and fluorescent microscopy imaging. The activation and degranulation of T cells induced by pAXL×CD3ε were evaluated by flow cytometry. The antitumor activity induced by pAXL×CD3ε in combination with trabectedin was also investigated. In vivo activity studies of pAXL×CD3ε were performed in immunocompromised mice (NSG), engrafted with human sarcoma cells and reconstituted with human peripheral blood mononuclear cells from healthy donors. Most sarcoma cells showed high expression of AXL. pAXL×CD3ε triggered T-lymphocyte activation and induced dose-dependent T-cell-mediated cytotoxicity. The combination of pAXL×CD3ε with trabectedin increased cytotoxicity. pAXL×CD3ε inhibited the in vivo growth of human sarcoma xenografts, increasing the survival of treated mice. Our data demonstrate the antitumor efficacy of pAXL×CD3ε against sarcoma cells, providing a translational framework for the clinical development of pAXL×CD3ε in the treatment of human sarcomas, aggressive and still-incurable malignancies.

Keywords: sarcomas; AXL; pronectins™; bispecific T-cell engager; BTCE; immunotherapy; cancer

Citation: Polerà, N.; Mancuso, A.; Riillo, C.; Caracciolo, D.; Signorelli, S.; Grillone, K.; Ascrizzi, S.; Hokanson, C.A.; Conforti, F.; Staropoli, N.; et al. The First-In-Class Anti-AXL×CD3ε Pronectin™-Based Bispecific T-Cell Engager Is Active in Preclinical Models of Human Soft Tissue and Bone Sarcomas. *Cancers* **2023**, *15*, 1647. https://doi.org/10.3390/cancers15061647

Academic Editor: Catrin Sian Rutland

Received: 13 February 2023
Revised: 3 March 2023
Accepted: 5 March 2023
Published: 8 March 2023

1. Introduction

Sarcomas are a large group of heterogeneous malignancies of mesenchymal origin, commonly characterized by a poor prognosis, of which the onset may occur at any age [1,2]. Among them, soft tissue sarcomas (STSs) represent 80%, bone sarcomas 15% and gastrointestinal stromal tumors 5%. Because of their heterogeneity and common aggressive nature, they are resistant to available therapies and clinical management is still highly challenging [3–5]. Conventional treatment, including surgery, radiation therapy and chemotherapy (Doxorubicin, Ifosfamide, trabectedin, and others), differs from one subtype to another. Surgery is the first-line treatment for localized sarcomas, in combination with pre- or post-operative therapies [6], while chemotherapy is the standard treatment for metastatic disease. Unfortunately, the median survival for advanced disease is around 12 months [7]. In this scenario, targeted therapies which might overcome the limitations of current treatments are eagerly awaited [8]. Different signaling pathways involved in sarcoma genesis have been investigated so far. Targeting therapies involving (i) cell cycle progression, through cell cycle inhibitors (CDKIs) [9,10]; (ii) and growth receptors and pro-survival signaling molecules, through tyrosine kinase inhibitors (TKIs) [11], IGFR [12] and mTOR inhibitors [13], have shown efficacy against sarcomas, but only the VEGR inhibitor pazopanib has reached the prime time [14]. Inhibition of epigenetic regulators [15] and poly-ADP-ribose-polymerase (PARP) inhibitors have also demonstrated promising anti-cancer activity in preclinical and clinical studies [16].

Even if immunotherapy may be considered a new therapeutic path and some clinical trials based on the use of immune checkpoint inhibitors are currently ongoing [17,18], to date, it is not considered a valuable option for most sarcomas. Other clinical trials are investigating strategies based on endogenous, transgenic, or chimeric antigen receptor (CAR)-expressing T cells for the targeting of specific antigens, such as tyrosine-kinase-like orphan receptor 2, CD133, GD-2, Muc1 and CD117 (e.g., NCT03356782, NCT00902044, NCT04995003, NCT01953900) (https://clinicaltrials.gov/, accessed on 10 February 2023) [19]. Nevertheless, the effective treatment of advanced disease is an unmet clinical need, and most sarcomas can still be considered incurable. Novel strategies based on new therapeutic targets are highly desirable.

Recently, the AXL receptor has emerged as a promising candidate target for a variety of sarcomas [20–22]. The AXL gene is located on chromosome 19q13.2. It encodes for the protein called AXL (UFO, ARK, Tyro7, or JTK11), a member of the TAM family of tyrosine kinase receptors (RTKs) characterized by an extracellular, transmembrane, and intracellular domain [22]. The extracellular structure consists of two immunoglobulins (Ig-like) and two fibronectin type III (Fro III-like) chains, while the intracellular domain is important for auto-phosphorylation and signaling kinase activity [23]. In normal cells and tissues, AXL regulates cell survival, non-inflammatory clearance of apoptotic cells, natural killer cell differentiation and platelet aggregation. AXL is also expressed in cancer cells and microenvironmental immune cells, including dendritic cells, macrophages, and NK cells. It drives several cellular processes that are critical for the development, growth and spread of tumors, including proliferation, invasiveness and migration, epithelial–mesenchymal transition, angiogenesis, and immune resistance [24–26].

Different therapeutic agents targeting AXL have been recently developed, including: (i) small molecule inhibitors, which block AXL auto-phosphorylation and kinase activities, such as BGB324, presently investigated in phase I/II clinical trials [27]; (ii) anti-AXL monoclonal antibodies (mAbs), such as YW327.6S2, which bind both human and murine AXL [22]; (iii) nucleotide aptamers, such as the RNA aptamer GL21.T [28]; (iv) soluble AXL receptor that acts as a decoy receptor for the AXL ligand GAS6; (v) natural compounds, such as the *Viscum album* (L.) extract [29].

It was also demonstrated that AXL is over-expressed in Kaposi sarcoma and Kaposi sarcoma herpesvirus-transformed endothelial cells. MAbs generated to induce AXL degradation inhibited Kaposi-sarcoma-cell invasion in in vitro models and tumor growth in vivo [30]. A subsequent study identified AXL as a potential therapeutic target for Ewing sarcoma. AXL inhibitors were shown to affect the viability of Ewing sarcoma cells [31 In addition, a high expression of AXL gene was found in leiomyosarcoma, and its a tivity was suppressed by two different multi-tyrosine kinase inhibitors (Crizotinib an Foretinib) [20]. Finally, other studies have found that osteosarcoma cells highly expres AXL [32], of which the inhibition significantly reduces lung metastases [21]. Despite a these promising findings, an effective anti-AXL treatment is still not available for thes aggressive malignancies.

A novel class of non-immunoglobulin, single-domain therapeutic proteins (Pronectins™ based on "antibody mimics" technology, has been just developed with the aim of providin a novel platform for the treatment of various diseases, including cancer. Pronectins™ were isolated from synthetic human libraries, built upon the 14th domain of Fibronecti III (14FN3) scaffold, which is selected by a bioinformatic approach, and on advance complementarity-determining region (CDR) diversity of more than 25 billion loop se quences [33]. Since Pronectins™ mimic the natural human repertoire, they are poorl immunogenic [34]. Several pharmacological properties are associated to the fibronectin II scaffold, such as high stability, tissue penetration, and low cost of production. Furthermore they are smaller than a conventional mAb, representing a favorable feature for the loca delivery to the solid tumor mass [35]. Starting from Pronectins™, it is possible to generate multimers, fusion proteins, bispecifics or constructs with site-specific modifications fo tailored therapy [36]. Bispecific T-Cell Engagers (BTCEs) emerged as a novel promising strategy for hematologic malignancies, but their application in solid tumors is highly challenging, due to the paucity of selective tumor-associated antigens (TAAs) and the struggle in penetrating the solid tumor mass [37]. In this scenario, a Pronectin™-based BTCE (pBTCE) can help overcome these limitations.

Based on this rationale, we investigated the in vitro and in vivo activity of a first-in class pBTCE targeting AXL (pAXL×CD3ε) as a potential immunotherapeutic agent for the treatment of sarcomas.

2. Materials and Methods

2.1. Generation and Development of pAXL×CD3ε

A highly specific anti-AXL Pronectin™, a non-immunoglobulin and single-domain protein, has been isolated from synthetic libraries based on the human scaffold of the 14th domain of fibronectin III (14FN3), as previously described [38]. By bioinformatic analysis aimed to select the best candidate within the amino acid loop diversity and minimize or prevent immunogenicity, 6 Pronectins™ with a KD < 10 nM were identified and AXL54 was chosen for targeting purposes (KD = 8 nM). This Pronectin™ was used to develop a first-in-class BTCE (AXL54 (Pronectin™)-linker-scFV CD3, pAXL×CD3ε), for investigation as an anti-tumor novel agent. The linker is made of a single unit of Gly4–Ser (GGGGS) [38]

2.2. Cell Lines

CAL-72, ESS-I, HT-1080, SAOS-2 and Rh-30 were purchased by DSMZ. SW982 and RD-ES were purchased from ATCC. ESS-I (endometrial stromal sarcoma), SAOS-2 (osteogenic sarcoma), SW982 (synovial sarcoma) and RD-ES (Ewing's sarcoma) were grown in RPMI 1640 (Gibco®, Thermo Fisher Scientific, Waltham, MA, USA), supplemented with 20% fetal bovine serum (FBS) (Lonza Group Ltd., Basel, Switzerland), penicillin (100 U/mL) and streptomycin (100 µg/mL) (Gibco®, Thermo Fisher Scientific). Rh-30 (rhabdomyosarcoma) was cultured in RPMI 1640, supplemented with 10% FBS, 100 U/mL penicillin and 100 µg/mL streptomycin. CAL-72 (osteosarcoma) and HT-1080 (fibrosarcoma) cell lines were cultured in DMEM-GlutaMAX™ (Gibco®, Thermo Fisher Scientific), respectively, supplemented with 20% FBS and 10% FBS, 100 U/mL penicillin and 100 µg/mL streptomycin. Cell lines were maintained at 37 °C, in a humidified atmosphere with 5% CO_2.

2.3. Transduction of Sarcoma Cell Lines

Sarcoma cells were plated at 1×10^5 cells/mL in 6-well plate and incubated O/N. To obtain sarcoma cells stably expressing green fluorescent protein (GFP) transgene, a lentiviral GFP-encoding vector was added according to the manufacturer's instruction (SBI System Biosciences, Mountain View, CA, USA). Polybrene (Sigma-Aldrich, Saint Louis, MO, USA) was also used to a final concentration of 8 μg/mL. Two days after transduction, cells were selected using DMEM, supplemented with 20% FBS, containing 1 mg/mL puromycin (Sigma Aldrich). After antibiotic selection, puromycin-resistant transduced cells were assessed for the expression of GFP by flow cytometry, using Attune NxT Flow cytometer (Thermo Fisher Scientific) and microscopy (Thunder Imaging Systems, Leica Microsystems, Wetzlar, Germany).

2.4. Peripheral Blood Mononuclear Cell (PBMC) Isolation

Mononuclear cells were obtained from healthy donor buffy coats. Briefly, PBMCs were isolated by Ficoll-Paque Plus (Cytiva Europe GmbH, Buccinasco, Milan, Italy) density gradient centrifugation, according to the manufacturer's recommendations, and washed twice in the culture medium (RPMI-1640 supplemented with 10% FBS), as previously described [39,40].

2.5. Detection of AXL Expression and Target Quantification

AXL expression was analyzed on each sarcoma cell line by flow cytometry. Cells were incubated with FITC-conjugated AXL antibody (#MAB154-100, R&D Systems, Minneapolis, MN, USA) for 15 min at RT in the dark. The tubes were washed in PBS 1X and centrifuged $400\times g$ for 5 min, resuspended in 500 μL of PBS 1X and analyzed by a flow cytometer.

To quantify AXL expression on sarcoma cell lines, calibrated microspheres (Quantum Simply Cellular, Bangs Laboratories Inc., Fishers, Castenaso, BO, Italy) were used according to the manufacturer's protocol. Briefly, saturating amounts of FITC-conjugated AXL antibody were added to one drop of each microbead suspension, and the final mixes were incubated for 30 min at RT in the dark. Samples were washed twice using PBS 1X ($2500\times g$), resuspended in 500 μL of PBS 1X and analyzed by a flow cytometer. Simultaneously, each cell line was stained with FITC-conjugated AXL antibody, as previously described. The analysis was performed maintaining the same instrument setting used for QSC beads. A QuickCal® spreadsheet, provided by Bangs Laboratories, was used to convert the main fluorescence intensity (MFI) from microspheres to antibody-binding capacity (ABC) values.

2.6. Redirected T-Cell Cytotoxicity Assay

PBMCs were isolated from at least 3 donors and labeled with CellTrace™ Violet viable marker (Invitrogen, Waltham, MA, USA), according to the manufacturer's instructions, and co-cultured with sarcoma cell lines (CAL-72, ESS-I, HT-1080, SAOS-2, Rh-30, SW982 or RD-ES) at different effector-to-target-cell (E:T) ratio, in the presence of increasing concentrations of pAXL×CD3ε (0.1 μg/mL, 1 μg/mL and 2.5 μg/mL) or anti-B-cell maturation antigen (BCMA) Pronectin™-based BTCE, pBCMA×CD3ε (2.5 μg/mL), as a negative control. BCMA is in fact highly restricted to hematopoietic B cells and is not expressed by solid tumors, therefore representing a suitable negative control in our case. Cells were incubated for 72 h at 37 °C and 5% CO_2, and finally stained with 7-AAD (BD Biosciences, La Jolla, CA, USA). The cytotoxic effect on sarcoma cell lines was detected by flow cytometry and reported as the percentage of 7-AAD$^+$/CellTrace™ Violet$^-$ cells. The 10:1 E:T ratio was selected because it allowed for the highest toxicity.

Cells stably expressing GFP gene were co-cultured with PBMCs from at least 3 donors at 10:1 E:T ratio, in the presence of pAXL×CD3ε (2.5 μg/mL). Cells were incubated for 72 h at 37 °C and 5% CO_2. Cytotoxicity was assessed by flow cytometry monitoring MFI in GFP-positive cells.

For microscope imaging, co-cultured cells were plated on a round cover glass (Fisher Scientific) above 24 wells, and fixed using 4% paraformaldehyde (PFA) for 15 min. Sections were washed three times with PBS 1X, mounted in Vectashield with DAPI (Vector Lab, Newark, CA, USA) and analyzed using Thunder Imaging Systems (Leica, Wetzlar, Germany).

2.7. Cell Viability Assay

Cells were plated in 96 wells treated with different concentrations of pAXL×CD3ε, and cell viability was evaluated by Cell Titer-Glo Luminescent Assay (CTG; Promega, Madison, WI, USA), as previously reported [41].

2.8. Western Blot

Whole-cell protein extracts were obtained using NP40 lysis buffer containing Halt Protease and Phosphatase Inhibitor cocktail (Invitrogen, Thermo Fisher Scientific), separated using 4–12% Novex Bis-Tris SDS-acrylamide gels (Invitrogen), and transferred on nitrocellulose membranes (Bio-Rad, Hercules, CA, USA), as previously reported [42]. Nitrocellulose membranes were incubated O/N at 4 °C with primary antibody. In detail, anti-AXL (#4566) by Cell Signaling Technology (Danvers, MA, USA) and anti-GAPDH (sc-25778) by Santa Cruz (Dallas, TX, USA) were used for Western blotting (WB) procedures. The membrane was washed thrice with PBS-Tween and incubated with the secondary antibody (anti-rabbit IgG HRP-linked antibody #7074S, Cell Signaling Technology) for 1 h at RT. Chemiluminescence was recorded using SuperSignal West Pico PLUS Chemiluminescent Substrate (Thermo Scientific). Densitometric analysis of blots was performed using LI-COR Image Studio Digits Ver 5.0 (Bad Homburg, Germany).

2.9. RNA Isolation and Quantitative Real-Time PCR

The WizPrep™ Total RNA Mini Kit (Wizbiosolutions, Seongnam, South Korea) was used, according to the manufacturer's guidelines, to extract purified RNA from sarcoma cell lines. The RNA quantity and quality were assessed by NanoDrop® (ND-1000 Spectrophotometer). cDNA was obtained from the reverse transcription of total RNA, using the "high-capacity cDNA reverse transcription kit" (Applied Biosystems, Foster City, CA, USA). Taq-Man® assay (Life Technologies, Carlsbad, CA, USA) was used to detect and quantify AXL (Hs01064439_m1), and GAPDH (Hs03929097_g1) was considered to normalize the recorded threshold cycle values. qRT-PCR was performed in triplicate and relative expression was obtained through the comparative cross threshold method on a ViiA7 System (Thermo Fisher Scientific, Waltham, MA, USA).

2.10. T-Cell Activation

Sarcoma cells lines were co-cultured with PBMCs from at least 3 donors at selected 10:1 E:T ratio in the presence of increasing concentrations of pAXL×CD3ε (0.1 μg/mL, 1 μg/mL, and 2.5 μg/mL) or negative control, and were incubated for 72 h at 37 °C and 5% CO_2. T cells were stained using anti-human CD4 (SK3) FITC (#345768), CD8 (SK1) APC-Cy7 (#641400), CD25 APC (#555434), CD69 PE (#555531), CD3 (UCHT1) PerCP-Cy5.5 (#560835), CD45 (HI30) BV510 (#563204) and CD107a PE (#555801) (BD Biosciences), for 4 h at 37 °C and 5% CO_2. T cells were selected as CellTrace™ Violet-positive, gated for CD4-, CD8- or CD3-positive, and CD69-, CD25- or CD107a-positive cells. The intracellular production of cytokines and cytolytic enzymes was investigated adding brefeldin A 10 mg/mL. After 4 h, cells were incubated with surface antibodies and treated using FIX&PERM® kit (Nordic MUbio, Susteren, The Netherlands), according to the manufacturer's guidelines. Subsequently, cells were incubated with anti-TNFα PE-Cy™7 (Mab11) (#560678), anti-IFNγ PE (#559327) and anti-Granzyme B (AlexaFluor®647) (#560212) (BD Biosciences) for 15 min at RT in the dark. Samples were finally washed in PBS 1X and analyzed by a flow cytometer.

2.11. Analysis of the Activity of pAXL×CD3ε in Combination with Chemotherapeutic Drugs

SAOS-2 were plated in 24 wells and co-cultured at selected 10:1 E:T ratio with PBMCs from at least 3 donors, labelled with CellTrace™ Violet. Cells were treated with pAXL×CD3ε (1 μg/mL), trabectedin (0.2 nM) or their combination (pAXL×CD3ε + trabectedin). After 72 h of incubation, cells were stained using 7-AAD and analysis of positive cells were performed through flow cytometry.

2.12. In Vivo Studies

In vivo experiments were performed according to standard guidelines and approved protocols by the National and Institutional Animal Committee (483/2020-PR, 18 May 2020). Four-to-six-week-old male NSG (NOD.Cg-PrkdcscidIl2rgtm1Wjl/SzJ) mice were purchased from Charles River Laboratories (Wilmington, MA, USA). Animals were regularly monitored and euthanized when signs of disease-related symptoms or graft-versus-host disease (GvHD) developed.

To obtain a subcutaneous (sc) xenografted in vivo model, 10 mice were inoculated in the dorsal right flank with HT-1080 cells (3×10^6) resuspended in 100 μL of PBS 1X. On day 4, 10×10^6 PBMCs from healthy donors were intraperitoneally (ip) injected into each mouse. The same day, mice were randomized in 2 groups (5 mice for each group), and 0.1 mg/kg pAXL×CD3ε or vehicle were ip injected for 15 consecutive days. Tumor sizes were measured with a digital caliper. The tumor volume (tv) was calculated using the formula:

$$tv = (W^2 \times L)/2, \tag{1}$$

where W is the tumor width and L is the tumor length, as previously described [43].

Mice were sacrificed when the tv reached >2000 mm^3. At the time of sacrifice, blood samples were collected. Red blood cell lysis was performed, and cells were stained with anti-human CD45 BV510 and CD3 PerCP-Cy5.5 to evaluate PBMCs engraftment. Explanted tumors were analyzed by WB, as previously described, using anti-caspase-3 (#9668, Cell Signaling Technology) and anti-PARP (#9532, Cell Signaling Technology), and by immunohistochemistry (IHC) using anti-CD3 antibody (#GA503, Agilent Dako, Glostrup, Denmark).

2.13. Statistical Analysis

Statistical evaluations were carried out using a parametric Student's *t*-test by the GraphPad software (www.graphpad.com, accessed on 10 February 2023). Graphpad Prism version 6.0 was used to make graphs. Only results with a p value < 0.05 were accepted as statistically significant. Each value is reported as the mean of at least 2 experiments ± SD/SEM.

3. Results

3.1. Evaluation of AXL Expression on Sarcoma Cell Lines

To investigate the expression of AXL on sarcoma cells (Figure 1A), we collected a panel of seven human cell lines, including CAL-72 (osteosarcoma), ESS-I (endometrial stromal sarcoma), HT-1080 (fibrosarcoma), SAOS-2 (osteosarcoma), Rh-30 (rhabdomyosarcoma), SW982 (synovial sarcoma) and RD-ES (Ewing's sarcoma). Flow cytometry showed different AXL expression levels on the surface of tumor cells: high expression on CAL-72, ESS-I and HT-1080; intermediate expression on SAOS-2 and Rh-30; and low- or no-expression on SW982 and RD-ES cell lines, respectively (Figure 1B,C). This trend was confirmed performing quantitative analysis of antigen expression density for each cell line, using calibrated microspheres to assess the antibody-binding capacity (ABC). As reported in Figure 1D, AXL expression was in a range between 21,000 and 4200 antigen molecules on CAL-72 and SW982 cells, respectively. Through qRT-PCR analysis, we assessed the AXL mRNA expression in sarcoma cells (Figure 1E). Our findings revealed a different pattern of target expression, which was in accordance with data retrievable by cBioPortal for the Cancer Genomics dataset (cbioportal.org, accessed on 10 February 2023) and Cancer Cell Line Encyclopedia (CCLE) dataset (https://depmap.org/portal/interactive, accessed on 10 February 2023). Further Western blot analyses were performed to investigate the expression of AXL protein in sarcoma cells, reporting a clear difference in the band intensity between various cell lines (Figure 1F). According to our results, AXL expression is not correlated to specific sarcoma sub-types.

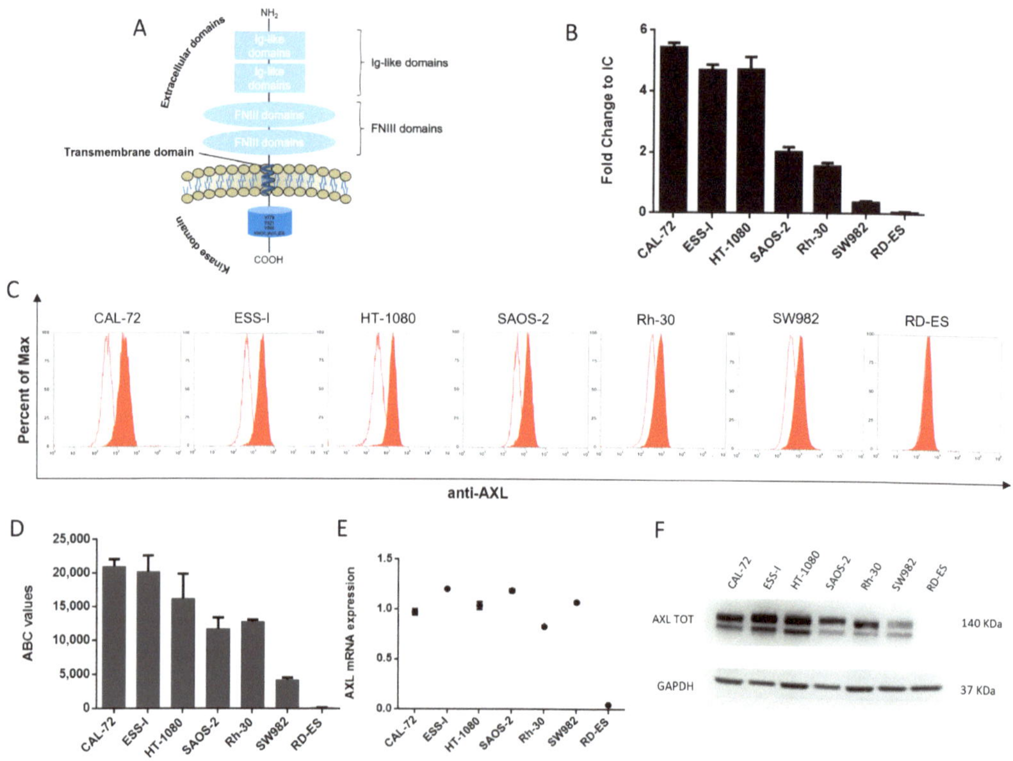

Figure 1. AXL expression. (**A**) Schematic representation of AXL receptor tyrosine kinase structure. It is composed of two immunoglobulin (Ig)-like domains that characterize extracellular domains, two fibronectin type III (FNIII) domains, a transmembrane domain and a kinase domain that is intracellular. (**B**) Flow cytometry analysis of AXL expression of sarcoma cells. (**C**) Representative FACS overlays between unstained (empty) and stained (full red) sample of each cell line. (**D**) Quantification of antibody-binding capacity (ABC) assay. (**E**) AXL-relative mRNA level determined by qRT-PCR and normalized on GAPDH housekeeping. (**F**) Western blot of AXL total form reported in a collection panel of sarcoma cells. The uncropped blots are shown in Figure S3.

Our data demonstrate that AXL is highly expressed in sarcoma cells, therefore representing a potential mean for selective targeting. The interaction of pAXL×CD3ε with sarcoma cells was assessed through indirect staining, using an anti-human IgG secondary antibody (Jackson ImmunoResearch, West Grove, PA, USA) (Figure S1).

3.2. T-Cell Mediated Cytotoxicity Is Induced by pAXL×CD3ε In Vitro

To assess the activity of pAXL×CD3ε, sarcoma cells with different expression levels of AXL were co-cultured with purified human T cells (E:T ratio selected at 10:1) from healthy donors, in the presence of three different concentrations of pAXL×CD3ε (0.1 μg/mL, 1 μg/mL, and 2.5 μg/mL) for 72 h. Increasing concentrations of pAXL×CD3ε produced T-cell-mediated cytotoxicity on sarcoma cells except for the RD-ES because of its low expression of the target antigen (Figure 2A,B). In particular, the cytotoxic activity of pAXL×CD3ε followed a binding-response effect leading to around 50% of cell death on CAL-72, Rh-30, and HT-1080, 40% on SAOS-2, 30% and 20% in SW982 and ESS-I, respectively, at 2.5 μg/mL. As a negative control, we performed a cytotoxicity assay using a Pronectin™-based BTCE binding a different target expressed by B cells only, the BCMA, not expressed by sarcoma cells (pBCMA×CD3ε).

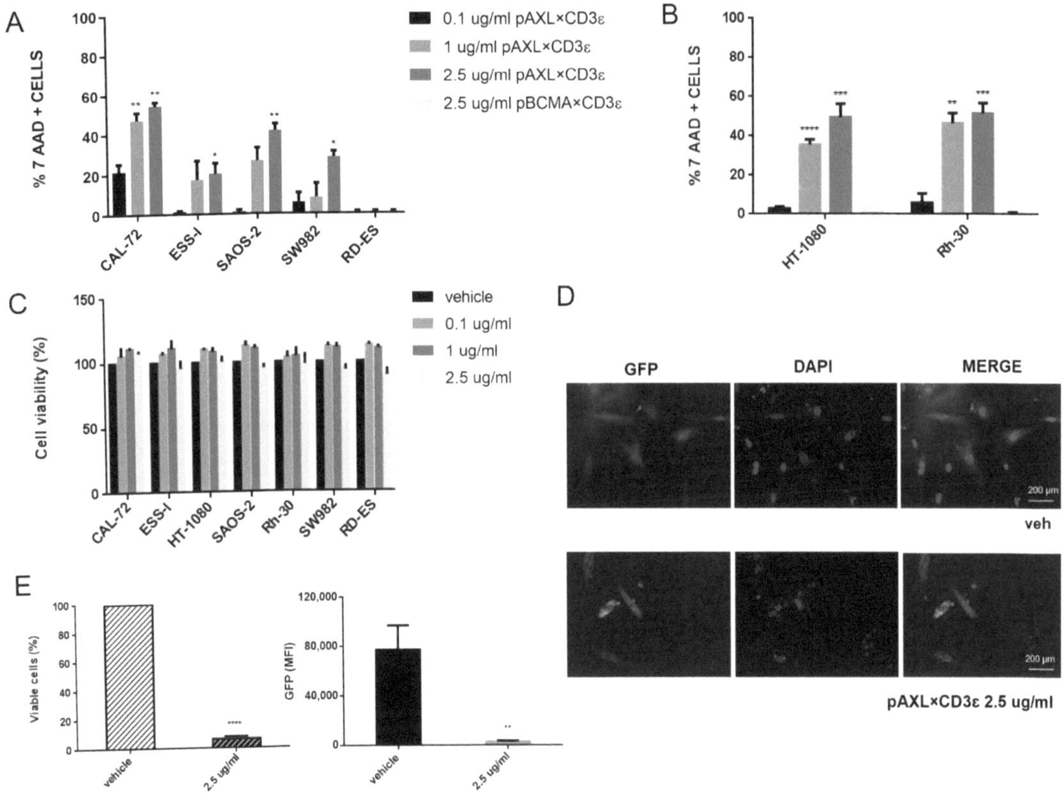

Figure 2. Redirected T-lymphocyte cytotoxicity by pAXL×CD3ε in sarcoma cell lines. (**A**) FACS analysis of 7-AAD(7-amino-actinomycin D)-positive cells. (**B**) HT-1080 and Rh-30 treated with increasing concentrations of pAXL×CD3ε (0.1 µg/mL, 1 µg/mL and 2.5 µg/mL), and as a negative control 2.5 µg/mL of pBCMA×CD3ε after 72 h. Each group has been compared to the 0.1 µg/mL group for statistical analysis. Results are normalized to the recorded vehicle values. (**C**) Cell Titer-Glo luminescent cell viability (%) assay performed on sarcoma cell lines without effector cells, following pAXL×CD3ε 72 h treatment. (**C**) Positive cells (relative percentage %) of sarcoma cell lines co-cultured with peripheral blood mononuclear cells (PBMCs) and treated with different concentrations of pAXL×CD3ε or vehicle for 72 h. Each group has been compared to the 0.1 µg/mL group for statistical analysis. Results are normalized to the recorded vehicle values. (**D**) Imaging of CAL-72 stably expressing green fluorescent protein (GFP) in untreated cells (vehicle) and 2.5 µg/mL of pAXL×CD3ε-treated cells after 72 h. Nuclei were stained with DAPI and microscopies were performed at 10-fold magnification. (**E**) Percentage of stably expressing CAL-72 GFP viable cells and median fluorescence intensity (MFI) of GFP analyzed by flow cytometry. PBMCs were obtained from 3 healthy donors and results are expressed as the mean value of triplicate experiments from each donor. * $p < 0.0332$; ** $p < 0.0021$; *** $p < 0.0002$; **** $p < 0.0001$.

As shown in Figure 2B, 2.5 µg/mL of pBCMA×CD3ε did not induce cytotoxicity in two different sarcoma cell lines, such as HT-1080 and Rh-30, when it was used at a higher dose. To exclude direct cytotoxicity of pAXL×CD3ε on sarcoma cells, we performed a cell viability assay that allows the evaluation of metabolically active cells, in the absence of effector cells. We found that different concentrations of pAXL×CD3ε did not alter cell growth capability of cancer cells (Figure 2C), indicating that T lymphocytes are indeed required to induce the redirected cytotoxicity of sarcoma cell lines. Furthermore, we performed co-culture experiments on sarcoma cells stably expressing GFP in the presence of 2.5 µg/mL of pAXL×CD3ε. As expected, we observed a strong reduction of GFP signal via imaging analysis, confirming the cytotoxic activity of pAXL×CD3ε observed in our experimental models (Figrues 2D and S2A). Moreover, the percentage of viable cells and fluorescence quantification evaluated by flow cytometry led to the same result (Figrues 2E and S2B).

Taken together, our findings demonstrate that pAXL×CD3ε has an antitumor effect through the recruitment of cytotoxic T lymphocytes.

3.3. pAXL×CD3ε Triggers T-Lymphocyte Activation against Sarcoma Cells

Functional effects on PBMCs co-cultured with sarcoma cells at 10:1 E:T ratio, in the presence of increasing concentrations of pAXL×CD3ε or vehicle, were also evaluated after 72 h of treatment. As shown in Figure 3, we observed the upregulation of early and late T-cell surface activation markers (CD69 and CD25) in experiments performed in three different sarcoma cell lines, such as Rh-30, HT-1080 and CAL-72. Additionally, pAXL×CD3ε induced the release of inflammatory cytokine Interferon-γ (IFN-γ) and cytolytic enzyme, Granzyme B.

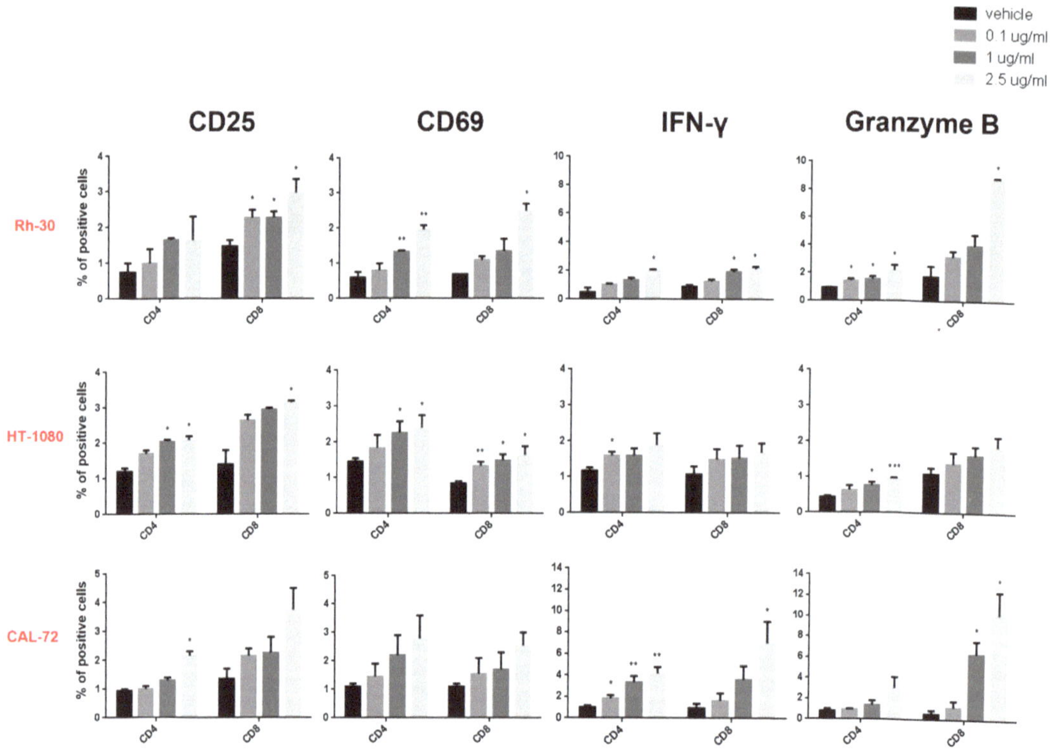

Figure 3. Functional experiments on CD4–CD8 gated T cells. Surface early and late activation markers (CD69 and CD25), cytokine release (IFN-γ) and cytolytic enzyme (Granzyme B) on CD4 and CD8-positive T lymphocytes from at least 3 donors co-cultured with Rh-30, HT-1080 and CAL-72 sarcoma cell lines at 10:1 E:T ratio, in the presence of different concentrations of pAXL×CD3ε. Each result is expressed as the mean value of triplicate experiments obtained from each donor. * $p < 0.0332$; ** $p < 0.0021$; *** $p < 0.0002$.

Consistent with their cytotoxic function, T lymphocytes were also positive for CD107a degranulation marker (Figure 4A,B).

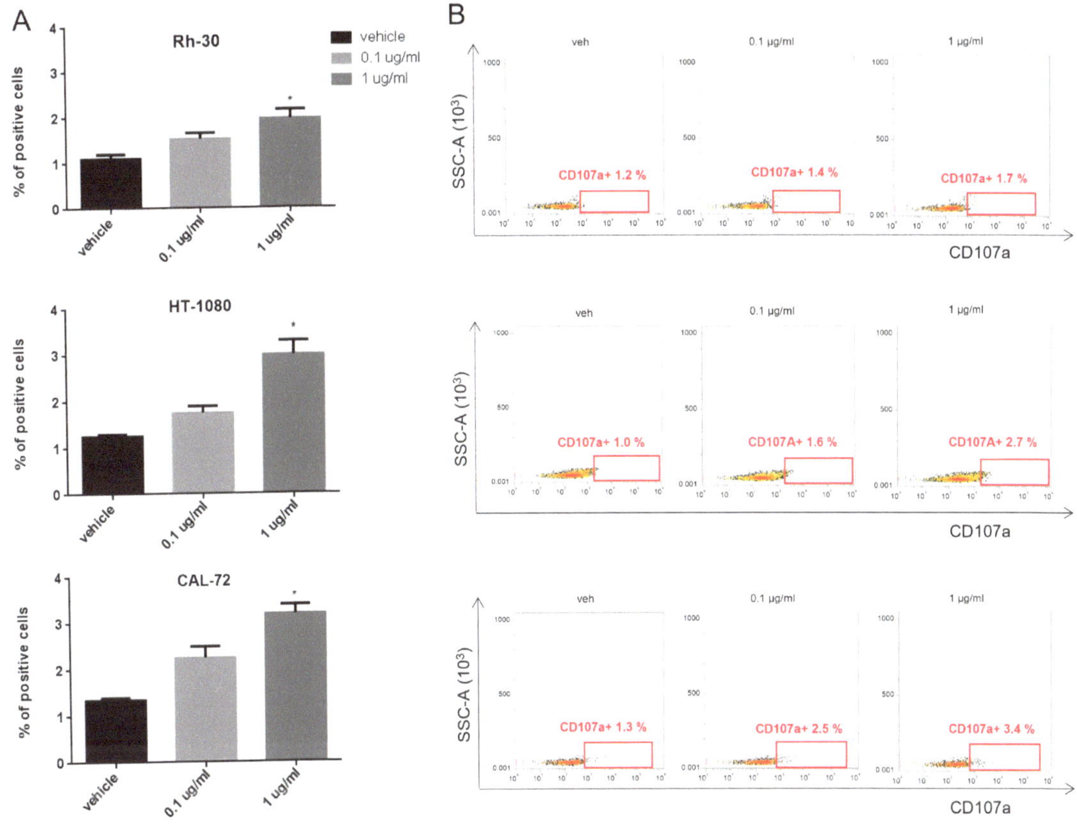

Figure 4. Degranulation assay on activated T lymphocytes. (**A**) Histogram quantification of CD107a-positive T cells from at least 3 donors, co-cultured with Rh-30, HT-1080 and CAL-72 sarcoma cell lines and treated with 0.1 µg/mL and 1 µg/mL of pAXL×CD3ε for 72 h. (**B**) Representative dot plots of CD107a-positive T lymphocytes analyzed by flow cytometry. Each result is expressed as the mean value of triplicate experiments obtained from each donor. * $p < 0.0332$.

These data indicate that pAXL×CD3ε produces a dose-dependent activation of T lymphocytes against AXL-positive sarcoma cells.

3.4. pAXL×CD3ε Increases Cytotoxicity Induced by Trabectedin

To verify if pAXL×CD3ε could make tumor cells more sensitive to conventional chemotherapeutic drugs, SAOS-2 was selected as the cell model to investigate the effect of redirected T-cell toxicity. Cells were co-treated with pAXL×CD3ε (1 µg/mL) and trabectedin (0.2 nM). After 72 h of treatment, an enhanced cytotoxic effect was observed for pAXL×CD3ε plus trabectedin, as compared to the effect induced by the single agents. In detail, pAXL×CD3ε increased cell death >20% in SAOS-2 cells compared to the effect induced by trabectedin alone (Figure 5A). The dot plots in Figure 5B provide a graphical overview of the reduction in cell viability (%). These data suggest a potential advantage induced by the combination of pAXL×CD3ε with chemotherapeutics commonly used for sarcoma therapy.

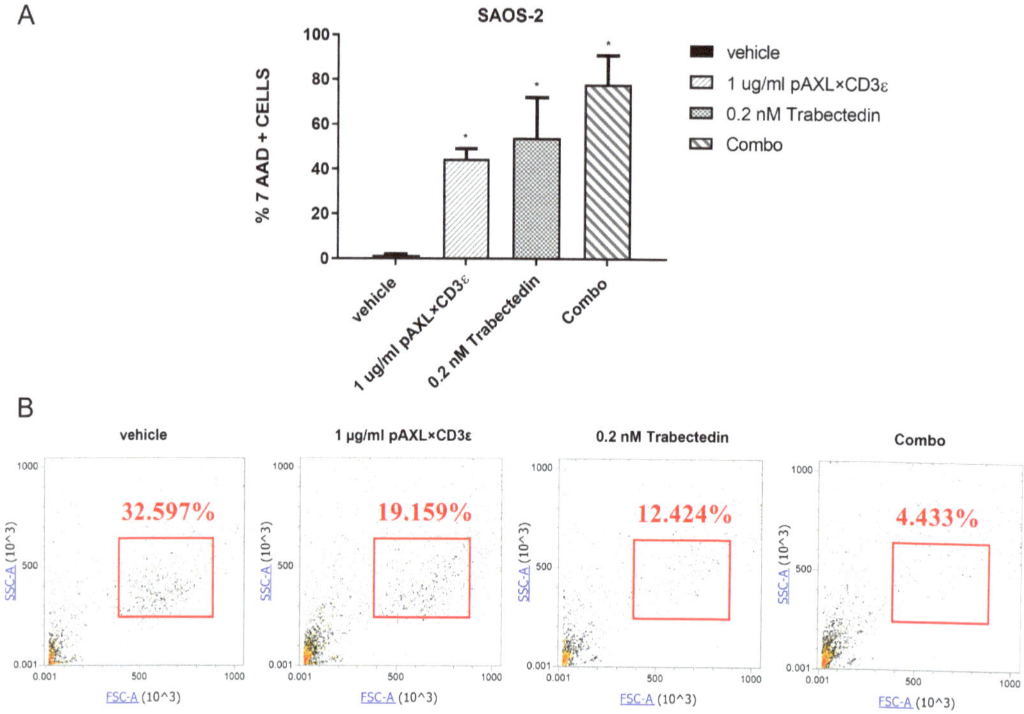

Figure 5. pAXL × CD3ε sensitizes chemotherapeutic drug activity in vitro. (**A**) Redirected cytotoxicity T-cell of pAXL × CD3ε in SAOS-2, analyzed in co-culture experiments using 1 μg/mL of pAXL × CD3ε, 0.2 nM of trabectedin and their combination (Combo) for 72 h. (**B**) Representative dot plots of SAOS-2 viability reduction (%) after treatment with pAXL × CD3ε and trabectedin alone or in combination. PBMCs were obtained from 3 healthy donors and results are expressed as the mean value of triplicate experiments from each donor. * $p < 0.0332$.

3.5. *pAXL × CD3ε In Vivo Activity*

The in vivo antitumor efficacy of pAXL × CD3ε was validated against human HT-1080 cell xenografts in NSG-immunocompromised mice (Figure 6A). A total of 10 xenografted mice were randomized to receive pAXL×CD3ε (0.1 mg/kg, five mice) or the vehicle alone (VEH, five mice) as the control group. A significant reduction of tumor growth was observed in NSG mice treated with pAXL × CD3ε as compared to VEH (Figure 6B). After 20 days from the cell engraftment, mice treated with pAXL × CD3ε showed a tumor volume of about 630 mm^3 versus 1200 mm^3 in the VEH-only group. This effect translated into a prolonged survival of treated animals (Figure 6C). To demonstrate the engraftment of human T-lymphocytes in these immunocompromised mice, flow cytometry analyses were performed on peripheral blood samples collected from mice on the day of sacrifice. An anti-CD3-fluorochrome-conjugated antibody was used for the staining, and T-cell engraftment was confirmed both in pAXL × CD3ε and VEH groups (Figure 6D). Retrieved xenografts from mice were homogenized and WB analysis was performed on the whole-cell protein extracts. The analysis revealed the induction of apoptotic processes, which was demonstrated by the increase of cleaved PARP and cleaved caspase-3 in treated mice as compared to VEH (Figure 6E). IHC analyses also highlighted the infiltration of CD3+ cells in tumor xenografts from mice treated with pAXL × CD3ε, thus demonstrating the effective engagement of T-lymphocytes at the tumor site (Figure 6F).

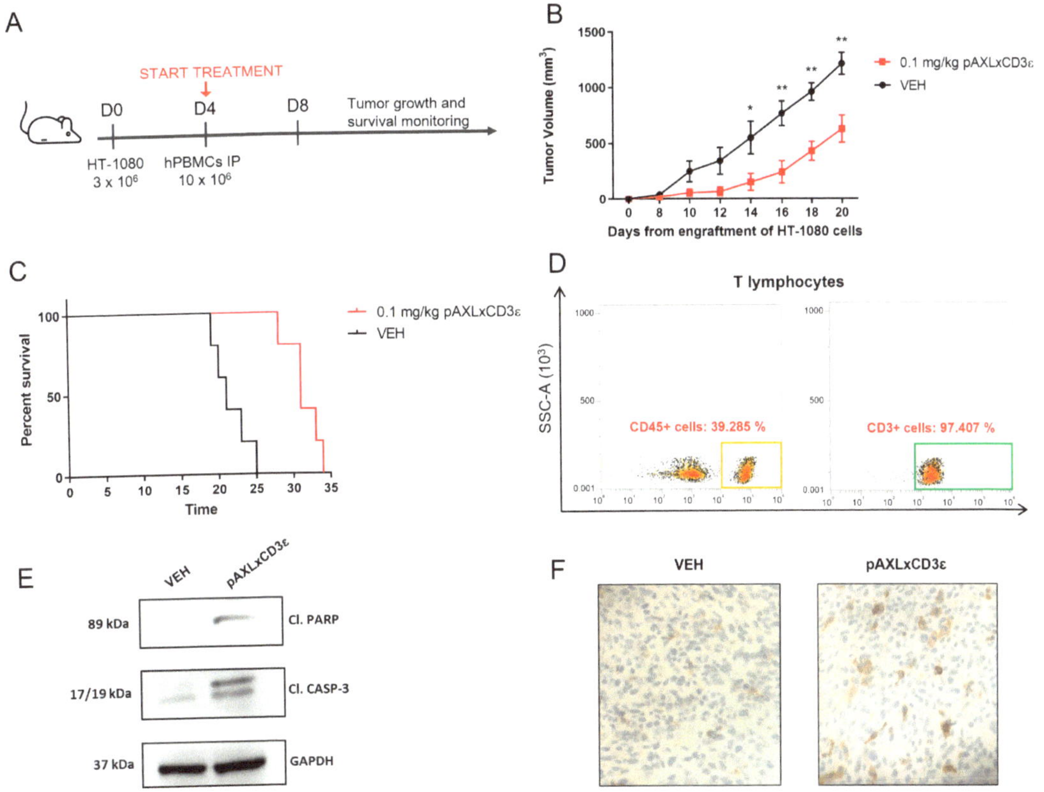

Figure 6. pAXL×CD3ε in vivo activity. (**A**) Experimental timeline of in vivo study on a sarcoma xenograft model. (**B**) Tumor volume curve of mice treated with pAXL×CD3ε 0.1 mg/kg or vehicle alone. Results obtained from each group have been compared at the same timepoint for statistical purposes. (**C**) Survival curves (Kaplan–Meier) of mice treated with pAXL×CD3ε or vehicle. (**D**) Representative FACS dot plots of T-cell engraftment evaluated on the day of sacrifice on intracardiac blood samples collected from mice. CD3-positive cells were evaluated on gated CD45-positive lymphocytes. (**E**) WB analysis of cleaved caspase-3 and cleaved PARP in whole-cell protein extracts from representative retrieved xenografts. GAPDH was used as a loading control. The uncropped blots are shown in Figure S3. (**F**) IHC staining of CD3 lymphocytes performed on tumors explanted from mice treated with vehicle or pAXL×CD3ε 0.1 mg/kg at 20-fold magnification. * $p < 0.0332$; ** $p < 0.0021$.

Based on these findings, pAXL×CD3ε demonstrates promising antitumor activity against sarcoma xenografts in vivo.

4. Discussion

Cancer immunotherapy based on T-cell engagement is a valuable therapeutical option and is in an advanced phase of clinical evaluation for different hematological malignancies [44–46]. While conventional mAbs bind the same antigen with both fragment antigen-binding (Fab) arms [47], BTCEs simultaneously bind a TAA on cancer cells and the epsilon (ε) subunit of CD3 on the T lymphocytes and, therefore, can efficiently trigger redirected T-cell cytotoxicity in an MHC-independent fashion [48,49]. This simultaneous engagement of the antigen on tumor cells and effector cells leads to an immunological synapse, resulting in T-cell activation and subsequent release of inflammatory cytokines and cytolytic molecules that lead to the killing of cancer cells [39,40,50].

Despite their demonstrated efficacy in patients with hematological malignancies, no BTCEs have been approved so far for the treatment of solid tumors [51]. There are, in fact, some main hurdles that can hamper the use of BTCEs in solid tumors: (i) on-target off-tumor toxicities due to the absence of specific TAAs; (ii) impaired anti-cancer activity due to the hostile and immunosuppressive tumor microenvironment (TME) that antagonizes T-cell infiltration into the tumor mass; (iii) reduced bioavailability and scarce penetration within a solid tumor mass [52,53]. In this scenario, the TME may play a relevant role in cancer progression and can influence the clinical management of these diseases [54–56], since it includes immune cells and stromal cells interacting with malignant cells through contact mechanisms or cytokines and subcellular structures, inducing both pro-tumor and antitumor activity. Among them, CD4+ and CD8+ T cells, together with NK cells, dendritic cells, and M1 tumor-associated macrophages (TAMs), promotes cell killing [57–59]. Novel approaches would aim to restore the immune function overcoming cancer suppressive effects [60]. In this light, different strategies are emerging by innovative protein-based scaffolds and novel targets [39,40,61].

Here, we assessed that AXL is highly expressed among a variety of sarcomas, confirming previous data performed on primary sarcoma samples [20,22,31,32,62,63] and representing a promising target for the development of innovative immunotherapeutic approaches, especially in chemo-refractory disease. Previous studies, using mAbs against AXL, have reported activity and manageable toxicity in sarcoma patients, suggesting its targeting potential also in the clinical setting [22,64]. Recently, different strategies based on mAbs and CAR-T cells showed encouraging results against AXL-expressing sarcomas and some of them are currently under clinical investigation. The safety and tolerability of CCT301-38 CAR-modified autologous T cells are being investigated in subjects with r/r sarcomas (NCT05128786). Patients with AXL gene alterations were also recruited for another phase I study to determine the safety, tolerability, pharmacokinetics, and antitumor effects induced by Mipasetamab Uzoptirine (ADCT-601) alone, or in combination with other anti-cancer drugs (NCT05389462). The immunogenicity and antitumor efficacy of BA3011, a conditionally active biologic (CAB) AXL-targeted antibody drug conjugate (CAB-AXL-ADC), is being investigated in a phase I/II study in different sarcoma subtypes, in monotherapy or combined with a PD-1 inhibitor (NCT03425279). Finally, a trial has been completed on different tumor types, including sarcoma, to investigate the safety and efficacy of Enapotamab Vedotin (HuMax-AXL-ADC), an AXL-specific antibody drug conjugate (NCT02988817).

On these premises, we focused on AXL as an immunotherapeutic target to be exploited in sarcoma treatment by BTCE-based strategy. We used an emerging protein therapeutic class, called Pronectins™ [36], taking advantage of their small size and low molecular weight to reach a higher concentration within the tumor tissue. Consistently, here we demonstrated that pAXL×CD3ε indeed redirects T cells toward AXL-expressing sarcoma cells, leading to a dose-dependent T-cell activation, with a consequent release of inflammatory cytokines and cytolytic molecules. Our results are in accordance with data recently reported, showing that pAXL×CD3ε exhibits cytotoxic effects on AXL-positive MDA-MB-231 cells and minimal cytotoxicity on AXL-negative CHO cells [38]. Moreover, we found enhanced cytotoxic effects as a function of increased concentrations of pAXL×CD3ε, which was highly promising taking into account the presence of immune cells in the TME of sarcomas. Furthermore, we demonstrated that the combination of pAXL×CD3ε with trabectedin, a conventional active chemotherapeutic drug, improved the cytotoxicity of sarcoma cells. Even if the use of conventional chemotherapeutics is under interindividual variability term of efficacy and toxicity [65–67], our data suggest the feasibility of combinatorial treatments and are consistent with preliminary reports, which showed anti-sarcoma activity of immunotherapy/chemotherapy combination [68]. Importantly, in our in vivo model, pAXL×CD3ε guaranteed the recruitment of T lymphocytes in the tumor site and significantly inhibited the growth of sarcoma xenografts, suggesting that this strategy has the potential to also control the fast-growing tumor cells in patients. These findings are of translational relevance, since conventional approaches are still largely unsuccessful, and the only favorable strategy at the present, for the treatment of this incurable disease, is represented by surgery in combination with pre- or post-surgery therapies.

Overall, we demonstrate that the first-in-class pAXL×CD3ε-based immunotherapy exerts significant anti-sarcoma activity in vitro and in vivo, and therefore represents a promising tool to be developed in the clinical setting, offering a novel opportunity to overcome the unmet need of long-term control of drug refractory disease.

5. Conclusions

Despite the identification of molecular mechanisms driving sarcoma genesis, as well as the discovery of key transcription factors, sarcoma treatment still represents a great challenge. The variability in response to current therapies can be ascribed to their heterogeneity and aggressive behavior. Taken together, our results indicate that AXL-targeting by the Pronectin™-based BTCE platform may represent a new-generation strategy for the treatment of this still-incurable disease.

Supplementary Materials: The following supporting information can be downloaded at: https://www.mdpi.com/article/10.3390/cancers15061647/s1, Figure S1: Interaction of pAXL×CD3ε with sarcoma cells; Figure S2: Redirected T-lymphocyte cytotoxicity by pAXL×CD3ε in sarcoma cell lines; Figure S3: Original Western blots.

Author Contributions: N.P., A.M., C.R., D.C., S.S., K.G., S.A. and G.J. performed experiments and/or analyzed the data. C.A.H. and R.C. developed the Pronectin™-based Bispecific T-Cell Engager. F.C. performed IHC analysis. N.S., L.G., M.T.D.M., M.A., G.N. and R.C. provided critical evaluation of experimental data of the manuscript. N.P., A.M., C.R., G.J., P.T. (Pierosandro Tagliaferri) and P.T. (Pierfrancesco Tassone) conceived the study and wrote the manuscript. P.T. (Pierosandro Tagliaferri) and P.T. (Pierfrancesco Tassone) supervised the study. All authors have read and agreed to the published version of the manuscript.

Funding: This manuscript has been supported by Institutional funds at Department of Experimental and Clinical Medicine (T27), University Magna Graecia of Catanzaro.

Institutional Review Board Statement: The animal study protocol was approved by the Institutional Ethics Committee of Magna Graecia University and all the procedures were performed according to the Institutional Animal-Care-approved protocols and guidelines (483/2020-PR, 18 May 2020).

Informed Consent Statement: Not applicable.

Data Availability Statement: The data presented in this study are available in this article and Supplementary Materials.

Acknowledgments: The authors are grateful to Leonardo Migale for his support in flow cytometry analyses.

Conflicts of Interest: R.C. and C.A.H. are associated with Protelica, Inc. that is the owner of Pronectins™ patent. The other authors declare no conflict of interest.

References

1. Agnoletto, C.; Caruso, C.; Garofalo, C. Heterogeneous circulating tumor cells in sarcoma: Implication for clinical practice. *Cancers* **2021**, *13*, 2189. [CrossRef] [PubMed]
2. Siegel, R.L.; Miller, K.D.; Wagle, N.S.; Jemal, A. Cancer statistics, 2023. *CA A Cancer J. Clin.* **2023**, *73*, 17–48. [CrossRef]
3. Damerell, V.; Pepper, M.S.; Prince, S. Molecular mechanisms underpinning sarcomas and implications for current and future therapy. *Signal Transduct. Target. Ther.* **2021**, *6*, 246. [CrossRef] [PubMed]
4. Petrella, A.; Storey, L.; Hulbert-Williams, N.J.; Fern, L.A.; Lawal, M.; Gerrand, C.; Windsor, R.; Woodford, J.; Bradley, J.; O'Sullivan, H. Fear of Cancer Recurrence in Patients with Sarcoma in the United Kingdom. *Cancers* **2023**, *15*, 956. [CrossRef]
5. Pushpam, D.; Garg, V.; Ganguly, S.; Biswas, B. Management of refractory pediatric sarcoma: Current challenges and future prospects. *OncoTargets Ther.* **2020**, *13*, 5093. [CrossRef] [PubMed]
6. Grimer, R.; Judson, I.; Peake, D.; Seddon, B. Guidelines for the management of soft tissue sarcomas. *Sarcoma* **2010**, *2010*, 506182. [CrossRef] [PubMed]

7. Italiano, A.; Mathoulin-Pelissier, S.; Cesne, A.L.; Terrier, P.; Bonvalot, S.; Collin, F.; Michels, J.J.; Blay, J.Y.; Coindre, J.M.; Bui, B. Trends in survival for patients with metastatic soft-tissue sarcoma. *Cancer* **2011**, *117*, 1049–1054. [CrossRef]
8. Evdokimova, V.; Gassmann, H.; Radvanyi, L.; Burdach, S.E. Current State of Immunotherapy and Mechanisms of Immune Evasion in Ewing Sarcoma and Osteosarcoma. *Cancers* **2023**, *15*, 272. [CrossRef]
9. Kohlmeyer, J.L.; Gordon, D.J.; Tanas, M.R.; Monga, V.; Dodd, R.D.; Quelle, D.E. CDKs in sarcoma: Mediators of disease and emerging therapeutic targets. *Int. J. Mol. Sci.* **2020**, *21*, 3018. [CrossRef]
10. Thiel, J.T.; Daigeler, A.; Kolbenschlag, J.; Rachunek, K.; Hoffmann, S. The Role of CDK Pathway Dysregulation and Its Therapeutic Potential in Soft Tissue Sarcoma. *Cancers* **2022**, *14*, 3380. [CrossRef]
11. Wilding, C.P.; Elms, M.L.; Judson, I.; Tan, A.-C.; Jones, R.L.; Huang, P.H. The landscape of tyrosine kinase inhibitors in sarcomas: Looking beyond pazopanib. *Expert Rev. Anticancer. Ther.* **2019**, *19*, 971–991. [CrossRef] [PubMed]
12. Guenther, L.M.; Dharia, N.V.; Ross, L.; Conway, A.; Robichaud, A.L.; Catlett, J.L.; Wechsler, C.S.; Frank, E.S.; Goodale, A.; Church, A.J. A Combination CDK4/6 and IGF1R Inhibitor Strategy for Ewing SarcomaCDK4/6 and IGF1R Inhibitors Are Synergistic in Ewing Sarcoma. *Clin. Cancer Res.* **2019**, *25*, 1343–1357. [CrossRef]
13. Lim, H.J.; Wang, X.; Crowe, P.; Goldstein, D.; Yang, J.-L. Targeting the PI3K/PTEN/AKT/mTOR pathway in treatment of sarcoma cell lines. *Anticancer. Res.* **2016**, *36*, 5765–5771. [CrossRef] [PubMed]
14. Bramwell, V.H. Pazopanib and the treatment palette for soft-tissue sarcoma. *Lancet* **2012**, *379*, 1854–1856. [CrossRef]
15. Sanaei, M.; Kavoosi, F. Histone deacetylases and histone deacetylase inhibitors: Molecular mechanisms of action in various cancers. *Adv. Biomed. Res.* **2019**, *8*, 63. [PubMed]
16. Grignani, G.; D'Ambrosio, L.; Pignochino, Y.; Palmerini, E.; Zucchetti, M.; Boccone, P.; Aliberti, S.; Stacchiotti, S.; Bertulli, R.; Piana, R. Trabectedin and olaparib in patients with advanced and non-resectable bone and soft-tissue sarcomas (TOMAS): An open-label, phase 1b study from the Italian Sarcoma Group. *Lancet Oncol.* **2018**, *19*, 1360–1371. [CrossRef] [PubMed]
17. Nathenson, M.J.; Conley, A.P.; Sausville, E. Immunotherapy: A new (and old) approach to treatment of soft tissue and bone sarcomas. *Oncologist* **2018**, *23*, 71–83. [CrossRef]
18. Peterson, C.; Denlinger, N.; Yang, Y. Recent advances and challenges in cancer immunotherapy. *Cancers* **2022**, *14*, 3972. [CrossRef]
19. Hattinger, C.M.; Patrizio, M.P.; Magagnoli, F.; Luppi, S.; Serra, M. An update on emerging drugs in osteosarcoma: Towards tailored therapies? *Expert Opin. Emerg. Drugs* **2019**, *24*, 153–171. [CrossRef]
20. Dantas-Barbosa, C.; Lesluyes, T.; Loarer, F.L.; Chibon, F.; Treilleux, I.; Coindre, J.-M.; Meeus, P.; Brahmi, M.; Bally, O.; Ray-Coquard, I. Expression and role of TYRO3 and AXL as potential therapeutical targets in leiomyosarcoma. *Br. J. Cancer* **2017**, *117*, 1787–1797. [CrossRef]
21. Lamhamedi-Cherradi, S.-E.; Mohiuddin, S.; Mishra, D.K.; Krishnan, S.; Velasco, A.R.; Vetter, A.M.; Pence, K.; McCall, D.; Truong, D.D.; Cuglievan, B. Transcriptional activators YAP/TAZ and AXL orchestrate dedifferentiation, cell fate, and metastasis in human osteosarcoma. *Cancer Gene Ther.* **2021**, *28*, 1325–1338. [CrossRef] [PubMed]
22. Zhu, C.; Wei, Y.; Wei, X. AXL receptor tyrosine kinase as a promising anti-cancer approach: Functions, molecular mechanisms and clinical applications. *Mol. Cancer* **2019**, *18*, 153. [CrossRef] [PubMed]
23. Kimani, S.G.; Kumar, S.; Bansal, N.; Singh, K.; Kholodovych, V.; Comollo, T.; Peng, Y.; Kotenko, S.V.; Sarafianos, S.G.; Bertino, J.R. Small molecule inhibitors block Gas6-inducible TAM activation and tumorigenicity. *Sci. Rep.* **2017**, *7*, 43908. [CrossRef] [PubMed]
24. Goyette, M.-A.; Côté, J.-F. AXL receptor tyrosine kinase as a promising therapeutic target directing multiple aspects of cancer progression and metastasis. *Cancers* **2022**, *14*, 466. [CrossRef] [PubMed]
25. Auyez, A.; Sayan, A.E.; Kriajevska, M.; Tulchinsky, E. AXL receptor in cancer metastasis and drug resistance: When normal functions go askew. *Cancers* **2021**, *13*, 4864. [CrossRef]
26. Wium, M.; Ajayi-Smith, A.F.; Paccez, J.D.; Zerbini, L.F. The role of the receptor tyrosine kinase Axl in carcinogenesis and development of therapeutic resistance: An overview of molecular mechanisms and future applications. *Cancers* **2021**, *13*, 1521. [CrossRef]
27. Wu, X.; Liu, X.; Koul, S.; Lee, C.Y.; Zhang, Z.; Halmos, B. AXL kinase as a novel target for cancer therapy. *Oncotarget* **2014**, *5*, 9546. [CrossRef]
28. Cerchia, L.; Esposito, C.L.; Camorani, S.; Rienzo, A.; Stasio, L.; Insabato, L.; Affuso, A.; De Franciscis, V. Targeting Axl with an high-affinity inhibitory aptamer. *Mol. Ther.* **2012**, *20*, 2291–2303. [CrossRef]
29. Kim, S.; Kim, K.-C.; Lee, C. Mistletoe (*Viscum album*) extract targets Axl to suppress cell proliferation and overcome cisplatin-and erlotinib-resistance in non-small cell lung cancer cells. *Phytomedicine* **2017**, *36*, 183–193. [CrossRef]
30. Liu, R.; Gong, M.; Li, X.; Zhou, Y.; Gao, W.; Tulpule, A.; Chaudhary, P.M.; Jung, J.; Gill, P.S. Induction, regulation, and biologic function of Axl receptor tyrosine kinase in Kaposi sarcoma. *Blood J. Am. Soc. Hematol.* **2010**, *116*, 297–305. [CrossRef]
31. Fleuren, E.D.; Hillebrandt-Roeffen, M.H.; Flucke, U.E.; Te Loo, D.M.W.; Boerman, O.C.; van der Graaf, W.T.; Versleijen-Jonkers, Y.M. The role of AXL and the in vitro activity of the receptor tyrosine kinase inhibitor BGB324 in Ewing sarcoma. *Oncotarget* **2014**, *5*, 12753. [CrossRef] [PubMed]
32. Han, J.; Tian, R.; Yong, B.; Luo, C.; Tan, P.; Shen, J.; Peng, T. Gas6/Axl mediates tumor cell apoptosis, migration and invasion and predicts the clinical outcome of osteosarcoma patients. *Biochem. Biophys. Res. Commun.* **2013**, *435*, 493–500. [CrossRef] [PubMed]
33. Cappuccilli, G.; Crea, R.; Shen, R.; Hokanson, C.A.; Kirk, P.B.; Liston, D.R. Universal Fibronectin Type III Binding-Domain Libraries. U.S. Patent US20090176654A1, 9 July 2009.

34. Chandler, P.G.; Buckle, A.M. Development and differentiation in monobodies based on the fibronectin type 3 domain. *Cells* **2020**, *9*, 610. [CrossRef] [PubMed]
35. Lipovšek, D.; Lippow, S.M.; Hackel, B.J.; Gregson, M.W.; Cheng, P.; Kapila, A.; Wittrup, K.D. Evolution of an interloop disulfide bond in high-affinity antibody mimics based on fibronectin type III domain and selected by yeast surface display: Molecular convergence with single-domain camelid and shark antibodies. *J. Mol. Biol.* **2007**, *368*, 1024–1041. [CrossRef] [PubMed]
36. Mintz, C.S.; Crea, R. Protein scaffolds: The next generation of protein therapeutics? *Bioprocess Int.* **2013**, *11*, 40–48.
37. Wu, Y.; Yi, M.; Zhu, S.; Wang, H.; Wu, K. Recent advances and challenges of bispecific antibodies in solid tumors. *Exp. Hematol. Oncol.* **2021**, *10*, 56. [CrossRef]
38. Hokanson, C.A.; Zacco, E.; Cappuccilli, G.; Odineca, T.; Crea, R. AXL-Receptor Targeted 14FN3 Based Single Domain Proteins (Pronectins™) from 3 Synthetic Human Libraries as Components for Exploring Novel Bispecific Constructs against Solid Tumors. *Biomedicines* **2022**, *10*, 3184. [CrossRef]
39. Caracciolo, D.; Riillo, C.; Ballerini, A.; Gaipa, G.; Lhermitte, L.; Rossi, M.; Botta, C.; Duroyon, E.; Grillone, K.; Cantafio, M.E.G. Therapeutic afucosylated monoclonal antibody and bispecific T-cell engagers for T-cell acute lymphoblastic leukemia. *J. Immunother. Cancer* **2021**, *9*, e002026. [CrossRef]
40. Riillo, C.; Caracciolo, D.; Grillone, K.; Polerà, N.; Tuccillo, F.M.; Bonelli, P.; Juli, G.; Ascrizzi, S.; Scionti, F.; Arbitrio, M. A Novel Bispecific T-Cell Engager (CD1a x CD3ε) BTCE Is Effective against Cortical-Derived T Cell Acute Lymphoblastic Leukemia (T-ALL) Cells. *Cancers* **2022**, *14*, 2886. [CrossRef]
41. Grillone, K.; Riillo, C.; Rocca, R.; Ascrizzi, S.; Spanò, V.; Scionti, F.; Polerà, N.; Maruca, A.; Barreca, M.; Juli, G. The New Microtubule-Targeting Agent SIX2G Induces Immunogenic Cell Death in Multiple Myeloma. *Int. J. Mol. Sci.* **2022**, *23*, 10222. [CrossRef]
42. Neri, P.; Tagliaferri, P.; Di Martino, M.T.; Calimeri, T.; Amodio, N.; Bulotta, A.; Ventura, M.; Eramo, P.O.; Viscomi, C.; Arbitrio, M. In vivo anti-myeloma activity and modulation of gene expression profile induced by valproic acid, a histone deacetylase inhibitor. *Br. J. Haematol.* **2008**, *143*, 520–531.
43. Cosco, D.; Bulotta, A.; Ventura, M.; Celia, C.; Calimeri, T.; Perri, G.; Paolino, D.; Costa, N.; Neri, P.; Tagliaferri, P. In vivo activity of gemcitabine-loaded PEGylated small unilamellar liposomes against pancreatic cancer. *Cancer Chemother. Pharmacol.* **2009**, *64*, 1009–1020. [CrossRef]
44. Li, L.; Wang, Y. Recent updates for antibody therapy for acute lymphoblastic leukemia. *Exp. Hematol. Oncol.* **2020**, *9*, 33. [CrossRef] [PubMed]
45. Topp, M.; Stelljes, M.; Zugmaier, G.; Barnette, P.; Heffner, L.; Trippett, T.; Duell, J.; Bargou, R.; Holland, C.; Benjamin, J. Blinatumomab retreatment after relapse in patients with relapsed/refractory B-precursor acute lymphoblastic leukemia. *Leukemia* **2018**, *32*, 562–565. [CrossRef]
46. Waldman, A.D.; Fritz, J.M.; Lenardo, M.J. A guide to cancer immunotherapy: From T cell basic science to clinical practice. *Nat. Rev. Immunol.* **2020**, *20*, 651–668. [CrossRef]
47. Shin, C.; Kim, S.S.; Jo, Y.H. Extending traditional antibody therapies: Novel discoveries in immunotherapy and clinical applications. *Mol. Ther.-Oncolytics* **2021**, *22*, 166–179. [CrossRef]
48. Slaney, C.Y.; Wang, P.; Darcy, P.K.; Kershaw, M.H. CARs versus BiTEs: A Comparison between T Cell–Redirection Strategies for Cancer TreatmentCAR T-cell and BiTE Therapies for Cancers. *Cancer Discov.* **2018**, *8*, 924–934. [CrossRef]
49. Yu, F.; Gao, Y.; Wu, Y.; Dai, A.; Wang, X.; Zhang, X., Liu, G.; Xu, Q.; Chen, D. Combination of a Novel Fusion Protein CD3εζ28 and Bispecific T Cell Engager Enhances the Persistance and Anti-Cancer Effects of T Cells. *Cancers* **2022**, *14*, 4947. [CrossRef] [PubMed]
50. Wu, Z.; Cheung, N. T cell engaging bispecific antibody (T-BsAb): From technology to therapeutics. *Pharmacol. Ther.* **2018**, *182*, 161–175. [CrossRef] [PubMed]
51. Clynes, R.A.; Desjarlais, J.R. Redirected T cell cytotoxicity in cancer therapy. *Annu. Rev. Med.* **2019**, *70*, 437–450. [CrossRef] [PubMed]
52. Middelburg, J.; Kemper, K.; Engelberts, P.; Labrijn, A.F.; Schuurman, J.; van Hall, T. Overcoming challenges for CD3-bispecific antibody therapy in solid tumors. *Cancers* **2021**, *13*, 287. [CrossRef] [PubMed]
53. Grünewald, T.G.; Alonso, M.; Avnet, S.; Banito, A.; Burdach, S.; Cidre-Aranaz, F.; Di Pompo, G.; Distel, M.; Dorado-Garcia, H.; Garcia-Castro, J. Sarcoma treatment in the era of molecular medicine. *EMBO Mol. Med.* **2020**, *12*, e11131. [CrossRef] [PubMed]
54. Deng, J.; Zeng, W.; Kong, W.; Shi, Y.; Mou, X. The study of sarcoma microenvironment heterogeneity associated with prognosis based on an immunogenomic landscape analysis. *Front. Bioeng. Biotechnol.* **2020**, *8*, 1003. [CrossRef] [PubMed]
55. Maman, S.; Witz, I.P. A history of exploring cancer in context. *Nat. Rev. Cancer* **2018**, *18*, 359–376. [CrossRef] [PubMed]
56. Quail, D.F.; Joyce, J.A. Microenvironmental regulation of tumor progression and metastasis. *Nat. Med.* **2013**, *19*, 1423–1437. [CrossRef]
57. Ye, H.; Hu, X.; Wen, Y.; Tu, C.; Hornicek, F.; Duan, Z.; Min, L. Exosomes in the tumor microenvironment of sarcoma: From biological functions to clinical applications. *J. Nanobiotechnol.* **2022**, *20*, 403. [CrossRef]
58. Zhu, S.; Yi, M.; Wu, Y.; Dong, B.; Wu, K. Roles of tumor-associated macrophages in tumor progression: Implications on therapeutic strategies. *Exp. Hematol. Oncol.* **2021**, *10*, 60. [CrossRef]
59. Mikolajczyk, A.; Mitula, F.; Popiel, D.; Kaminska, B.; Wieczorek, M.; Pieczykolan, J. Two-Front War on Cancer—Targeting TAM Receptors in Solid Tumour Therapy. *Cancers* **2022**, *14*, 2488. [CrossRef]

60. Clemente, O.; Ottaiano, A.; Di Lorenzo, G.; Bracigliano, A.; Lamia, S.; Cannella, L.; Pizzolorusso, A.; Di Marzo, M.; Santorsola, M.; De Chiara, A. Is immunotherapy in the future of therapeutic management of sarcomas? *J. Transl. Med.* **2021**, *19*, 173. [CrossRef]
61. Gebauer, M.; Skerra, A. Engineered protein scaffolds as next-generation antibody therapeutics. *Curr. Opin. Chem. Biol.* **2009**, *13*, 245–255. [CrossRef]
62. Van Renterghem, B.; Wozniak, A.; Castro, P.G.; Franken, P.; Pencheva, N.; Sciot, R.; Schöffski, P. Enapotamab Vedotin, an AXL-Specific Antibody-Drug Conjugate, Demonstrates Antitumor Efficacy in Patient-Derived Xenograft Models of Soft Tissue Sarcoma. *Int. J. Mol. Sci.* **2022**, *23*, 7493. [CrossRef] [PubMed]
63. May, C.D.; Garnett, J.; Ma, X.; Landers, S.M.; Ingram, D.R.; Demicco, E.G.; Al Sannaa, G.A.; Vu, T.; Han, L.; Zhang, Y. AXL is a potential therapeutic target in dedifferentiated and pleomorphic liposarcomas. *BMC Cancer* **2015**, *15*, 901. [CrossRef] [PubMed]
64. Rios-Doria, J.; Favata, M.; Lasky, K.; Feldman, P.; Lo, Y.; Yang, G.; Stevens, C.; Wen, X.; Sehra, S.; Katiyar, K. A potent and selective dual inhibitor of AXL and MERTK possesses both immunomodulatory and tumor-targeted activity. *Front. Oncol.* **2020**, *10*, 598477. [CrossRef] [PubMed]
65. Arbitrio, M.; Di Martino, M.T.; Scionti, F.; Barbieri, V.; Pensabene, L.; Tagliaferri, P. Pharmacogenomic profiling of ADME gene variants: Current challenges and validation perspectives. *High-Throughput* **2018**, *7*, 40. [CrossRef] [PubMed]
66. Arbitrio, M.; Scionti, F.; Di Martino, M.T.; Caracciolo, D.; Pensabene, L.; Tassone, P.; Tagliaferri, P. Pharmacogenomics biomarker discovery and validation for translation in clinical practice. *Clin. Transl. Sci.* **2021**, *14*, 113–119. [CrossRef] [PubMed]
67. DeRidder, L.; Rubinson, D.A.; Langer, R.; Traverso, G. The past, present, and future of chemotherapy with a focus on individualization of drug dosing. *J. Control. Release* **2022**, *352*, 840–860. [CrossRef]
68. Engelsen, A.S.; Lotsberg, M.L.; Abou Khouzam, R.; Thiery, J.-P.; Lorens, J.B.; Chouaib, S.; Terry, S. Dissecting the role of AXL in cancer immune escape and resistance to immune checkpoint inhibition. *Front. Immunol.* **2022**, *13*, 869676. [CrossRef]

cancers MDPI

Article

What Is the Significance of Indeterminate Pulmonary Nodules in High-Grade Soft Tissue Sarcomas? A Retrospective Cohort Study

Marcus J. Brookes [1,*], Corey D. Chan [1,2], Timothy P. Crowley [1], Maniram Ragbir [1], Thomas Beckingsale [1], Kanishka M. Ghosh [1] and Kenneth S. Rankin [1,2]

1 North of England Bone and Soft Tissue Tumour Service, Royal Victoria Infirmary, Queen Victoria Road, Newcastle upon Tyne NE1 4LP, UK; corey.chan@newcastle.ac.uk (C.D.C.); timothy.crowley1@nhs.net (T.P.C.); maniram.ragbir@nhs.net (M.R.); t.beckingsale@nhs.net (T.B.); kanishka.ghosh1@nhs.net (K.M.G.); kenneth.rankin@newcastle.ac.uk (K.S.R.)

2 Translational and Clinical Research Institute, Newcastle University, Newcastle upon Tyne NE1 7RU, UK

* Correspondence: marcus.brookes1@nhs.net

Simple Summary: Sarcomas are rare cancers; they can arise anywhere in the body and most often spread to the lungs. When patients are diagnosed, they have a scan of the chest to look for this. The scan often finds small nodules whereby we cannot be certain whether they are cancer or not; these are called indeterminate pulmonary nodules or IPNs. We do not yet understand what the presence of IPNs means for patients with high-grade sarcomas in their soft tissues, although we know that some of these reveal themselves later on as being a spreading of the cancer. Currently, patients with IPNs normally have repeat scans a number of months down the line to see whether they have changed in size, suggesting that they may be cancer. This study has identified a number of different characteristics that make these IPNs more likely to be cancer.

Abstract: Background: Sarcomas are rare, aggressive cancers which frequently metastasise to the lungs. Following diagnosis, patients typically undergo staging by means of a CT scan of their chest. This often identifies indeterminate pulmonary nodules (IPNs), but the significance of these in high-grade soft tissue sarcoma (STS) is unclear. Identifying whether these are benign or malignant is important for clinical decision making. This study analyses the clinical relevance of IPNs in high-grade STS. Methods: All patients treated at our centre for high-grade soft tissue sarcoma between 2010 and 2020 were identified from a prospective database. CT scans and their reports were reviewed, and survival data were collected from patient records. Results: 389 suitable patients were identified; 34.4% had IPNs on their CT staging scan and 20.1% progressed into lung metastases. Progression was more likely with IPNs ≥ 5 mm in diameter ($p = 0.006$), multiple IPNs ($p = 0.013$) or bilateral IPNs ($p = 0.022$), as well as in patients with primaries ≥ 5 cm ($p = 0.014$), grade 3 primaries ($p = 0.009$) or primaries arising deep to the fascia ($p = 0.041$). The median time to progression was 143 days. IPNs at diagnosis were associated with an increased risk of developing lung metastases and decreased OS in patients with grade 3 STS ($p = 0.0019$ and $p = 0.0016$, respectively); this was not observed in grade 2 patients. Conclusions: IPNs at diagnosis are associated with significantly worse OS in patients with grade 3 STS. It is crucial to consider the primary tumour as well as the IPNs when considering the risk of progression. Surveillance CT scans should be carried out within 6 months.

Keywords: indeterminate pulmonary nodules (IPNs); sarcoma; soft tissue sarcoma; survival; metastasis; metastases

Citation: Brookes, M.J.; Chan, C.D.; Crowley, T.P.; Ragbir, M.; Beckingsale, T.; Ghosh, K.M.; Rankin, K.S. What Is the Significance of Indeterminate Pulmonary Nodules in High-Grade Soft Tissue Sarcomas? A Retrospective Cohort Study. *Cancers* **2023**, *15*, 3531. https://doi.org/10.3390/cancers15133531

Academic Editor: Catrin Sian Rutland

Received: 13 April 2023
Revised: 29 June 2023
Accepted: 5 July 2023
Published: 7 July 2023

1. Introduction

Sarcomas are rare [1], aggressive tumours arising from mesenchymal cells which most commonly metastasise to the lungs [2]; up to 30% of soft tissue sarcoma (STS) patients present with synchronous metastases [3–6]. Following diagnosis, guidelines dictate that patients should undergo staging by means of a CT scan of their chest [7,8]. Whilst CT scanning has good utility for the identification of metastases, allowing important treatment decisions to be made, they often identify indeterminate pulmonary nodules (IPNs), but the significance of these is currently unclear.

The prevalence of IPNs varies widely in the literature; Rissing et al. reported IPNs at diagnosis in 21% of the 331 sarcoma patients whom they followed prospectively [5], whilst a retrospective review by Saifuddin et al. identified IPNs in up to 49.5% of 200 STS patients [6]. Whilst there is variation in CT scanning modalities, differences in technology and variation in radiological reporting that may explain some of this; there remains uncertainty surrounding IPNs. There is no agreed definition of what an IPN is and what metastatic disease is based on CT imaging, and as such, there is a high degree of variability; non-calcified nodules either <5 mm or <10 mm in size are frequently used [3,5,6]. Arguably more importantly, with regard to the significance of IPNs in STS, Rissing et al. demonstrated that 28% of IPNs progressed into overt lung metastases, with IPNs > 5 mm in size associated with decreased disease-free survival at 3 years [5]. Nakamura et al. analysed the factors associated with an increased likelihood of nodules being malignant rather than benign, finding that larger nodules, $n > 1$, bilateral distribution and first detection during follow up rather than at screening were more likely to prove malignant [3].

Differentiation between benign and metastatic lung nodules is of high clinical importance, as it helps to guide treatment decisions which have proven impacts on survival; Billingsley et al. demonstrated the complete resection of metastatic disease as being the most important prognostic factor in STS patients with pulmonary metastases [4]. Not only is this important for clinical decision making, but the detection of IPNs can add significant stress to patients who may not actually have metastatic disease. Whilst new technologies such as positron emission tomography (PET) show promise in the detection of metastatic disease [9], a study by Fortes et al. (including sarcoma patients) reported a 30% false negative rate in the detection of metastatic pulmonary nodules [10]. As such, there is an unmet clinical need both for the detection and interpretation of pulmonary nodules in patients with high-grade STS.

Whilst IPN rates have previously been reported in STS, we aimed to perform the largest, most in-depth analysis to date of IPNs, their progression and detection, as well as looking at their effect on patient overall survival (OS), focusing specifically on high-grade STS.

2. Materials and Methods

A retrospective cohort study of consecutive patients treated for high-grade soft tissue sarcoma in the North of England Bone and Soft Tissue Tumour Service between 1 January 2010 and 1 May 2020 was performed, following identification from a prospectively maintained database. This study was registered with the local institutional review board (number 13952). Low-grade sarcomas, as well as tumours arising from visceral, retroperitoneal and intracranial locations, were excluded from the study. Patients with no available staging scans for review were also excluded. All grading and classification of tumour subtypes were conducted by expert sarcoma pathologists, according to the WHO classification of bone and soft tissue tumours [11]. Tumours were considered to be high-grade if they were scored as being grade 2 or 3 using the French Fédération Nationale des Centres de Lutte Contre le Cancer (FNCLCC) grading system [12]. As the FNCLCC grading system does not apply to all sarcoma subtypes, sarcomas reported as being morphologically high-grade in the pathology report were also included.

Patients underwent staging as part of their initial work up; this was usually performed at our centre using a Scanner Somatom Definition AS by Siemens, Erlangen, Germany—3 mm slices, although this was occasionally performed at local hospitals due to the logistics of travel. The scans and reports were reviewed by the lead author (MJB), whilst blinded to the outcomes at this point. Nodules were classified as metastatic, benign or indeterminate, with indeterminate pulmonary nodules defined as non-calcified nodules <10 mm in maximum diameter [5]. Follow up imaging was reviewed to determine whether these nodules remained unchanged or progressed, as well as to monitor for the development of a new disease. Survival data were also collected by reviewing patients' clinical notes and collecting information including age, gender, tumour location, histological subtype, grade, size and depth relative to the fascia.

Differences between groups were compared using independent T-tests and Fisher's exact test, accordingly, using SPSS statistics (Version 28.0, IBM Corp, Armonk, NY, USA). The time taken to develop lung metastases in different groups was analysed using Kaplan–Meier plots and log rank tests to calculate *p* values. The influence of other risk factors on the progression of IPNs and the development of metastases was analysed as part of both univariate and multivariate models using the Cox proportional hazards model. Survival analysis was conducted using R statistics (version 4.2.1, R Foundation for Statistical Computing, Vienna, Austria).

3. Results

A total of 389 patients were identified as being suitable for the study and their basic clinical information is displayed in Table 1. Patients had a mean age of 61.9 years (range 2–97), with a male predominance of 62.4% of the cohort. A wide range of histological subtypes were included, with undifferentiated pleomorphic sarcoma (UPS) (25.7%) and myxofibrosarcoma (25.7%) being the most frequent subtypes. Only high-grade sarcomas were included, with 71.5% being grade 3, and the remaining 28.5% being grade 2.

After reviewing the staging CT scans and reports, 222 (57.1%) patients had no evidence of lung metastases or IPNs, 134 (34.4%) had IPNs and 33 (8.5%) had synchronous lung metastases (Figure 1). Of the patients with no IPNs or lung metastases upon CT staging, 62 (27.9%) went on to develop metastases at a later date, with a median time to development of lung metastases of 448 days (range 87–1998). A greater percentage of patients with IPNs at diagnosis went on to develop lung metastases, with 48 (35.8%) developing metastatic disease after a median of 249 days (range 13–1676). Patients with IPNs at diagnosis appeared to be at a higher risk of developing lung metastases than patients without, although this did not reach significance (Figure 2A, KM $p = 0.14$, HR 1.33, HR $p = 0.14$). When patients with grade 2 primaries were excluded, this became significant (Figure 2B, KM $p = 0.019$, HR 1.62, HR $p = 0.021$).

Of those patients with IPNs from the initial CT staging developing metastases in the future, 27 (56.3%) had progression of these IPNs, whereas the remaining 11 (43.7%) had new lesions, suggesting that only 20.1% of IPNs progressed. Of those with grade 3 primaries, progression occurred in 27.2%. The progression of known IPNs occurred sooner than the development of new lesions, with a median time to progression of 143 days (range 13–557) compared to 409 days (range 96–1676) ($p < 0.001$). Table 2 displays the distribution of risk factors amongst IPNs which did and did not progress; the progression group contained a significantly higher proportion of multiple IPNs ($p = 0.010$), IPNs ≥ 5 mm in diameter ($p = 0.008$), bilateral IPNs ($p = 0.029$), primaries ≥ 5 cm ($p = 0.010$), grade 3 primaries ($p = 0.002$) and primaries arising deep to the fascia ($p = 0.032$). When the above risk factors for IPN progression were analysed using a Cox regression model, all demonstrated a significant increase in the risk of progression to lung metastases (Table 3). When analysed at the multivariate level, IPNs ≥ 5 mm in diameter at diagnosis and grade 3 primaries retained significance (HR = 2.37, $p = 0.03$ and HR = 6.07, $p = 0.015$, respectively). Supplementary Table S1 shows the clinical details of all of the patients with IPNs that progressed into metastases; only two patients had a grade 2 sarcoma, with the rest having grade 3 primaries, whilst only three patients had a primary sarcoma <5 cm. The median age of patients whose IPNs progressed was 68 years (range 11–87) and there was no significant difference in the average age between the group of patients whose IPNs progressed and those whose remained stable ($p = 0.531$). Of these 27 patients, the progression of IPNs was detected on interval CT scans in 9 patients, and six were detected upon routine surveillance chest X-ray (CXR) prior to confirmation with CT scanning of the chest; the remainder were detected on scans carried out for patients who were acutely unwell or during restaging following the detection of metastatic disease elsewhere.

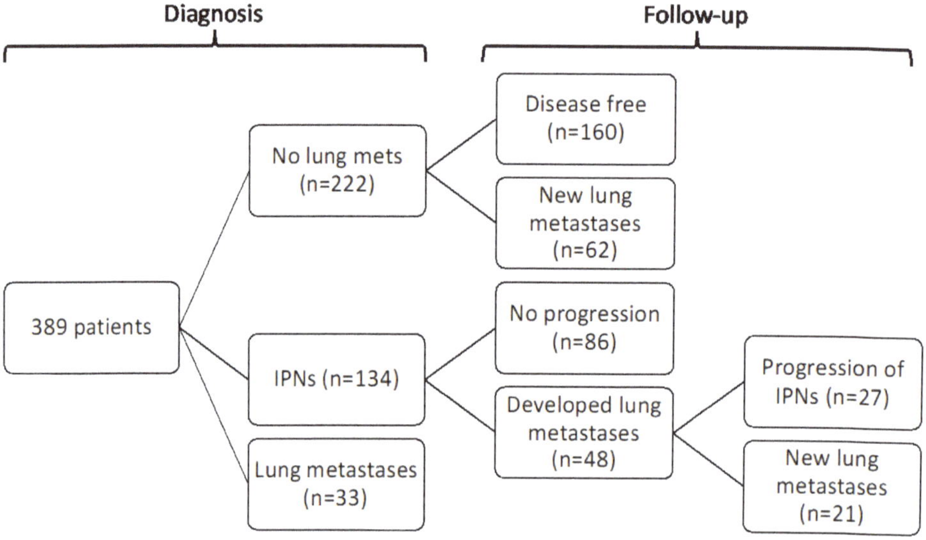

Figure 1. Flowchart depicting patients' IPN statuses at diagnosis and their progression during follow up.

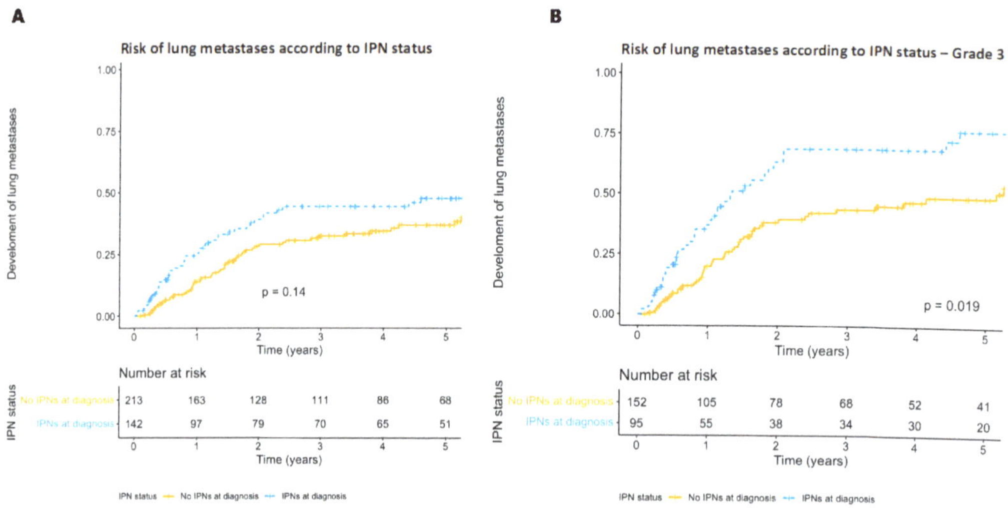

Figure 2. (**A**) Kaplan–Meier graph depicting the cumulative hazard of the development of lung metastases in patients with and without IPNs on the CT staging scan. Patients with IPNs at diagnosis appeared to be at higher risk of developing lung metastases, but this did not reach significance (p = 0.14, HR 1.33, HR p = 0.14). (**B**) Kaplan–Meier graph depicting the risk of lung metastases according to IPN status at diagnosis after patients with grade 2 primaries were excluded. Patients with IPNs at diagnosis are at a significantly higher risk of developing lung metastases (p = 0.019, HR 1.62, HR p = 0.021).

Table 1. Summary of demographic and basic clinical information.

Characteristic	
Mean age, years (range)	61.9 (2–97)
Gender, number (%)	
Male	243 (62.4%)
Female	146 (37.6%)
Location, number (%)	
Lower limb	223 (57.3%)
Upper limb	69 (17.7%)
Trunk	87 (22.4%)
Head and neck	10 (2.6%)
Histological subtype, number (%)	
Angiosarcoma	38 (9.8%)
Extra-skeletal Ewing sarcoma	9 (2.3%)
Leiomyosarcoma	42 (10.8%)
Liposarcoma	32 (8.2%)
MPNST	0 (2.6%)
Myxofibrosarcoma	100 (25.7%)
Rhabdomyosarcoma	19 (4.9%)
Synovial sarcoma	28 (7.2%)
Undifferentiated pleomorphic sarcoma	100 (25.7%)
Other	11 (3.8%)
FNCLCC grade, number (%)	
Grade 2	111 (28.5%)
Grade 3	278 (71.5%)
Size	
<5 cm	121 (31.1%)
≥5 cm	268 (68.9%)
Depth relative to fascia	
Superficial	179 (46.0%)
Deep	210 (54.0%)

Table 2. Distribution of risk factors amongst patients with IPNs that did and did not progress, *p* value calculated using Fisher's exact test. * indicates significance.

		Clinical Outcome of IPNs (n = 134)		
		Stable (n = 107)	**Progressed (n = 27)**	***p* Value**
IPN number	1	66	9	**0.010** *
	>1	41	18	
IPN size ≥5 mm	1	74	11	**0.008** *
	>1	33	16	
Bilateral IPNs	No	83	15	**0.029** *
	Yes	24	12	
Primary size ≥5 cm	No	40	3	**0.010** *

Table 2. *Cont.*

		Clinical Outcome of IPNs (n = 134)		
		Stable (n = 107)	**Progressed (n = 27)**	**p Value**
	Yes	67	24	
Primary grade	2	40	2	**0.002 ***
	3	67	25	
Primary depth	Superficial	53	7	**0.032 ***
	Deep	54	20	

Table 3. Analysis of IPN and tumour characteristics on the progression of IPNs to metastatic disease, calculated using the Cox regression model. * indicates significance.

	Progression of IPNs			
	Univariate		**Multivariate**	
	HR	**p Value**	**HR**	**p Value**
IPN number ≥1	2.76	**0.013 ***	2.29	0.087
IPN size ≥5 mm	2.96	**0.006 ***	2.37	**0.030 ***
Bilateral IPNs	2.43	**0.022 ***	1.66	0.282
Primary size ≥5 cm	4.55	**0.014 ***	2.58	0.160
Primary grade	6.79	**0.009 ***	6.07	**0.015 ***
Primary depth	2.46	**0.041 ***	1.71	0.280

Figure 3A displays the OS of the three patient groups, demonstrating a significant difference in the OS between the three groups ($p < 0.001$). When those presenting with frank metastases were removed, a trend towards poorer OS (Figure 3B) was seen in those with IPNs upon staging CT, although this did not reach significance ($p = 0.19$, HR = 1.23, HR $p = 0.190$). This remained insignificant at the multivariate level when analysed with known prognostic risk factors of tumour size (<5 cm or ≥5 cm), depth relative to the fascia and grade (Table 4). When patients with grade 2 primaries were excluded, worse OS was seen in patients presenting with IPNs at diagnosis ($p = 0.016$, HR 1.50, HR $p = 0.017$) (Figure 3C). All but one patient with IPNs that progressed are now deceased, with a median OS of 248 days (range 23–840).

The percentage of patients developing lung metastases varied between histological subtypes; during follow up, 53.6% of synovial sarcoma patients, 46% of UPS patients, 40.5% of leiomyosarcoma patients, 33.3% of liposarcoma patients, 28.9% of angiosarcoma patients and 20% of myxofibrosarcoma patients developed lung metastases. The cohort included 100 UPSs and 100 myxofibrosarcomas; as such, these were analysed independently as a subanalysis. Of the 100 patients with UPSs, 53 (53%) had no lung metastasis, 35 (35%) had IPNs and 12 (12%) had synchronous lung metastases. Of those with no lung metastases or IPNs at diagnosis, 18 (33.9%) developed lung metastases at a later date. Of those presenting with IPNs, 12 (34.2%) progressed into lung metastases, 4 (11.4%) developed new lung metastases and 18 (51.4%) remained stable, with no metastases developing elsewhere in the lungs. The presence of IPNs did not increase the likelihood of developing lung metastases ($p = 0.45$, Figure 4A), nor did it confer worse OS ($p = 0.64$, Figure 4B). Of the 100 patients included with myxofibrosarcoma, 66 (66%) presented with no lung metastases or IPNs, 32 (32%) presented with IPNs and 2 (2%) presented with lung metastases. Only 12 (18.2%) of the patients presenting without metastases or IPNs went on to develop metastases. Of those presenting with IPNs, 0 progressed and 6 (18.8%) patients developed lung metastases in different areas of the lung. There was no difference in the likelihood of developing lung metastases between those presenting with IPNs and those presenting with no signs of IPNs or lung metastases ($p = 0.68$, Figure 4C), and no difference in overall survival ($p = 0.46$, Figure 4D).

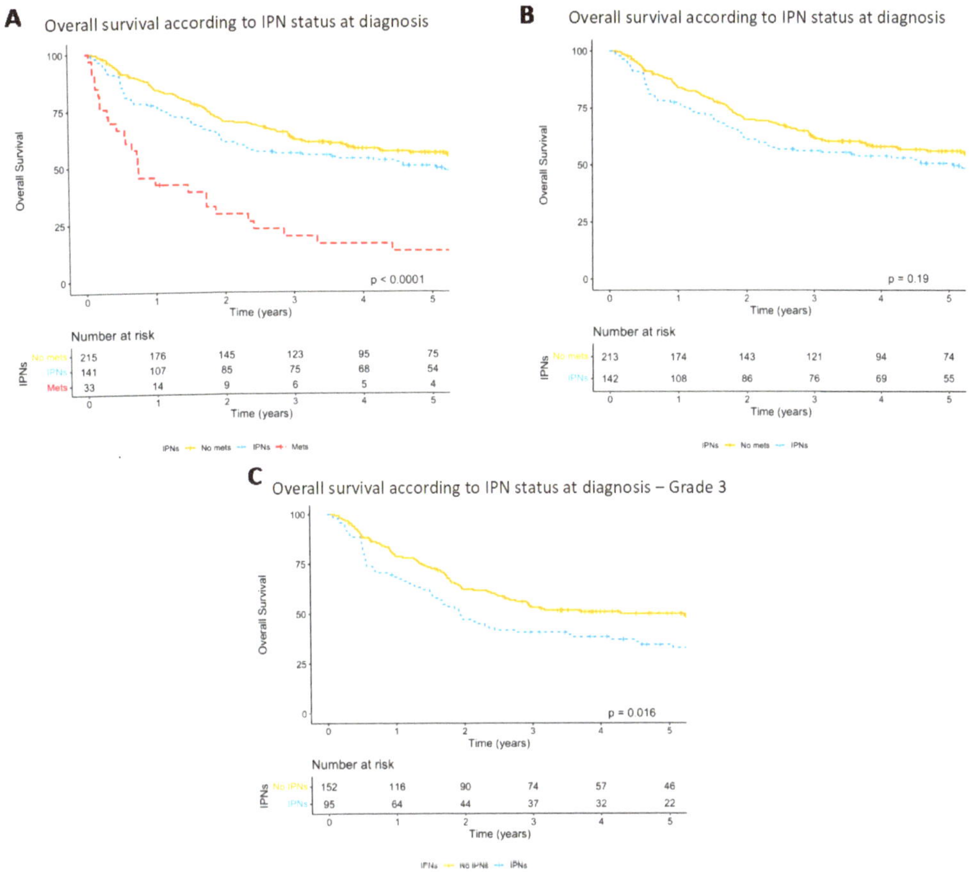

Figure 3. (**A**) Kaplan–Meier graph comparing survival in patients with no IPNs or lung metastases at diagnosis, IPNs and overt lung metastases, with a significant difference detected between groups ($p < 0.001$). (**B**) The same Kaplan–Meier graph with the lung metastases at presentation group removed, demonstrating a trend to decreased OS in patients with IPNs at diagnosis, although this did not reach significance ($p = 0.19$). (**C**) OS according to IPN status once patients with grade 2 primaries are excluded, demonstrating significantly decreased OS in patients with IPNs at diagnosis ($p = 0.016$).

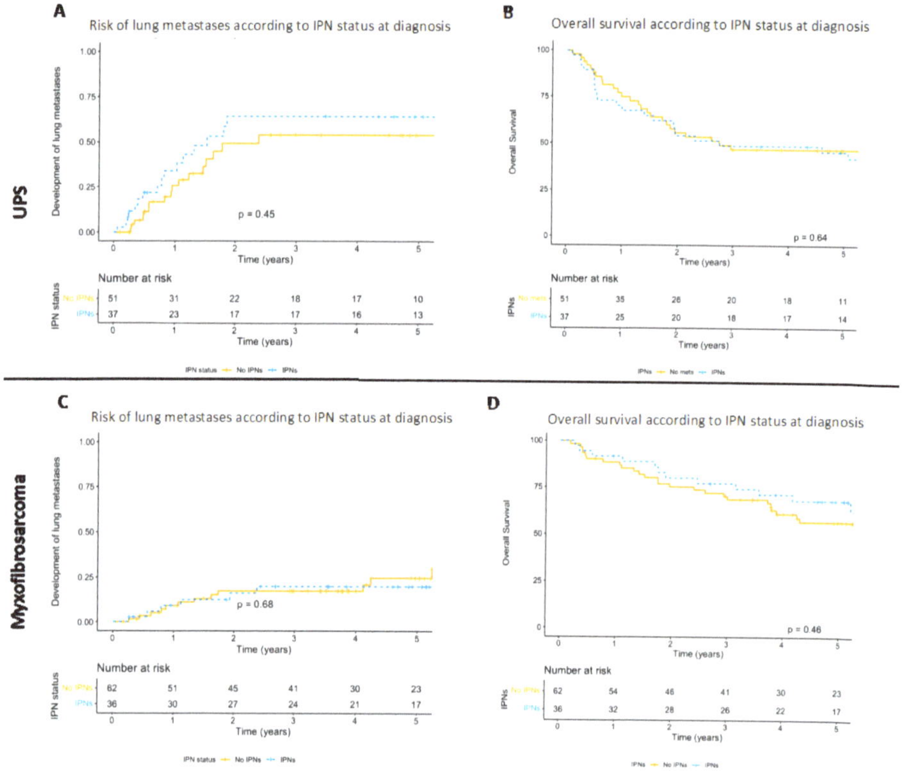

Figure 4. (**A**) Kaplan–Meier graph depicting the cumulative hazard of the development of lung metastases in UPS patients with and without IPNs on the CT staging scan. Patients with IPNs at diagnosis were not at increased risk of developing lung metastases (p = 0.45). (**B**) Kaplan–Meier graph comparing survival in UPS patients with no IPNs and IPNs detected upon CT staging, with no significant difference between groups seen (p = 0.64). (**C**) Kaplan–Meier graph depicting the cumulative hazard of the development of lung metastases in myxofibrosarcoma patients with and without IPNs on the CT staging scan. Patients with IPNs at diagnosis did not have an increased risk of developing lung metastases (p = 0.68). (**D**) Kaplan–Meier graph comparing survival in myxofibrosarcoma patients with no IPNs and IPNs detected upon CT staging, with no significant difference seen between groups (p = 0.46).

Table 4. Analysis of tumour characteristics and IPN status (IPNs or no IPNs/lung metastases) at diagnosis on the progression of IPNs to metastatic disease, calculated using the Cox regression model. * indicates significance.

	Overall Survival			
	Univariate		Multivariate	
	HR	*p* Value	HR	*p* Value
Primary size ≥5 cm	3.60	**<0.001 ***	3.03	**<0.001 ***
Primary grade	2.74	**<0.001 ***	2.25	**<0.001 ***
Primary depth	1.48	**0.011 ***	1.01	0.947
IPN status	1.23	0.190	1.31	0.080

4. Discussion

This study provides the largest and most in-depth analysis of IPNs in patients presenting with high-grade STS and provides new data on an area which remains poorly understood. We included 389 patients and demonstrated that the presence of IPNs is associated with an increased risk of developing lung metastases and poorer OS in patients with grade 3 STS, but not grade 2 STS. In this cohort, 34.4% of patients had IPNs at diagnosis, and of which, 20.1% progressed after a median time of 143 days.

This study builds on work conducted by Saifuddin et al. [6], with similar inclusion criteria, but with a longer period of follow up and a significantly larger sample, allowing for greater study of the progression of IPNs, rather than frequency at staging, and focusing solely on high-grade STS. We had similar issues as these authors in terms of inconsistency in follow up scans, meaning the rate of detection of progression may be lower than the true value; the length of follow up here reduces the chances of this however. Our results demonstrated that grade 3 STS patients presenting with IPNs have significantly poorer survival than patients presenting with no metastatic disease, although this was not observed in patients with grade 2 primaries. Previous studies in other types of sarcoma have produced various results; Tsoi et al. demonstrated worse OS in patients with IPNs in osteosarcoma [13], whilst Ghosh et al. found no difference [14]. Tsoi et al. also studied the relevance of IPNs in Ewing sarcoma, finding no difference in OS [15]. It is important to note that Tsoi et al.'s sample size for their osteosarcoma study was significantly larger than the other two studies mentioned, allowing for much greater power to detect a difference.

This study has once again reinforced the clinical conundrum raised by IPNs and highlights an unmet clinical need. This is particularly important given that pulmonary nodules have been shown to have a higher risk of being malignant in sarcoma patients [16]. We analysed the factors associated with the progression of IPNs to frank metastatic disease; this occurred in 27 out of 134 patients (20.1%) with IPNs at diagnosis. The results demonstrated that IPNs $\geq$ 5 mm, multiple IPNs and a bilateral IPN distribution were more likely to be indicators of metastatic disease, as shown previously [3,6,16]. Interestingly, the two factors with the highest HR for progression actually related to the primary rather than the IPNs, with a primary size of $\geq$5 cm and a grade 3 primary having HRs of 4.55 and 6.79, respectively. Of the 27 patients with IPNs that progressed, only 3 had a primary <5 cm and only 2 had a grade 2 primary. As suggested previously by Mayo et al. [17], this suggests that it is important to consider the characteristics of the primary as well as the IPN when considering the risk of progression; this is logical given that larger, higher-grade sarcomas are associated with higher rates of metastasis in general [18]. As such, even single IPNs < 5 mm warrant close surveillance in patients with large grade 3 primaries. The small number of IPNs progressing prohibited examining prognostic factors separately in different subtypes. The results demonstrated variation between the percentage of patients developing lung metastases during follow up however, with synovial sarcoma and UPS having particularly high rates; it would be logical that IPNs in these patients also had a higher risk of progression.

There are currently no established guidelines to guide the follow up of IPNs on CT staging scans in STS patients. Often, an interval scan at 3, 6 or 12 months is recommended by radiologists to look for changes in the nodules. Our results would suggest that this is insufficient however; 5 of the 27 IPNs that progressed did so after more than 12 months. Given that the median time to progression was 143 days, it seems reasonable that the initial interval scan should be performed within 6 months and be repeated after 12. Interestingly, progression was detected via surveillance CXR in 6 out of 27 patients who progressed. This highlights the importance of regular chest surveillance in STS patients, something which the SAFETY trial is currently investigating [19]. Gamboa et al. previously investigated surveillance methods in high-grade STS patients, comparing CT surveillance to CXR surveillance, finding no improvement in detection and intervention rates in the CT arm [20].

It is important that imaging technologies advance in order to aid clinical decision making in the management of these complex patients. The early identification of malignant nodules followed by rapid intervention with metastasectomy or stereotactic radiotherapy may increase the cure rate in these patients. Relatively new technologies such as FDG PET are still insufficient, with a false negative rate of 30% [10]. One possible solution to this is the development of targeted agents specific to surface proteins expressed on sarcoma cells, such as MT1-MMP [21–24]. Pringle et al. recently published preclinical data of a targeted MT1-MMP antibody labelled with both IRDye800 and Zr-DFO [25]. Zr-DFO emits Cerenkov luminescence and can be imaged via PET scans [26], offering the potential for more accurate pre-operative local imaging and theoretically a differentiation between IPNs and metastatic sarcoma deposits in the lungs. Furthermore, IRDye800 fluoresces in a similar spectrum to indocyanine green, which is currently under investigation for its utility in fluorescence-guided sarcoma surgery [27,28], meaning it could also be used as a targeted fluorescent dye for intra-operative guidance using current camera systems. Given that surgical resection is the current cornerstone of curative STS management, this dual purpose is particularly enticing. In combination with the ever-evolving fields of machine learning and artificial intelligence, which have already been suggested to be equal in efficacy at identifying pulmonary nodules to consultant radiologists, huge progress could be made in the distinction of malignant and benign pulmonary nodules over the coming years [29,30]. Ideally, this will remove the concept of IPNs, being able to accurately distinguish between metastatic disease and benign nodules.

5. Conclusions

In conclusion, this study demonstrates that patients with grade 3 STS presenting with IPNs have significantly worse survival than those without. It also highlights the importance of the consideration of factors related to the primary tumour itself when evaluating the risk of IPN progression in patients with high-grade STS; IPNs in patients with larger, higher-grade primaries arising deep in the fascia are associated with an increased risk of progression, as are IPNs $\geq$ 5 mm in diameter, multiple IPNs and a bilateral distribution. In order to monitor for progression, we recommend that IPNs are followed up with CT scans at 6 and 12 months. Further study in a larger cohort of people with high-grade STS is required, particularly to allow for the analysis of the role of subtype on risk of progression.

Supplementary Materials: The following supporting information can be downloaded at https://www.mdpi.com/article/10.3390/cancers15133531/s1: Table S1: Clinical details of patients with IPNs at diagnosis progressing to lung metastases.

Author Contributions: Conceptualisation, M.J.B. and K.S.R.; methodology, M.J.B. and K.S.R.; formal analysis, M.J.B.; investigation, M.J.B., C.D.C., T.P.C., M.R., T.B., K.M.G. and K.S.R.; data curation, M.J.B.; writing—original draft, M.J.B. and C.D.C.; writing—review and editing, M.J.B., C.D.C., T.P.C., M.R., T.B., K.M.G. and K.S.R.; visualisation, M.J.B.; supervision, K.M.G. and K.S.R. All authors have read and agreed to the published version of the manuscript.

Funding: This research received no external funding.

Institutional Review Board Statement: The study was conducted according to the guidelines of the Declaration of Helsinki, and approved by the Institutional Review Board of Newcastle upon Tyne Hospitals NHS Foundation Trust (number 13952, 20 March 2023).

Informed Consent Statement: Patient consent was waived as this study only used anonymised, retrospective data, with no intervention.

Data Availability Statement: Further data are available upon request from the corresponding author.

Acknowledgments: We would like to thank Luke Wigney for his help with maintaining the sarcoma database.

Conflicts of Interest: The authors declare no conflict of interest.

References

1. Siegel, R.; Naishadham, D.; Jemal, A. Cancer statistics, 2013. *CA A Cancer J. Clin.* **2013**, *63*, 11–30. [CrossRef]
2. Cool, P.; Grimer, R.; Rees, R. Surveillance in patients with sarcoma of the extremities. *Eur. J. Surg. Oncol.* **2005**, *31*, 1020–1024. [CrossRef] [PubMed]
3. Nakamura, T.; Matsumine, A.; Matsusaka, M.; Mizumoto, K.; Mori, M.; Yoshizaki, T.; Matsubara, T.; Asanuma, K.; Sudo, A. Analysis of pulmonary nodules in patients with high-grade soft tissue sarcomas. *PLoS ONE* **2017**, *12*, e0172148. [CrossRef] [PubMed]
4. Billingsley, K.G.; Burt, M.E.; Jara, E.; Ginsberg, R.J.; Woodruff, J.M.; Leung, D.H.; Brennan, M.F. Pulmonary metastases from soft tissue sarcoma: Analysis of patterns of diseases and postmetastasis survival. *Ann. Surg.* **1999**, *229*, 602–610, Discussion 610–602. [CrossRef] [PubMed]
5. Rissing, S.; Rougraff, B.T.; Davis, K. Indeterminate pulmonary nodules in patients with sarcoma affect survival. *Clin. Orthop. Relat. Res.* **2007**, *459*, 118–121. [CrossRef]
6. Saifuddin, A.; Shafiq, H.; Rajakulasingam, R.; Tan, A.; O'Donnell, P.; Khoo, M. A review of staging chest CT in trunk and extremity soft tissue sarcoma. *Br. J. Radiol.* **2021**, *94*, 20201109. [CrossRef]
7. Dangoor, A.; Seddon, B.; Gerrand, C.; Grimer, R.; Whelan, J.; Judson, I. UK guidelines for the management of soft tissue sarcomas. *Clin. Sarcoma Res.* **2016**, *6*, 20. [CrossRef]
8. Casali, P.G.; Abecassis, N.; Aro, H.T.; Bauer, S.; Biagini, R.; Bielack, S.; Bonvalot, S.; Boukovinas, I.; Bovee, J.; Brodowicz, T.; et al. Soft tissue and visceral sarcomas: ESMO-EURACAN Clinical Practice Guidelines for diagnosis, treatment and follow-up. *Ann. Oncol.* **2018**, *29*, iv51–iv67. [CrossRef]
9. Lim, H.J.; Johnny Ong, C.-A.; Tan, J.W.-S.; Ching Teo, M.C. Utility of positron emission tomography/computed tomography (PET/CT) imaging in the evaluation of sarcomas: A systematic review. *Crit. Rev. Oncol./Hematol.* **2019**, *143*, 1–13. [CrossRef]
10. Fortes, D.L.; Allen, M.S.; Lowe, V.J.; Shen, K.R.; Wigle, D.A.; Cassivi, S.D.; Nichols, F.C.; Deschamps, C. The sensitivity of 18F-fluorodeoxyglucose positron emission tomography in the evaluation of metastatic pulmonary nodules. *Eur. J. Cardio-Thorac. Surg.* **2008**, *34*, 1223–1227. [CrossRef]
11. Fletcher, C.; Bridge, J.A.; Hogendoorn, P.C.W.; Mertens, F. *WHO Classification of Tumours of Soft Tissue and Bone: WHO Classification of Tumours*; World Health Organization: Geneva, Switzerland, 2013; Volume 5.
12. Coindre, J.M.; Nguyen, B.B.; Bonichon, F.; De Mascarel, I.; Trojani, M. Histopathologic grading in spindle cell soft tissue sarcomas. *Cancer* **1988**, *61*, 2305–2309. [CrossRef]
13. Tsoi, K.M.; Lowe, M.; Tsuda, Y.; Lex, J.R.; Fujiwara, T.; Almeer, G.; Gregory, J.; Stevenson, J.; Evans, S.E.; Botchu, R.; et al. How Are Indeterminate Pulmonary Nodules at Diagnosis Associated with Survival in Patients with High-Grade Osteosarcoma? *Clin. Orthop. Relat. Res.* **2021**, *479*, 298–308. [CrossRef]
14. Ghosh, K.M.; Lee, L.H.; Beckingsale, T.B.; Gerrand, C.H.; Rankin, K.S. Indeterminate nodules in osteosarcoma: What's the follow-up? *Br. J. Cancer* **2018**, *118*, 634–638. [CrossRef]
15. Tsoi, K.M.; Tan, D.; Stevenson, J.; Evans, S.; Jeys, L.M.; Botchu, R. Indeterminate pulmonary nodules are not associated with worse overall survival in Ewing Sarcoma. *J. Clin. Orthop. Trauma* **2021**, *16*, 58–64. [CrossRef] [PubMed]
16. Hanamiya, M.; Aoki, T.; Yamashita, Y.; Kawanami, S.; Korogi, Y. Frequency and significance of pulmonary nodules on thin-section CT in patients with extrapulmonary malignant neoplasms. *Eur. J. Radiol.* **2012**, *81*, 152–157. [CrossRef]
17. Mayo, Z.; Kennedy, S.; Gao, Y.; Miller, B.J. What Is the Clinical Importance of Incidental Findings on Staging CT Scans in Patients With Sarcoma? *Clin. Orthop. Relat. Res.* **2019**, *477*, 730–737. [CrossRef]
18. Stojadinovic, A.; Leung, D.H.Y.; Hoos, A.; Jaques, D.P.; Lewis, J.J.; Brennan, M.F. Analysis of the prognostic significance of microscopic margins in 2,084 localized primary adult soft tissue sarcomas. *Ann. Surg.* **2002**, *235*, 424–434. [CrossRef]
19. The SAFETY Investigators. The Surveillance After Extremity Tumor Surgery (SAFETY) trial: Protocol for a pilot study to determine the feasibility of a multi-centre randomised controlled trial. *BMJ Open* **2019**, *9*, e029054. [CrossRef]
20. Gamboa, A.C.; Ethun, C.G.; Switchenko, J.M.; Lipscomb, J.; Poultsides, G.A.; Grignol, V.; Howard, J.H.; Gamblin, T.C.; Roggin, K.K.; Votanopoulos, K.; et al. Lung Surveillance Strategy for High-Grade Soft Tissue Sarcomas: Chest X-Ray or CT Scan? *J. Am. Coll Surg.* **2019**, *229*, 449–457. [CrossRef] [PubMed]
21. Brookes, M.J.; Roundhill, E.A.; Jeys, L.; Parry, M.; Burchill, S.A.; Rankin, K.S. Membrane-type 1 matrix metalloproteinase as predictor of survival and candidate therapeutic target in Ewing sarcoma. *Pediatr. Blood Cancer* **2022**, *69*, e29959. [CrossRef] [PubMed]
22. Gonzalez-Molina, J.; Gramolelli, S.; Liao, Z.; Carlson, J.W.; Ojala, P.M.; Lehti, K. MMP14 in Sarcoma: A Regulator of Tumor Microenvironment Communication in Connective Tissues. *Cells* **2019**, *8*, 991. [CrossRef] [PubMed]
23. Uchibori, M.; Nishida, Y.; Nagasaka, T.; Yamada, Y.; Nakanishi, K.; Ishiguro, N. Increased expression of membrane-type matrix metalloproteinase-1 is correlated with poor prognosis in patients with osteosarcoma. *Int. J. Oncol.* **2006**, *28*, 33–42. [CrossRef] [PubMed]
24. Liu, M.; Qi, Y.; Zhao, L.; Chen, D.; Zhou, Y.; Zhou, H.; Lv, Y.; Zhang, L.; Jin, S.; Li, S.; et al. Matrix metalloproteinase-14 induces epithelial-to-mesenchymal transition in synovial sarcoma. *Hum. Pathol.* **2018**, *80*, 201–209. [CrossRef]
25. Pringle, T.A.; Chan, C.D.; Luli, S.; Blair, H.J.; Rankin, K.S.; Knight, J.C. Synthesis and In Vivo Evaluation of a Site-specifically Labeled Radioimmunoconjugate for Dual-Modal (PET/NIRF) Imaging of MT1-MMP in Sarcomas. *Bioconjug. Chem.* **2022**, *33*, 1564–1573. [CrossRef]

26. Manafi-Farid, R.; Ataeinia, B.; Ranjbar, S.; Jamshidi Araghi, Z.; Moradi, M.M.; Pirich, C.; Beheshti, M. ImmunoPET: Antibody-Based PET Imaging in Solid Tumors. *Front. Med.* **2022**, *9*, 916693. [CrossRef] [PubMed]
27. Brookes, M.J.; Chan, C.D.; Nicoli, F.; Crowley, T.P.; Ghosh, K.M.; Beckingsale, T.; Saleh, D.; Dildey, P.; Gupta, S.; Ragbir, M.; et al. Intraoperative Near-Infrared Fluorescence Guided Surgery Using Indocyanine Green (ICG) for the Resection of Sarcomas May Reduce the Positive Margin Rate: An Extended Case Series. *Cancers* **2021**, *13*, 6284. [CrossRef]
28. Gong, M.F.; Li, W.T.; Bhogal, S.; Royes, B.; Heim, T.; Silvaggio, M.; Malek, M.; Dhupar, R.; Lee, S.J.; McGough, R.L.; et al. Intraoperative Evaluation of Soft Tissue Sarcoma Surgical Margins with Indocyanine Green Fluorescence Imaging. *Cancers* **2023**, *15*, 582. [CrossRef]
29. Schreuder, A.; Scholten, E.T.; Van Ginneken, B.; Jacobs, C. Artificial intelligence for detection and characterization of pulmonary nodules in lung cancer CT screening: Ready for practice? *Transl. Lung Cancer Res.* **2021**, *10*, 2378–2388. [CrossRef]
30. Koh, D.M.; Papanikolaou, N.; Bick, U.; Illing, R.; Kahn, C.E., Jr.; Kalpathi-Cramer, J.; Matos, C.; Martí-Bonmatí, L.; Miles, A.; Mun, S.K.; et al. Artificial intelligence and machine learning in cancer imaging. *Commun. Med.* **2022**, *2*, 133. [CrossRef]

cancers

MDPI

Article

Rapid Classification of Sarcomas Using Methylation Fingerprint: A Pilot Study

Aviel Iluz [1,2], Myriam Maoz [3], Nir Lavi [1,2,4], Hanna Charbit [1,2], Omer Or [5], Noam Olshinka [5], Jonathan Abraham Demma [6], Mohammad Adileh [6], Marc Wygoda [7], Philip Blumenfeld [7], Masha Gliner-Ron [8], Yusef Azraq [8], Joshua Moss [3], Tamar Peretz [3], Amir Eden [9], Aviad Zick [3,*,†] and Iris Lavon [1,2,*,†]

1 Leslie and Michael Gaffin Center for Neuro-Oncology, Hadassah Medical Center and Faculty of Medicine, The Hebrew University of Jerusalem, Jerusalem 9190501, Israel
2 Agnes Ginges Center for Human Neurogenetics, Department of Neurology, Hadassah Medical Center and Faculty of Medicine, The Hebrew University of Jerusalem, Jerusalem 9190501, Israel
3 Oncology Department, Sharett Institute of Oncology, Hadassah Medical Organization and Faculty of Medicine, The Hebrew University of Jerusalem, Jerusalem 9190501, Israel
4 Department of Military Medicine and "Tzameret", Faculty of Medicine, The Hebrew University of Jerusalem, Jerusalem 9190501, Israel
5 Orthopedic Department, Hadassah Medical Organization and Faculty of Medicine, The Hebrew University of Jerusalem, Jerusalem 9190501, Israel
6 Surgical Department, Hadassah Medical Organization and Faculty of Medicine, The Hebrew University of Jerusalem, Jerusalem 9190501, Israel
7 Radiotherapy Institute, Sharett Institute of Oncology, Hadassah Medical Organization and Faculty of Medicine, The Hebrew University of Jerusalem, Jerusalem 9190501, Israel
8 Radiology Department, Hadassah Medical Organization and Faculty of Medicine, The Hebrew University of Jerusalem, Jerusalem 9190501, Israel
9 Department of Genetics, The Hebrew University of Jerusalem, Jerusalem 9190501, Israel
* Correspondence: aviadz@hadassah.org.il (A.Z.); irisl@hadassah.org.il (I.L.); Tel.: +972-50-4048024 (A.Z.); +972-50-8573933 (I.L.)
† These authors contributed equally to this work.

Simple Summary: Sarcomas encompass a diverse range of cancers, resulting in intricate classification that contributes to treatment delays. The aim of this pilot study, conducted within a specific subset of sarcoma types, is to demonstrates the feasibility of methylation and copy-number variation data obtained from low-coverage whole-genome sequencing using Oxford Nanopore for rapid point-of-care sarcoma classification. Oxford Nanopore sequencers are relatively affordable for laboratories, unlike other technologies used in previous studies for methylation-based sarcoma classification. Our findings indicate that this method attained an overall correct classification rate of 78%. This study could serve as the foundation for a rapid point-of-care sarcoma classification test, facilitating timely and efficient care across diverse clinical settings.

Citation: Iluz, A.; Maoz, M.; Lavi, N.; Charbit, H.; Or, O.; Olshinka, N.; Demma, J.A.; Adileh, M.; Wygoda, M.; Blumenfeld, P.; et al. Rapid Classification of Sarcomas Using Methylation Fingerprint: A Pilot Study. *Cancers* **2023**, *15*, 4168. https://doi.org/10.3390/cancers15164168

Academic Editor: Catrin Sian Rutland

Received: 25 July 2023
Revised: 14 August 2023
Accepted: 16 August 2023
Published: 18 August 2023

Abstract: Sarcoma classification is challenging and can lead to treatment delays. Previous studies used DNA aberrations and machine-learning classifiers based on methylation profiles for diagnosis. We aimed to classify sarcomas by analyzing methylation signatures obtained from low-coverage whole-genome sequencing, which also identifies copy-number alterations. DNA was extracted from 23 suspected sarcoma samples and sequenced on an Oxford Nanopore sequencer. The methylation-based classifier, applied in the nanoDx pipeline, was customized using a reference set based on processed Illumina-based methylation data. Classification analysis utilized the Random Forest algorithm and t-distributed stochastic neighbor embedding, while copy-number alterations were detected using a designated R package. Out of the 23 samples encompassing a restricted range of sarcoma types, 20 were successfully sequenced, but two did not contain tumor tissue, according to the pathologist. Among the 18 tumor samples, 14 were classified as reported in the pathology results. Four classifications were discordant with the pathological report, with one compatible and three showing discrepancies. Improving tissue handling, DNA extraction methods, and detecting point mutations and translocations could enhance accuracy. We envision that rapid, accurate, point-of-care sarcoma classification using nanopore sequencing could be achieved through additional validation in a diverse tumor cohort and the integration of methylation-based classification and other DNA aberrations.

Keywords: sarcoma; nanopore; methylation; copy-number; classification; machine learning

1. Introduction

Sarcoma is a cancer that originates from connective tissue [1]. Sarcomas are classified based on tissue and cell type and are typically divided into two major groups: bone sarcomas and soft-tissue sarcomas (STS) [2].

Sarcoma often presents as a painless mass that grows over months or years. Some types are more likely to affect children, while others affect mainly adults. Sarcomas can occur anywhere in the body, but the most common types occur in the arms, legs, and abdomen [3]. Generally, the cancer grade refers to its aggressiveness and the likelihood of spreading to other body parts. Low-grade sarcomas have a better prognosis than higher-grade sarcomas and are usually treated surgically, although sometimes radiation therapy or chemotherapy are used. Intermediate- and high-grade sarcomas are more frequently treated with surgery, chemotherapy, and radiation therapy. The treatment varies according to the exact type of sarcoma [4].

Diagnosis of bone sarcomas and soft-tissue sarcomas begins with a history, physical examination, and imaging studies [5]. Definitive diagnosis requires a tumor biopsy with extensive pathological review [4]. There is high inter-observer variability among pathologists. Using current pathological methods, up to 80–85% of sarcoma cases are classified, while the remaining cases remain unclassified [6]. Institutions with access to fluorescence in situ hybridization (FISH), Sanger sequencing, massively parallel DNA sequencing, and methylation-based arrays can gain a more accurate diagnosis by detecting point mutations, translocations, copy-number alterations [6,7], and methylation patterns [8].

Copy-number alterations in sarcomas are relatively uncommon, except for MDM2 amplification. The MDM2 gene is located on chromosome 12q13-15 and encodes the MDM2 protein. MDM2 amplification involves the presence of multiple copies of the MDM2 gene, which leads to elevated levels of MDM2 protein expression [9,10]. This amplification has been associated with heightened MDM2 protein expression and is linked to the process of de-differentiation in liposarcomas [11]. In de-differentiated liposarcoma, MDM2 is amplified in all tumors while in other tumors such as extraskeletal osteosarcoma MDM2 amplification is found in about 40% of the tumors.

The identification of MDM2 amplification employs techniques such as fluorescence in situ hybridization (FISH) and immunohistochemistry (IHC) to detect MDM2 overexpression, serving as the gold standard methods [12].

The DNA of normal and tumor cell types in the body carries unique methylation marks correlating with its gene-expression profile, representing a fundamental aspect of tissue identity [13]. Numerous independent studies have shown that most central nervous system tumor types can be reliably identified based on their epigenetic DNA methylation. This layer of molecular information in neuropathological practice has increased accuracy and reduced the error rate in classifying CNS tumors [14]. A similar tool was developed for 54 histological types of sarcomas [8]. Most studies have used Illumina-based arrays for DNA methylation analysis, more specifically, the HumanMethylation450 with 485K CpGs or the MethylationEPIC with 850K CpGs. These arrays are based on DNA that undergoes bisulfite treatment that introduces specific changes in the DNA sequence that depend on the methylation status of individual cytosine residues and thus yield single-nucleotide resolution information about the methylation status [15].

The Oxford Nanopore sequencer can directly detect methylated base pairs (bp) without bisulfite modification. The sequencing of methylated bp is achieved by differentiating between the ionic current changes produced by unmethylated cytosine vs. 5-methylated cytosine [16]. Bisulfite-converted sequencing, which is the basis for Illumina Array, is a widely used method for detecting DNA methylation. Nonetheless, this approach has drawbacks including DNA degradation, limited specificity, and the production of short reads with low sequence diversity. In comparison, nanopore sequencing technology enables the direct detection of base modifications in native DNA, without requiring harsh chemical treatment as in bisulfite sequencing [17]. Moreover, nanopore technology allows the sequencing of longer DNA fragments up to about 100 kbp, allowing tumor classification based on methylation patterns and chromosomal aberrations [18]. It has been demonstrated that accurate and reliable CNS tumor classification can be performed based on methylation signatures gained by nanopore sequencing. Studies have shown that nanopore sequencing of low-coverage whole-genome sequencing (lcWGS) yielding a minimum set of 1000 random CpG sites chosen from the 450K sites, is sufficient for reliable brain tumor classification [19,20].

In this study, we investigated the utility of nanopore sequencing in classifying sarcomas. We successfully implemented and customized a nanopore-based nanoDx pipeline [19] to classify a restricted range of sarcoma types. The pipeline employs machine-learning algorithms for methylation-based classification. In addition, we utilize copy-number alteration to validate the classification of specific sarcoma types.

2. Materials and Methods

2.1. Patients and DNA Isolation

23 Patients diagnosed with sarcoma between 2018 to 2023 who signed an informed consent form (0346-12) participated in this study. Surgically resected masses were freshly frozen, and a pathological report is available for all tumors with a molecular profile using Oncomine comprehensive panel (Thermo Fisher Scientific, Waltham, MA, USA) for some samples. Per the manufacturer's protocol, we extracted tumor DNA using a DNeasy blood and tissue kit (Qiagen, Hilden, Germany). DNA was quantified by Qubit (Thermo Fisher Scientific) or QuantiFluor (Promega, Madison, WI, USA) assays and quality controlled (260/280 ratio) (NanoDrop, Thermo Fisher Scientific).

2.2. Nanopore WGS

Between 200 and 400 ng of genomic tumor DNA of each sample is used for library preparation with barcode labeling using the Rapid Barcoding Kit (SQK-RBK004, Oxford Nanopore Technologies, Oxford, UK) according to the manufacturer's instructions. Low-coverage whole-genome sequencing (lcWGS) is performed on a Minion Mk1C device (OS ubuntu 18.04) using an R9.4.1 flow cell (FLO-MIN106D, Oxford Nanopore Technologies). Sequencing was performed until the recommended 100M bps (per correspondence with the nanoDx pipeline developer [19]). Output FAST5 files containing the raw signal data were generated by the manufacturer's software MinKNOW (v.22.12.5) and the equivalent FASTQ files. They were all transferred to high-performance computing (HPC) clusters for further analysis.

2.3. Data Analysis Pipeline

FAST5 and FASTQ files of the assigned barcode were processed on the HPC using the nanoDx pipeline (v.5.0.1) that uses snakemake [19] v5.4.0 workflow [19,20]. This pipeline was initially developed for nanopore methylation-based classification and used the Heidelberg reference cohort of brain tumor methylation profiles of CpG sites probed by Illumina BeadChip 450K array (Illumina, Cambridge, UK) [14]. The nanoDx pipeline for brain tumor classification converts the methylation data from the 450K CpG sites to match the nanopore methylation data type for analysis. We adapted the pipeline to use sarcoma tumor methylation profiles obtained by the same Illumina platform from the cohort of 1077 sarcomas tumors [8]. We downloaded the beta-value processed data from GEO (GSE140686) and adapted it to the requirement of the pipeline code as a sarcoma reference set.

2.3.1. Classification

1. Methylation-based Random Forest Classification

The processing of the FAST5 files calls the methylation status of genome-wide CpG sites of each sample using nanopolish software (v.0.13.2) [18]. The nanopolish software assigns a binary value of 1 or 0 to each detected CpG site, indicating methylation or unmethylation, respectively. This assignment is made through statistical analysis of the methylation detection algorithm [18]. The methylation frequency per site is then calculated by the fraction of reads classified as methylated. Given the fundamental differences between the Illumina Array-based methylation beta values and the nanopore methylation frequency values of the CpG sites, both data sources are subjected to binarization using a cutoff value of 0.6, consistent with previous nanoDx implementations [19,20] and are compatible with the nanopore methylation data type format. This enables the classification of each CpG site as either methylated (>0.6) or unmethylated. Subsequently, an ad-hoc Random Forest [20] classifier is trained using the most variable maximum of 100,000 overlapping sites within the sarcoma reference set.

The Random Forest classifier is built in Python using the RandomForestClassifier function from the scikit-learn package v.1.0.2 [21]. The classifier is then used to predict the methylation class of each sample. The Random Forest-estimated class probabilities are rescaled to be more accurately interpreted as confidence levels or "confidence scores" by the CalibratedClassifierCV function from the scikit-learn package in Python, as previously described [20]. Based on previous research conducted on CNS tumors, a confidence score greater than the threshold value of 0.15 is regarded as a reliable classification (see Discussion Section 4) [20] (Figure 1B,C). The sarcoma reference set was generated in the 'HDF5' binary data format, adhering to the pipeline's specifications, using R/Bioconductor and the rhdf5 package [22].

2. Unsupervised Clustering

Additional unsupervised clustering analysis using t-SNE (t-distributed stochastic neighbor embedding) was performed on the 50,000 most variable CpG sites, and a final plot was generated using the R package Rtsne [23] (Figure 1D). Yet, t-SNE plots are meant for visual quality control, not classification. It can help validate the Random Forest classification results but must be interpreted carefully (per correspondence with the nanoDx pipeline developer [19]).

2.3.2. Copy-Number Analysis

Briefly (as of [19,20]), FASTQ files are aligned to the hg19 human reference genome (minimap2 v2.15) [24] for the generation of copy-number profile (Figure 1A) which is generated from the same sequencing run using R/Bioconductor and the QDNAseq package [21]. Reads with a minimum mapping quality of 20 were sorted into 1000 kbp bins and analyzed using public data from a single flow cell sequencing run (FAF04090) generated with NA12878 reference DNA [22] for pseudo-germline subtraction. The circular binary segmentation method, implemented in the PSCBS R package, was utilized for the analysis. Change points were accepted based on an alpha value < 0.05. Arm-level copy-number calls were made by calculating the segment length weighted mean log ratio per chromosome arm.

2.3.3. Reporting

All the analysis and classification results are reported in a PDF file. Extracts from a typical PDF report are depicted in Figure 1. The full report format of all cases is shown in File S1.

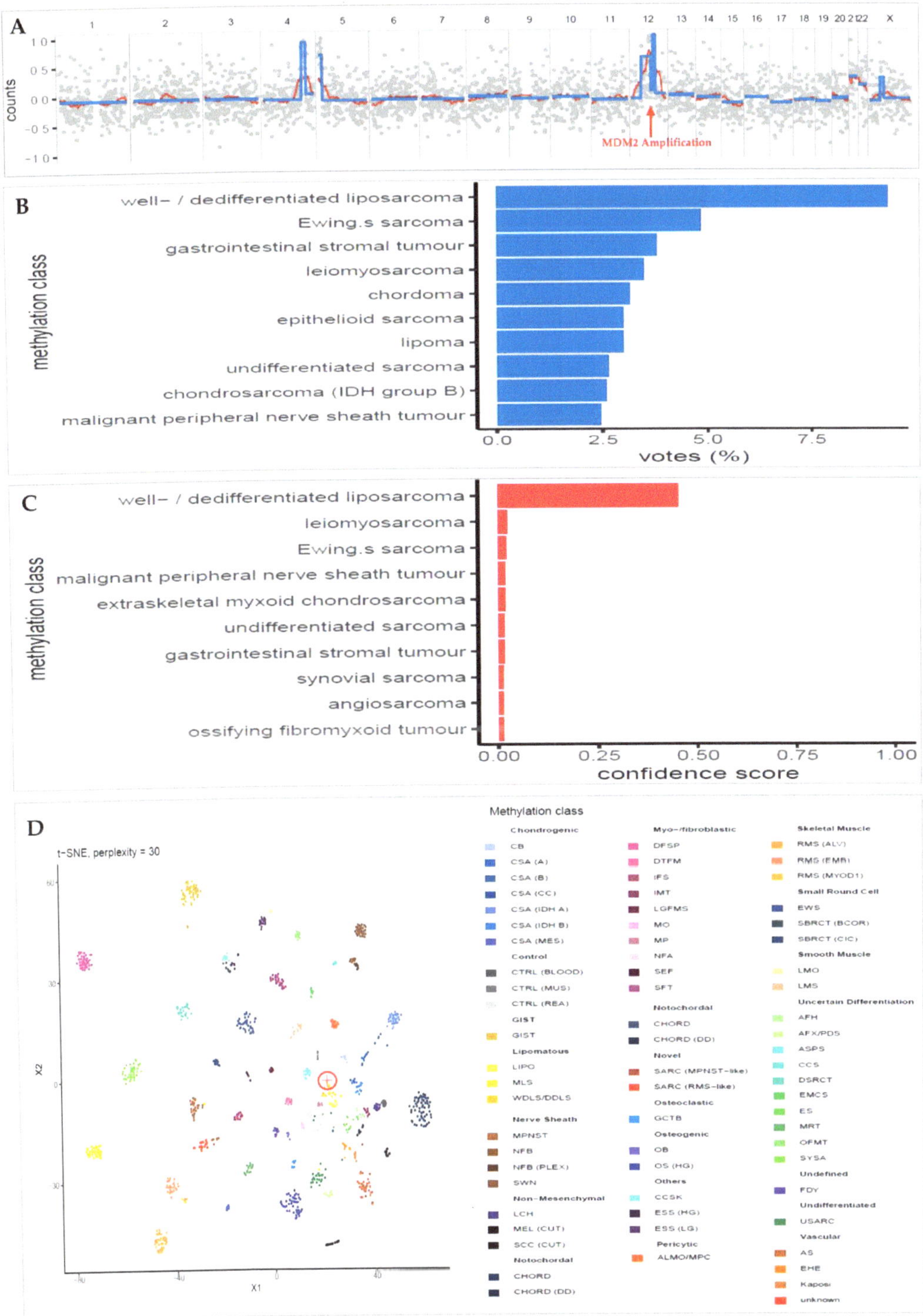

Figure 1. Extracts from an example of the final nanoDx analysis report for sample SARC-09 of a female patient with retroperitoneal MDM2 amplified well-differentiated liposarcoma. (**A**): Copy-number

profile. The x-axis is the chromosomal location, and the y-axis is the read counts (log2 transformation). The MDM2 amplification is shown in chromosome 12. (**B**): Bar plot of the Random Forest classification voting results. The category (y) axis shows this sample's 10 most frequent methylation classification votes and the percentage voting rate in the x-axis. (**C**): Bar plot of the confidence score of the voted Random Forest classification (in B). The most confident classification of this sample is "well/dedifferentiated liposarcoma" (WDLS/DDLS), with a confidence score of 0.45. This confidence score validates the top classification in the voting results (B) as a correct one. (**D**): t-SNE plot shows the clustering of the methylation pattern of the specific sample (circled cross sign) among the other methylation patterns of the sarcomas in the reference set. It shows that the sample clusters very close to the WDLS/DDLS group (dark yellow), as classified by the Random Forest classifier. Abbreviation: AFH, angiomatoid fibrous histiocytoma; AFX/PDS, atypical fibroxanthoma/pleomorphic dermal sarcoma; ALMO/MPC, angioleiomyoma/myopericytoma; AS, angiosarcoma; ASPS, alveolar soft part sarcoma; CB, chondroblastoma; CCS, clear cell sarcoma of soft parts; CCSK, clear cell sarcoma of the kidney; CHORD, chordoma; CHORD (DD), chordoma (dedifferentiated); CSA (A), chondrosarcoma (group A); CSA (B), chondrosarcoma (group B); CSA (CC), chondrosarcoma (clear cell); CSA (IDH A), chondrosarcoma (IDH group A); CSA (IDH B), chondrosarcoma (IDH group B); CSA [25], chondrosarcoma (mesenchymal); CTRL (BLOOD), control (blood); CTRL [26], control (muscle tissue); CTRL (REA), control (reactive tissue); DFSP, dermatofibrosarcoma protuberans; DSRCT, desmoplastic small round cell tumor; DTFM, desmoid-type fibromatosis; EHE, epithelioid hemangioendothelioma; EMCS, extraskeletal myxoid chondrosarcoma; ES, epithelioid sarcoma; ESS (HG), endometrial stromal sarcoma (high grade); ESS [27], endometrial stromal sarcoma (low grade); EWS, Ewing's sarcoma; FDY, fibrous dysplasia; GCTB, giant cell tumor of bone; GIST, gastrointestinal stromal tumor; IFS, infantile fibrosarcoma; IMT, inflammatory myofibroblastic tumor; Kaposi, Kaposi sarcoma; LCH, Langerhans cell histiocytosis; LGFMS, low-grade fibromyxoid sarcoma; LIPO, lipoma; LMO, leiomyoma; LMS, leiomyosarcoma; MEL (CUT), melanoma (cutaneous); MLS, myxoid liposarcoma; MO, myositis ossificans; MP, myositis proliferans; MPNST, malignant peripheral nerve sheath tumor; MRT, malignant rhabdoid tumor; NFA, nodular fasciitis; NFB, neurofibroma; NFB (PLEX), neurofibroma (plexiform); OB, osteoblastoma; OFMT, ossifying fibromyxoid tumor; OS (HG), osteosarcoma (high grade); RMS [4], rhabdomyosarcoma (alveolar); RMS [28], rhabdomyosarcoma (embryonal); RMS (MYOD1), rhabdomyosarcoma (MYOD1); SARC (MPNST-like), sarcoma (MPNST-like); SARC (RMS-like), sarcoma (RMS-like); SBRCT (BCOR), small blue round cell tumor with BCOR alteration; SBRCT [29], small blue round cell tumor with CIC alteration; SCC (CUT), squamous cell carcinoma (cutaneous); SEF, sclerosing epithelioid fibrosarcoma; SFT, solitary fibrous tumor; SWN, schwannoma; SYSA, synovial sarcoma; USARC, undifferentiated sarcoma; WDLS/DDLS, well/dedifferentiated liposarcoma; t-SNE, t-distributed stochastic neighbor embedding.

3. Results

3.1. DNA Extracted from Sarcoma Surgical Samples Are Successfully Utilized for Nanopore Sequencing

Out of the 23 samples, 20 were successfully run using nanopore and met the minimum sequencing coverage required to be processed by the nanoDx pipeline (see Methods). These include 18 tumors representing a limited range of 11 pathologically identified sarcoma types and 2 masses with no tumor; hence, we included in the statistical analysis the 18 tumor samples (Table 1). The included samples had a mean read length of 3966 bp (range 1310–7078), the mean number of CpG sites covered is 27594 (range 6539–100,000, Table S1), and the mean coverage is 0.53X (range 0.03X–6.4X).

The 2 non-tumor samples (not shown in Table 1) were analyzed in the pipeline. They have a mean read length of 1262 and 5769 bp, several CpG sites covered 7436 and 20,895, and coverage of 0.04X and 0.08X, respectively (Table S1).

Table 1. Results of the comparisons conducted between the nanoDx methylation-based classification and pathology diagnosis for the study samples (n = 18). See Table S1 for more details.

Sample	Max Calibrated Meth. Class	Pathology	Pathology Meth. Class	Concordance	Max Confidence Score	Mean Read Length	Mean Coverage	MDM2 Ampl.	t-SNE Agreed Cluster
SARC-01	WDLS/DDLS	Well differentiated liposarcoma	WDLS/DDLS	C	0.09	5192	0.43	Y	Y
SARC-02	MLS	myxoid liposarcoma	MLS	C	0.30	5038	6.4		Y
SARC-03	LMS	leiomyosarcoma	LMS	C	0.21	5679	1.78		Y
SARC-04	USARC	Undifferentiated small round spindle cell sarcoma	USARC	C	0.08	1957	0.08		
SARC-05	EMCS	Extraskeletal myxoid chondrosarcoma	EMCS	C	0.88	4537	0.11		Y
SARC-06	WDLS/DDLS	Dedifferentiated liposarcoma	WDLS/DDLS	C	0.08	1310	0.05	Y	Y
SARC-08	WDLS/DDLS	Well differentiated liposarcoma	WDLS/DDLS	C	0.10	5121	0.03		
SARC-09	WDLS/DDLS	Highly suspicious for well differentiated liposarcoma	WDLS/DDLS	C	0.45	5564	0.07	Y	Y
SARC-10	EWS	Ewing´s sarcoma	EWS	C	0.14	3081	0.06		Y
SARC-11	WDLS/DDLS	Well differentiated liposarcoma	WDLS/DDLS	C	0.15	5244	0.04	Y	Y
SARC-17	MLS	Myxoid liposarcoma	MLS	C	0.36	5900	0.06		Y
SARC-18	CSA (A)	Chondrosarcoma	CSA (A)	C	0.53	2638	0.05		Y
SARC-19	WDLS/DDLS	Well differentiated liposarcoma	WDLS/DDLS	C	0.14	7078	0.05	Y	Y
SARC-21	SYSA	synovial sarcoma	SYSA	C	0.23	2623	0.04		Y
SARC-13	MPNST	chondrosarcoma	CHORD	D	0.04	1774	0.06		
SARC-12	WDLS/DDLS	Myxofibrosarcoma	USARC	D	0.07	3072	0.06		Y
SARC-22	AFH	Myxofibrosarcoma	USARC	D	0.04	2749	0.04		
SARC-07	EWS	Unclassified spindle-round cell sarcoma	SRBCS	D	0.91	2830	0.08		Y

Abbreviations: AFH, angiomatoid fibrous histiocytoma; CHORD, chordoma; CSA (A), chondrosarcoma (group A); EMCS, extraskeletal myxoid chondrosarcoma; EWS, Ewing's sarcoma; LMS, leiomyosarcoma; MLS, myxoid liposarcoma; MPNST, malignant peripheral nerve sheath tumor; SFT, solitary fibrous tumor; SYSA, synovial sarcoma; USARC, undifferentiated sarcoma; WDLS/DDLS, well/dedifferentiated liposarcoma; Meth, methylation; Ampl, amplification; Y, yes; C, concordant; D, discordant.

3.2. Low-Overage DNA Methylation Successfully Classifies Sarcoma

The classification of sarcoma samples using the nanoDx Random Forest exhibited a concordance rate of 78% (14/18) with the pathological report (Table 1; File S1). The median confidence score of the Random Forest voting of the concordant classifications is 0.18 (range 0.08–0.88), with half above the threshold (0.15) considered the correct classification of CNS tumors. The Random Forest mean voting rate of the most confident concordant classifications is 11.23% (4.4–27.6%, Table 1).

Of the 18 samples analyzed, 4 (22%) exhibit discordant classification with the pathology report. One (SARC-07) is classified as a Ewing sarcoma (EWS) with a confidence score of 0.91 and a Random Forest voting rate of 30%. According to the pathologist, it is a small round blue cell tumor with no *EWS* translocation detected using FISH.

The two Myxofibrosarcomas samples (SARC-12, SARC-22) were not classified correctly, as well as the chondrosarcoma sample (SARC-13). In these three samples, the confidence score is 0.04 to 0.07.

The two non-tumor samples were discordantly classified as undifferentiated sarcoma (USARC) and malignant peripheral nerve sheath tumor (MPNST) with confidence scores of 0.03 and 0.05, respectively (Table S1; File S1).

3.3. t-SNE Unsupervised Clustering Matches Concordant Classifications

Based on the methylation fingerprint, out of the 14 cases that exhibited concordant classification with the pathology report, 12 (86%) were also clustered by the t-SNE analysis in agreement with the Random Forest classification (Table 1). Of the 4 discordant Random Forest classifications, 2 were not clustered by t-SNE analysis in agreement with the classification. They both had a low-confidence score of 0.04.

Notably, there is a discordant case (SARC-07) in which the final pathological report disagreed with the Random Forest classification. Nevertheless, there is an agreement between the t-SNE clustering and the Random Forest classification. This case achieved the highest confidence score of 0.91 and the highest Random Forest voting rate of 30%.

All the cases in the cohort where the t-SNE clustering disagreed with the Random Forest classification had low-confidence scores (range 0.04–0.10).

The t-SNE clustering also disagreed with the classification of the 2 non-tumor samples (excluded from the cohort) that also achieved low-confidence scores (0.03, 0.05).

3.4. Copy-Number Analysis Detects Typical Sarcoma Alteration

In 5 samples (28%), copy-number analysis identified MDM2 amplification, as depicted in Figure 1A and Table 1. All samples with MDM2 amplifications were classified under the methylation class of 'well/dedifferentiated liposarcoma' (WDLS/DDLS), characterized by MDM2 amplification [7]. In these instances, the MDM2 amplification validates the classification results. In one well-differentiated liposarcoma sarcoma, MDM2 amplification is not identified. The analysis of copy-number variations did not produce any particular modifications linked to alternative subtypes of sarcoma.

4. Discussion

This pilot study presents several key findings. First, in a cohort of 23 surgically resected sarcoma tumors, 20 were successfully sequenced using an Oxford Nanopore device with low-coverage whole-genome sequencing (lcWGS). Out of the 20 samples, 2 did not contain tumor tissue, and 18 tumors were included in the study. The 18 tumors included a limited range of 11 pathologically identified sarcoma types. Among the 18 tumors, 14 were classified in agreement with the pathological report based on their methylation fingerprints. Copy-number alterations were also detected from the same sequencing data and were used to validate the classification. Specifically, MDM2 amplification is successfully identified in five out of six liposarcomas [7].

Our results were accomplished by tailoring the nanoDx classification pipeline specifically for sarcoma tumors [19,20]. We achieved a significant concordant classification rate of 78% (14/18) for the restricted sarcoma types by making a single effective adjustment to the pipeline incorporating our in-house-built sarcoma reference set. This reference set is generated using Illumina Array data [8]. Although Illumina Array has limitations, as discussed in the Introduction section [17], it currently stands as our sole data source for constructing the machine-learning sarcoma training set for methylation-based.

It's worth noting that no changes were made to parameters related to the training of the Random Forest, such as the minimum number of CpG sites required for training or other hyperparameters [20]. The successful implementation of nanopore methylation-based classification using the nanoDx pipeline in this study for sarcomas indicates its potential applicability to other cancer types. It implies that low-pass methylation data obtained through nanopore sequencing might be meaningful and adequate to achieve a satisfactory classification rate using the Random Forest classification approach in other cancers. This potentially can be achieved by making similar adaptations to the current pipeline. However, as Koelsche and von Deimling pointed out, applying a methylation-based approach in hematopoietic tumors may present greater challenges than CNS or mesenchymal-derived tumors. This is primarily due to the already well-established classification system in hematopoietic tumors, which heavily relies on specific mutational events. Since individual mutations do not influence cellular methylation patterns in most cases, further evidence is needed to demonstrate their contribution to the existing classification system of hematopoietic tumors [30].

In CNS nanopore methylation-based classification, a confidence score is implemented. A platform-specific threshold is determined to ensure a more precise interpretation of the classifier results in a clinical context. The classification above a confidence score of 0.15 is considered reliable [20]. This study's median confidence score for the concordant cases is 0.18 (range 0.08–0.87), comparable to the confidence score observed in CNS classification. However, a nanopore-specific confidence score threshold for accurate interpretation has not yet been determined for sarcomas. Establishing such a threshold will help ensure a precise interpretation of the nanopore methylation-based classification results in sarcoma cases.

Factors such as low tumor cell content and DNA quality can influence confidence score values [20]. The current nanopore sequencing method is primarily optimized for fresh tissue samples from biopsies or collected during surgical procedures [20,30]. Using only fresh tissues is a limitation that restricts the ability to select sample regions with high tumor purity. Consequently, this can result in a methylation pattern that significantly deviates from that of cancerous tissue leading to a non-valid classification that is indicated by a low-confidence score. Furthermore, the possibility that the sample belongs to unknown sarcoma entities or different tumor types not represented in the Random Forest classifier training set [8,23] can also contribute to lower confidence scores and potentially discordant classifications.

Genomic alterations detected by copy-number analysis can help achieve more reliable classification, particularly in low-confidence score cases. In our results, six cases with low-confidence scores (range 0.08–0.14) were also classified in agreement with the t-SNE clustering analysis and the pathology report. In three of them, which are classified as liposarcoma (WDLS/DDLS), we identified the MDM2 amplification. This emphasizes the added value of the copy-number profile in validating the methylation-based classification results, mainly when a low-confidence value is achieved. Thus, as pointed out in [8,30], developing additional classifiers combining methylation patterns with other molecular parameters such as sequencing data, proteomic signatures, and histology might increase diagnostic accuracy.

In four cases (22%), the pathology report disagrees with the Random Forest classifications. Three of these cases achieved the lowest confidence score in the cohort (range 0.04–0.07). Moreover, the two non-tumor samples excluded from the cohort achieved a similar low-confidence score (0.03–0.05). Overall, all the cases that achieved a confidence score below 0.08 disagreed with the Random Forest classifications. This might imply that a confidence score lower than 0.08 indicates a non-valid classification in sarcoma. Still, this hypothesis should be tested in further research.

Particular attention should be focused on a specific instance of discordant classification referred to as SARC-07. In this case, the EWS class obtained a confidence score of 0.91. Notably, this class's Random Forest voting rate was remarkably high, at 30%, the highest among the entire cohort. Furthermore, the t-SNE clustering result also aligned with the assigned classification. The pathology report for SARC-07 classifies it as a small round blue cell sarcoma (SRBCS) without *EWS* translocations through FISH analysis. EWS, which frequently manifests as SRBCS, typically involves the prototypical translocation of the *ESWR1* gene with genes from the ETS family [31]. An additional SRBCS type, based on distinct pathology, molecular analysis, and clinical observations indicating a highly aggressive clinical course [32,33], is a distinct entity termed "*CIC* rearranged sarcoma" [34]. Given these findings, we suspect SARC-07 might be a case of *CIC* rearranged sarcoma, warranting molecular reassessment. Further investigation could potentially result in the reclassification of this case, leading to its inclusion among the concordant cases.

The findings of this pilot study should be interpreted with caution, considering the following limitations:

First, the hyperparameters of the Random Forest machine-learning algorithm, such as the minimum number of CpG sites required for model training and confidence score threshold, were determined based on the analysis of CNS tumor methylation data obtained by nanopore sequencing [19,20]. These hyperparameters were not adjusted or explicitly recalibrated to analyze sarcoma nanopore methylation data. Consequently, there is potential to enhance the classification process by rescaling these parameters specifically for sarcoma data.

Second, this study's limited size and diversity of the sarcoma cohort do not adequately represent the wide range of sarcoma types. Therefore, it is impossible to definitively claim that this customized classification pipeline is suitable for reliably classifying all sarcoma subtypes based on nanopore lcWGS methylation data.

Last, it is essential to note that the current reference set used in this study comprised 62 sarcoma methylation classes [8]. The analysis of additional sarcoma samples will contribute to further improvement of this tool [30].

However, it is essential to underscore that achieving a more accurate sarcoma classification can be facilitated by integrating supplementary layers of sarcoma-related data into the statistical analysis. These layers involve transcriptomic and proteomic analyses and consider pathological features and metabolic characteristics. Additionally, incorporating supplementary genomic and molecular data, such as copy-number alterations and point mutations, should be explored in conjunction with the current methylation data.

Despite these limitations, the classification concordance rate is significant, relying only on nanopore methylation data and minimal pipeline adaptations for sarcomas.

5. Conclusions

Previous studies have established the validity of methylation-based classification using Illumina Array methylation data, particularly in sarcomas and CNS tumors [8,14]. Illumina Array has limitations such as GC bias, time-consuming procedures, and reliance on central high-volume laboratories. In contrast, nanopore sequencing devices offer compact, rapid, and accessible technology that detects methylation patterns, point mutations, translocations, and copy-number alterations [16,18]. In the future, we expect that upcoming studies involving nanopore methylation data could provide more suitable and accurate information, potentially presenting an alternative to the currently employed Illumina-based methylation data.

In this study, we successfully customized the nanopore methylation-based classification pipeline for a restricted range of 11 pathologically identified sarcoma tumor types. This highlights its potential for aiding in the timely diagnosis of sarcoma. However, further validation is necessary across a broader range of tumors and in different centers, along with appropriate statistical refinement tailored for sarcomas. A more elaborate classifier that combines methylation patterns with sarcoma-specific CNA, translocations, point mutations, and additional multi-omic data layers is expected to increase accuracy further.

With these advancements, this method can potentially add to sarcoma diagnosis, providing accurate classification in a faster, point-of-care manner. Furthermore, rapid detection of methylation patterns, copy-number alterations, and translocation might be used in the future to plan patient-specific cell-free DNA biomarkers and shed light on sarcoma biology.

Supplementary Materials: The following supporting information can be downloaded at: https://www.mdpi.com/article/10.3390/cancers15164168/s1, File S1: Final nanoDx reports of all 23 cases, including an extract of the pathology microscopic report; Table S1: Supplemental data including additional sequencing and classification information.

Author Contributions: Conceptualization, I.L., A.I., A.Z. and T.P.; methodology, I.L., A.I., A.Z., H.C., M.M., A.E. and J.M.; software, A.I. and N.L.; validation, I.L. and A.Z.; formal analysis, I.L., A.I. and A.Z.; investigation, I.L. and A.I.; resources, A.I., M.A., H.C., J.A.D., M.G.-R., M.W., N.L., N.O., O.O., P.B. and Y.A.; data curation, A.I., I.L., N.L. and H.C.; writing—original draft preparation, A.I.; writing—review and editing, I.L., A.I. and A.Z.; visualization, I.L., A.I. and A.Z.; supervision, I.L. and A.Z.; project administration, I.L. and H.C.; funding acquisition, A.Z. and A.E. All authors have read and agreed to the published version of the manuscript.

Funding: This research was funded by the Israel Science Foundation, grant numbers 2279/22 and 3099/22.

Institutional Review Board Statement: The study is approved by the Hadassah Medical Center institutional review board (# 0346-12).

Informed Consent Statement: Informed consent was obtained from all subjects involved in the study.

Data Availability Statement: Sarcoma reference set data (.h5) will be available upon reasonable request.

Acknowledgments: The nanoDx pipeline was developed by Philipp Euskirchen [19,20] and is available at https://gitlab.com/pesk/nanoDx (accessed on 2 November 2022). The source methylation data for the generation of the sarcoma reference set was prepared by Christian Koelsche et al. from the Andreas von Deimling group [8]. We are grateful for the generous donations of Leslie and Michael Gaffin.

Conflicts of Interest: The authors declare no conflict of interest.

References

1. Yang, J.; Ren, Z.; Du, X.; Hao, M.; Zhou, W. The role of mesenchymal stem/progenitor cells in sarcoma: Update and dispute. *Stem Cell Investig.* **2014**, *1*, 18. [CrossRef] [PubMed]
2. Amin, M.B.; Greene, F.L.; Edge, S.B.; Compton, C.C.; Gershenwald, J.E.; Brookland, R.K.; Meyer, L.; Gress, D.M.; Byrd, D.R.; Winchester, D.P. The Eighth Edition AJCC Cancer Staging Manual: Continuing to build a bridge from a population-based to a more "personalized" approach to cancer staging. *CA Cancer J. Clin.* **2017**, *67*, 93–99. [CrossRef] [PubMed]
3. den Hollander, D.; Van der Graaf, W.T.A.; Fiore, M.; Kasper, B.; Singer, S.; Desar, I.M.E.; Husson, O. Unravelling the heterogeneity of soft tissue and bone sarcoma patients' health-related quality of life: A systematic literature review with focus on tumour location. *ESMO Open* **2020**, *5*, e000914. [CrossRef]
4. Martinez-Trufero, J.; Cruz Jurado, J.; Gomez-Mateo, M.C.; Bernabeu, D.; Floria, L.J.; Lavernia, J.; Sebio, A.; Garcia Del Muro, X.; Alvarez, R.; Correa, R.; et al. Uncommon and peculiar soft tissue sarcomas: Multidisciplinary review and practical recommendations for diagnosis and treatment. Spanish group for Sarcoma research (GEIS—GROUP). Part I. *Cancer Treat. Rev.* **2021**, *99*, 102259. [CrossRef] [PubMed]
5. Hwa Yun, B.; Guo, J.; Bellamri, M.; Turesky, R.J. DNA adducts: Formation, biological effects, and new biospecimens for mass spectrometric measurements in humans. *Mass Spectrom. Rev.* **2020**, *39*, 55–82. [CrossRef]
6. Gounder, M.M.; Agaram, N.P.; Trabucco, S.E.; Robinson, V.; Ferraro, R.A.; Millis, S.Z.; Krishnan, A.; Lee, J.; Attia, S.; Abida, W.; et al. Clinical genomic profiling in the management of patients with soft tissue and bone sarcoma. *Nat. Commun.* **2022**, *13*, 3406. [CrossRef]
7. Nacev, B.A.; Sanchez-Vega, F.; Smith, S.A.; Antonescu, C.R.; Rosenbaum, E.; Shi, H.; Tang, C.; Socci, N.D.; Rana, S.; Gularte-Merida, R.; et al. Clinical sequencing of soft tissue and bone sarcomas delineates diverse genomic landscapes and potential therapeutic targets. *Nat. Commun.* **2022**, *13*, 3405. [CrossRef]
8. Koelsche, C.; Schrimpf, D.; Stichel, D.; Sill, M.; Sahm, F.; Reuss, D.E.; Blattner, M.; Worst, B.; Heilig, C.E.; Beck, K.; et al. Sarcoma classification by DNA methylation profiling. *Nat. Commun.* **2021**, *12*, 498. [CrossRef]
9. Forus, A.; Florenes, V.A.; Maelandsmo, G.M.; Meltzer, P.S.; Fodstad, O.; Myklebost, O. Mapping of amplification units in the q13-14 region of chromosome 12 in human sarcomas: Some amplica do not include MDM2. *Cell Growth Differ.* **1993**, *4*, 1065–1070.

10. Patterson, H.; Barnes, D.; Gill, S.; Spicer, J.; Fisher, C.; Thomas, M.; Grimer, R.; Fletcher, C.; Gusterson, B.; Cooper, C. Amplification and Over-Expression of the MDM2 Gene in Human Soft Tissue Tumours. *Sarcoma* **1997**, *1*, 17–22. [CrossRef]
11. Nakayama, T.; Toguchida, J.; Wadayama, B.; Kanoe, H.; Kotoura, Y.; Sasaki, M.S. MDM2 gene amplification in bone and soft-tissue tumors: Association with tumor progression in differentiated adipose-tissue tumors. *Int. J. Cancer* **1995**, *64*, 342–346. [CrossRef] [PubMed]
12. Fletcher, J.A. Cytogenetics and molecular biology of soft tissue tumors. *Monogr. Pathol.* **1996**, *38*, 37–64. [PubMed]
13. Dor, Y.; Cedar, H. Principles of DNA methylation and their implications for biology and medicine. *Lancet* **2018**, *392*, 777–786. [CrossRef]
14. Capper, D.; Jones, D.T.W.; Sill, M.; Hovestadt, V.; Schrimpf, D.; Sturm, D.; Koelsche, C.; Sahm, F.; Chavez, L.; Reuss, D.E.; et al. DNA methylation-based classification of central nervous system tumours. *Nature* **2018**, *555*, 469–474. [CrossRef] [PubMed]
15. Laird, P.W. The power and the promise of DNA methylation markers. *Nat. Rev. Cancer* **2003**, *3*, 253–266. [CrossRef]
16. Kono, N.; Arakawa, K. Nanopore sequencing: Review of potential applications in functional genomics. *Dev. Growth Differ.* **2019**, *61*, 316–326. [CrossRef] [PubMed]
17. Zhang, J.; Xie, S.; Xu, J.; Liu, H.; Wan, S. Cancer Biomarkers Discovery of Methylation Modification With Direct High-Throughput Nanopore Sequencing. *Front. Genet.* **2021**, *12*, 672804. [CrossRef]
18. Simpson, J.T.; Workman, R.E.; Zuzarte, P.C.; David, M.; Dursi, L.J.; Timp, W. Detecting DNA cytosine methylation using nanopore sequencing. *Nat. Methods* **2017**, *14*, 407–410. [CrossRef]
19. Euskirchen, P.; Bielle, F.; Labreche, K.; Kloosterman, W.P.; Rosenberg, S.; Daniau, M.; Schmitt, C.; Masliah-Planchon, J.; Bourdeaut, F.; Dehais, C.; et al. Same-day genomic and epigenomic diagnosis of brain tumors using real-time nanopore sequencing. *Acta Neuropathol.* **2017**, *134*, 691–703. [CrossRef]
20. Kuschel, L.P.; Hench, J.; Frank, S.; Hench, I.B.; Girard, E.; Blanluet, M.; Masliah-Planchon, J.; Misch, M.; Onken, J.; Czabanka, M.; et al. Robust methylation-based classification of brain tumors using nanopore sequencing. *Neuropathol. Appl. Neurobiol.* **2023**, *49*, e12856. [CrossRef]
21. Pedregosa, F. Scikit-learn: Machine Learning in Python. *J. Mach. Learn. Res.* **2011**, *12*, 2825–2830.
22. Fischer, B.; Smith, M.; Pau, G.; Morgan, M.; van Twisk, D. rhdf5: R Interface to HDF5, R Package version 2.42.0. 2022. Available online: https://github.com/grimbough/rhdf5 (accessed on 29 January 2023).
23. Krijthe, R.; van der Maaten, L.R.; Rtsne, L. Rtsne: T-Distributed Stochastic Neighbor Embedding using a Barnes-Hut Implementation. 2015. Available online: https://github.com/jkrijthe/Rtsne (accessed on 29 January 2023).
24. Li, H. Minimap2: Pairwise alignment for nucleotide sequences. *Bioinformatics* **2018**, *34*, 3094–3100. [CrossRef] [PubMed]
25. Ellison, D.W.; Kocak, M.; Figarella-Branger, D.; Felice, G.; Catherine, G.; Pietsch, T.; Frappaz, D.; Massimino, M.; Grill, J.; Boyett, J.M.; et al. Histopathological grading of pediatric ependymoma: Reproducibility and clinical relevance in European trial cohorts. *J. Negat. Results BioMed.* **2011**, *10*, 7. [CrossRef] [PubMed]
26. Muschkin, C.G.; Patterson, C.J. Aging trends—Puerto Rico. *J. Cross Cult. Gerontol.* **1997**, *12*, 373–385. [CrossRef]
27. Stupp, R.; Hegi, M.E.; Mason, W.P.; van den Bent, M.J.; Taphoorn, M.J.; Janzer, R.C.; Ludwin, S.K.; Allgeier, A.; Fisher, B.; Belanger, K.; et al. Effects of radiotherapy with concomitant and adjuvant temozolomide versus radiotherapy alone on survival in glioblastoma in a randomised phase III study: 5-year analysis of the EORTC-NCIC trial. *Lancet Oncol.* **2009**, *10*, 459–466. [CrossRef]
28. Pembaur, A.; Sallard, E.; Weil, P.P.; Ortelt, J.; Ahmad-Nejad, P.; Postberg, J. Simplified Point-of-Care Full SARS-CoV-2 Genome Sequencing Using Nanopore Technology. *Microorganisms* **2021**, *9*, 2598. [CrossRef]
29. Chavala, S.H.; Kim, Y.; Tudisco, L.; Cicatiello, V.; Milde, T.; Kerur, N.; Claros, N.; Yanni, S.; Guaiquil, V.H.; Hauswirth, W.W.; et al. Retinal angiogenesis suppression through small molecule activation of p53. *J. Clin. Invest.* **2013**, *123*, 4170–4181. [CrossRef]
30. Koelsche, C.; von Deimling, A. Methylation classifiers: Brain tumors, sarcomas, and what's next. *Genes Chromosomes Cancer* **2022**, *61*, 346–355. [CrossRef]
31. Sbaraglia, M.; Righi, A.; Gambarotti, M.; Dei Tos, A.P. Ewing sarcoma and Ewing-like tumors. *Virchows Arch.* **2020**, *476*, 109–119. [CrossRef]
32. Antonescu, C.R.; Owosho, A.A.; Zhang, L.; Chen, S.; Deniz, K.; Huryn, J.M.; Kao, Y.C.; Huang, S.C.; Singer, S.; Tap, W.; et al. Sarcomas With CIC-rearrangements Are a Distinct Pathologic Entity With Aggressive Outcome: A Clinicopathologic and Molecular Study of 115 Cases. *Am. J. Surg. Pathol.* **2017**, *41*, 941–949. [CrossRef]
33. Yoshida, A.; Goto, K.; Kodaira, M.; Kobayashi, E.; Kawamoto, H.; Mori, T.; Yoshimoto, S.; Endo, O.; Kodama, N.; Kushima, R.; et al. CIC-rearranged Sarcomas: A Study of 20 Cases and Comparisons With Ewing Sarcomas. *Am. J. Surg. Pathol.* **2016**, *40*, 313–323. [CrossRef] [PubMed]
34. Sbaraglia, M.; Bellan, E.; Dei Tos, A.P. The 2020 WHO Classification of Soft Tissue Tumours: News and perspectives. *Pathologica* **2021**, *113*, 70–84. [CrossRef] [PubMed]

cancers

MDPI

Article

High-Grade Pleomorphic Sarcomas Treated with Immune Checkpoint Blockade: The MD Anderson Cancer Center Experience

Lewis F. Nasr [1,†], Marianne Zoghbi [1,†], Rossana Lazcano [2], Michael Nakazawa [3], Andrew J. Bishop [4], Ahsan Farooqi [4], Devarati Mitra [4], Beverly Ashleigh Guadagnolo [4], Robert Benjamin [3], Shreyaskumar Patel [3], Vinod Ravi [3], Dejka M. Araujo [3], Andrew Livingston [3], Maria A. Zarzour [3], Anthony P. Conley [3], Ravin Ratan [3], Neeta Somaiah [3], Alexander J. Lazar [2,5], Christina Roland [6], Emily Z. Keung [6] and Elise F. Nassif Haddad [3,7,*]

1 Department of Leukemia, University of Texas MD Anderson Cancer Center, Houston, TX 77030, USA; lfnasr@mdanderson.org (L.F.N.)
2 Department of Translational Molecular Pathology, University of Texas MD Anderson Cancer Center, Houston, TX 77030, USA
3 Department of Sarcoma Medical Oncology, University of Texas MD Anderson Cancer Center, Houston, TX 77030, USA; rbenjami@mdanderson.org (R.B.); maflores@mdanderson.org (M.A.Z.)
4 Department of Radiation Oncology, University of Texas MD Anderson Cancer Center, Houston, TX 77030, USA; abishop2@mdanderson.org (A.J.B.); afarooqi@mdanderson.org (A.F.)
5 Department of Genomic Medicine, University of Texas MD Anderson Cancer Center, Houston, TX 77030, USA
6 Department of Surgical Oncology, University of Texas MD Anderson Cancer Center, Houston, TX 77030, USA; clroland@mdanderson.org (C.R.)
7 Department of Investigational Cancer Therapeutics, University of Texas MD Anderson Cancer Center, Houston, TX 77030, USA
* Correspondence: efnassif@mdanderson.org
† These authors contributed equally to this work.

Simple Summary: Undifferentiated pleomorphic sarcomas (UPSs) represent 10–20% of all soft tissue sarcomas (STSs) and have quickly emerged as one of the more immune-sensitive types. There are few real-world data on the use of immune checkpoint blockade (ICB) in UPS patients and those with other high-grade pleomorphic STSs. This is a retrospective, observational study of all patients with metastatic high-grade pleomorphic sarcomas treated with FDA-approved ICB at MD Anderson Cancer Center intended to describe the efficacy and toxicity of ICB in this particular group of patients. We find that our outcomes are comparable to those in the published literature and pose a question regarding the need to further evaluate the optimal sequencing of radiotherapy and prior lines of systemic therapy.

Abstract: Background: Undifferentiated pleomorphic sarcomas (UPSs) are amongst the most common subtypes of soft-tissue sarcomas. Few real-world data on the use of immune checkpoint blockade (ICB) in UPS patients and other high-grade pleomorphic STS patients are available. Purpose: The purpose of our study is to describe the efficacy and toxicity of ICB in patients with advanced UPSs and other high-grade pleomorphic sarcomas treated at our institution. Methods: This is a retrospective, observational study of all patients with metastatic high-grade pleomorphic sarcomas treated with FDA-approved ICB at MD Anderson Cancer Center between 1 January 2015 and 1 January 2023. Patients included in trials for which results are not yet published were excluded. Results: Thirty-six patients with advanced/metastatic pleomorphic sarcomas were included. The median age was 52 years. A total of 26 patients (72%) had UPSs and 10 patients (28%) had other high-grade pleomorphic sarcomas. The median follow-up time was 8.8 months. The median PFS was 2.9 months. The 3-month PFS and 6-month PFS were 46% and 32%, respectively. The median OS was 12.9 months. The 12-month OS and 24-month OS were 53% and 29%, respectively. The best response, previous RT, and type of ICB treatment were significantly and independently associated with shorter PFS ($p = 0.0012$, $p = 0.0019$ and $p = 0.036$, respectively). No new safety signal was identified, and the toxicity was overall manageable with no toxic deaths and only four patients (11%) stopping treatment

Citation: Nasr, L.F.; Zoghbi, M.; Lazcano, R.; Nakazawa, M.; Bishop, A.J.; Farooqi, A.; Mitra, D.; Guadagnolo, B.A.; Benjamin, R.; Patel, S.; et al. High-Grade Pleomorphic Sarcomas Treated with Immune Checkpoint Blockade: The MD Anderson Cancer Center Experience. *Cancers* **2024**, *16*, 1763. https://doi.org/10.3390/cancers16091763

Received: 19 March 2024
Revised: 24 April 2024
Accepted: 29 April 2024
Published: 1 May 2024

due to toxicity. Conclusions: Real-world retrospective data are consistent with the published literature, with a promising 6-month PFS of 32%. Partial or stable responders to ICB treatment have significantly improved PFS compared to progressors.

Keywords: immunotherapy; anti-PD1; real world; sarcoma; undifferentiated pleomorphic sarcomas; survival; clinical

1. Introduction

Undifferentiated pleomorphic sarcomas (UPSs) are amongst the most common subtypes of soft-tissue sarcomas (STSs), representing 10–20% of all STSs [1,2]. Morphologically, UPS consists of pleomorphic spindle cells with no specific line of differentiation [3] and is a specific entity distinct from other high-grade pleomorphic sarcomas not otherwise specified [4,5]. The mainstay of advanced/metastatic treatment for STSs in general and UPSs in particular consists of systemic therapies, with doxorubicin-based [6,7] and gemcitabine-based treatments used as front-line treatment, with a median progression-free survival (PFS) of around 6 months and a median overall survival (OS) of around 18–20 months [2,8–12].

UPS has quickly emerged as one of the more immune-sensitive types of STS, which was identified during the first immune checkpoint blockade (ICB) clinical trials in patients with advanced and pretreated STS with an objective response rate (ORR) of 20–40% and a median PFS of 3 months [13–19], whereas other types of sarcomas such as leiomyosarcomas or synovial sarcomas have an ORR < 10% with ICB. Building upon this, several clinical trials have looked at the role of combination therapies with ICB across several STS types, including UPSs, and demonstrated increased efficacy of combination treatments over single agents [20–23]. These combinations are now moving to the earlier setting including first-line advanced [19] and peri-operative settings [24].

While there is an increasing number of ICB-based clinical trials with UPS patients, few real-world data on the use of ICB in UPS patients and other high-grade pleomorphic STS patients are available. Thus, our aim was to describe the efficacy and toxicity of ICB in patients with advanced UPSs and other high-grade pleomorphic sarcomas treated at our institution.

2. Materials and Methods

2.1. Study Design

This is a retrospective, observational study of all patients with high-grade pleomorphic sarcomas treated at MD Anderson Cancer Center between 1 January 2015 and 1 January 2023 with anti-PDL1 and anti-CTLA4 ICB. Patients were identified through our MD Anderson Cancer Center's pharmacy database using the following molecule names: ipilimumab, nivolumab, durvalumab, tremelimumab, atezolizumab, and pembrolizumab. Patients receiving ICB outside of our institution were not identified and, thus, not included. Patients with at least one month of follow-up after initiation of ICB were included. To keep a homogeneous cohort, patients with localized disease receiving ICB treatment in the neoadjuvant setting were excluded. Patients included in trials for which results are not yet published were excluded.

Clinical variables recorded included demographic characteristics such as sex, age, BMI, race and ethnicity, and European Cooperative Oncology Group (ECOG) performance status at initiation of ICB treatment. Disease-associated variables collected included the site of the primary tumor, stage at ICB treatment (locally recurrent/advanced or metastatic), location of metastasis if present, tumor size (biggest dimension, evaluated per RECIST criteria) [25] at ICB, and histologic type. Prior treatment modalities were recorded, including systemic therapies, radiation therapy (RT), and surgical resections. For patients with multiple surgical resections, treatment modalities (chemotherapy regimens and RT) between each surgical resection and any eventual recurrence were recorded. Variables pertaining to ICB treatment included the type of ICB treatment, whether it was administered as a standalone or in combination with another type of systemic therapy or RT, the best radiographic response by Response Evaluation Criteria In Solid Tumors version 1.1 (RECIST1.1) [25], any toxicities experienced, time on treatment, time to progression, and the reason for treatment discontinuation. The last known status of each patient was censored as of 5 April 2023.

Pathology was evaluated by experienced pathologists (RN and AL) in soft-tissue tumors to differentiate between UPSs and other high-grade pleomorphic sarcomas.

ORR is defined as the percentage of patients who achieve a response, whether complete response (complete disappearance of lesions) or partial response (reduction in the sum of maximal tumor diameters by at least 30% or more) per RECIST1.1. Clinical benefit rate (CBR) is defined as the percentage of advanced cancer patients who achieve complete response, partial response, or stable disease for at least 6 months as a result of therapy [25].

This retrospective study of patients treated with sarcomas was approved by the Institutional Review Board.

2.2. Statistical Considerations

Categorical variables were reported as percentages and continuous variables as medians and interquartile ranges (IQRs). Comparisons between categorical variables were conducted using Fisher's exact test. Progression-free survival (PFS) was defined as the time from initiation of ICB treatment to radiographic or clinical progression, death of any cause, or last follow-up, whichever occurred first. Overall survival (OS) was defined as the time from initiation of ICB treatment to death of any cause or last follow-up. The median PFS and OS were calculated using the Kaplan–Meier method, and a 95% confidence interval (95% CI) was estimated. We used the log-rank method to compare the significance of differences between survival curves. The association of clinical factors (disease stage, location of primary, sex, age, ECOG performance status (PS), presence of lung or liver metastasis, number of prior systemic lines, prior receipt of RT, type of ICB received and combination or single agent, and radiographic best response RECIST1.1) with PFS and OS was assessed using Cox univariate and multivariable proportional hazard models. Only those variables that were associated with survival at $p < 0.1$ in univariate models were included in the multivariable model. All analyses were performed using GraphPad Prism version 9 and IBM SPSS version 26.

3. Results

3.1. Patient Characteristics

Thirty-six patients with advanced/metastatic pleomorphic sarcomas were included in this study. Patient demographics and disease characteristics are outlined in Table 1. The median age of the patients at the time of initiation of ICB treatment was 52 years (range: 22–79), and 66% were male (n = 24/36). Most patients had an ECOG performance status of 1 (n = 23; 64%). The median body mass index (BMI) at the time of ICB treatment was 30.3 (range: 20–52). The disease histology was UPS in 72% of patients (n = 26/36). The median size of the biggest tumor at the start of ICB treatment was 6 cm (range: 1.3–25). The site of the primary tumor was divided between extremities (n = 15; 42%), trunk (n = 15; 42%), and other (n = 6; 16%), respectively. Lung and liver metastases were seen in 72% (n = 26) and 11% (n = 4) of patients, respectively.

Regarding treatment modalities prior to ICB treatment, 61% (n = 22/36) of patients had prior RT, including neoadjuvant or adjuvant RT for primary sarcoma in 36% (n = 8/22) of cases and palliative intent RT in 64% (n = 14/22) of cases. The median time between the last RT treatment and initiation of ICB was 12 months (IQR: 1.9–49.7). Eight patients (22%) received subsequent courses of RT at any time after the start of ICB with a median of two RT treatments (IQR: 1–3). The median number of surgical resections prior to ICB treatment was one (range: 0–8), and the median number of lines of systemic therapy prior to ICB treatment was two (range: 0–10). Patients received prior anthracycline-based and gemcitabine-based chemotherapy in 72% (n = 26/36) and 75% (n = 27/36) of cases, respectively. As 53%

(n = 19/36) of patients received their first line of chemotherapy in the peri-operative setting (neoadjuvant and/or adjuvant), the median number of prior systemic lines in the metastatic setting was 1.5 (range 0–6).

Table 1. Patient and disease characteristics.

Characteristic	Category	N (%)/Median [Range]		
		Whole Cohort N = 36	UPS N = 26	Other N = 10
Age at first ICB (years)		52 [22–79]	52 [22–79]	53 [30–79]
Sex	Male Female	24 (66) 12 (33)	18 (69)	6 (60)
Race	Caucasian African American	31 (86) 5 (14)	22 (85)	9 (90)
ECOG performance status	0 1 2	11 (30) 23 (64) 2 (6)	10 (38)	1 (10)
BMI		30.3 [20–52]	31 [21–50]	26 [20–52]
Histology	UPS Other unclassified pleomorphic	26 (72) 10 (28)	26 (100) 0	0 10 (100%)
Biggest tumor diameter at start of ICB (cm)		6 [1.3–25]	5.3 [1.3–22]	6.7 [1.5–25]
Lung metastasis		26 (72)	18 (69)	8 (80)
Liver metastasis		4 (11)	2 (8)	2 (20)
Site of primary tumor	Extremities Trunk Other *	15 (42) 15 (42) 6 (16)	12 (46) 10 (38) 4 (16)	3 (30) 5 (50) 2 (20)
Previous RT prior to ICB	Yes	22 (61)	17 (65)	5 (50)
Number of surgical resections prior to ICB		1 [0–8]	1 [0–6]	2 [1–8]
Number of lines of systemic therapy prior to ICB		2 [0–10]	2 [0–5]	4 [1–10]
Type of ICB ** treatment	Standalone ICB Combination of ICB + ICB Combination of ICB + RT Combination of ICB + chemo or antiangiogenic	15 (42) 16 (44) 3 (8) 2 (6)	11 (42) 10 (38) 3 (12) 2 (8)	4 (40) 6 (60) 0 0
Received ICB as part of a clinical trial		25 (69)	18 (69)	7 (70)
Best response to ICB	Partial/complete response Stable disease Progressive disease Unknown	3 (8) 9 (25) 21 (58) 3 (9)	3 (12) 6 (23) 15 (58) 2 (7)	0 3 (30) 6 (60) 1 (10)

* Other sites of disease include abdomen, heart, and lumbar spine, ** ICB drugs used: atezolizumab, durvalumab, ipilimumab, nivolumab, pembrolizumab, tremelimumab. Abbreviations: ICB, immune checkpoint blockade; ECOG, Eastern Cooperative Oncology Group; BMI, Body Mass Index; UPS, undifferentiated pleomorphic sarcoma; RT, radiotherapy.

The ICB treatments received were atezolizumab (n = 2), durvalumab (n = 11), ipilimumab (n = 2), nivolumab (n = 5), pembrolizumab (n = 14), tremelimumab (n = 11), and other PDL1 inhibitors (n = 3). ICB treatment was given either as single-agent standalone therapy (n = 15; 42%), in combination with another ICB agent (n= 16; 44%), in combination with RT (n= 3; 8%), or in combination with chemotherapy or an antiangiogenic (n = 2; 6%). Twenty-five patients (69%) received ICB as part of a clinical trial.

3.2. Responses to ICB

The best response of patients to treatment was progression of disease (n = 21; 58%), stable disease (n = 9; 25%), partial response (n = 2; 6%), or complete response (n = 1; 3%). One patient with UPS achieved CR after seven cycles of pembrolizumab, having received prior RT (56 months prior to ICB) and one prior line of gemcitabine-based systemic therapy. Three patients did not have an evaluable response due to logistical issues related to insurance coverage that prevented imaging at MD Anderson after ICB start. Responses are illustrated in the waterfall plot shown in Figure 1.

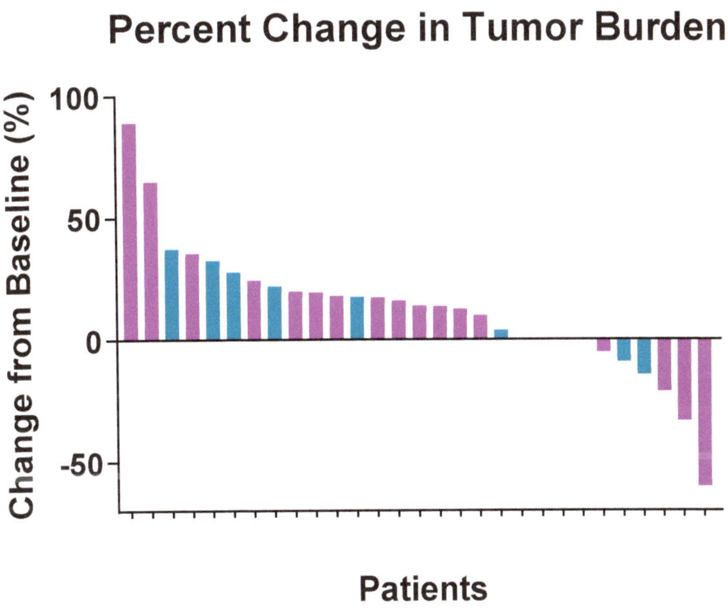

Figure 1. Waterfall plot by histology: purple represents UPS, and blue represents other pleomorphic sarcomas.

In the 33 patients with an evaluable response by imaging, the ORR was 9.1% (n = 3/33) and the CBR was 30.3% (n = 10/33). Table 2 shows the response rates according to the studied clinical characteristics. Previous RT, sex, BMI, histology, presence of liver or lung metastasis, number of previous lines of systemic therapy, and type of ICB treatment did not statistically significantly affect CBR. None of the aforementioned characteristics had a significant impact on ORR either.

Table 2. Response rates according to clinical characteristics.

Variable N (%)	Number of Patients *	Objective Response Rate (ORR)	ORR *p*-Value	Clinical Benefit Rate (CBR)	CBR *p*-Value
All cohort	33	3 (9.1)	-	10 (30.3)	-
Histology		-		-	
UPS	24	3 (12.5)	0.54	7 (29.2)	1
Other	9	0 (0)		3 (33.3)	

Table 2. *Cont.*

Variable N (%)	Number of Patients *	Objective Response Rate (ORR)	ORR p-Value	Clinical Benefit Rate (CBR)	CBR p-Value
Previous Radiotherapy		-		-	
Yes	20	2 (10)	0.54	4 (20)	0.14
No	13	1 (7.7)		6 (46)	
Previous Radiotherapy Intent					
Peri-op	12	1 (8)	1	2 (17)	1
Palliative	8	1 (12.5)		2 (25)	
Number of Previous Systemic Therapies		-		-	
≤2	18	3 (16.7)	0.233	7 (38.9)	0.28
>2	15	0 (0)		3 (20)	
Lung Metastasis		-		-	
Yes	24	3 (12.5)	0.54	9 (37.5)	0.22
No	9	0 (0)		1 (11.1)	
Liver Metastasis		-		-	
Yes	3	0 (0)	1	0 (0)	0.54
No	30	3 (10)		9 (30)	
ICB Combination Type		-		-	
Standalone	13	2 (15.4)	0.55	5 (38.5)	0.46
Combination	20	1 (5)		5 (25)	
Sex		-		-	
Male	21	2 (9.5)	1	6 (28.6)	1
Female	12	1 (9.1)		3 (27.3)	
BMI		-		-	
<25	8	0 (0)	0.56	1 (12.5)	0.38
>25	25	3 (12)		9 (36)	

* A total of 33 patients were evaluable for response.

3.3. Progression-Free Survival with ICB

With a median follow-up time of 8.8 months, the median PFS was 2.9 months, as seen in Figure 2A. The 3-month PFS and 6-month PFS were 46% and 32%, respectively.

The median PFS was 2.9 months and 3.8 months in the UPS group and in the other high-grade pleomorphic sarcoma group, respectively.

In univariate analyses of PFS (Table 3; Figure 3), ICB combination was associated with significantly shorter PFS (combination: 2.3 months vs. no combination 9.2 months, $p = 0.021$) while previous RT, sex, histology, age, presence of liver or lung metastasis, and number of previous lines of systemic therapy did not. Sarcoma histology (UPSs vs. other high-grade pleomorphic sarcomas) did not have a statistically significant impact on PFS ($p = 0.93$) (Figure 4A). The previous number of systemic lines of therapies (>2 vs. 1 or 2) was not significantly associated with PFS despite an HR of 2.06 ($p = 0.053$). Likewise, previous RT was not significantly associated with PFS despite an HR of 0.49 ($p = 0.054$).

The best response to ICB treatment was significantly associated with PFS ($p = 0.0012$) (Figure 5A): the median PFS was 40.7 months and 5.6 months in patients whose tumors responded (PR/CR) and stabilized (SD) per RECIST 1.1, respectively, compared to 2.2 months in non-responders.

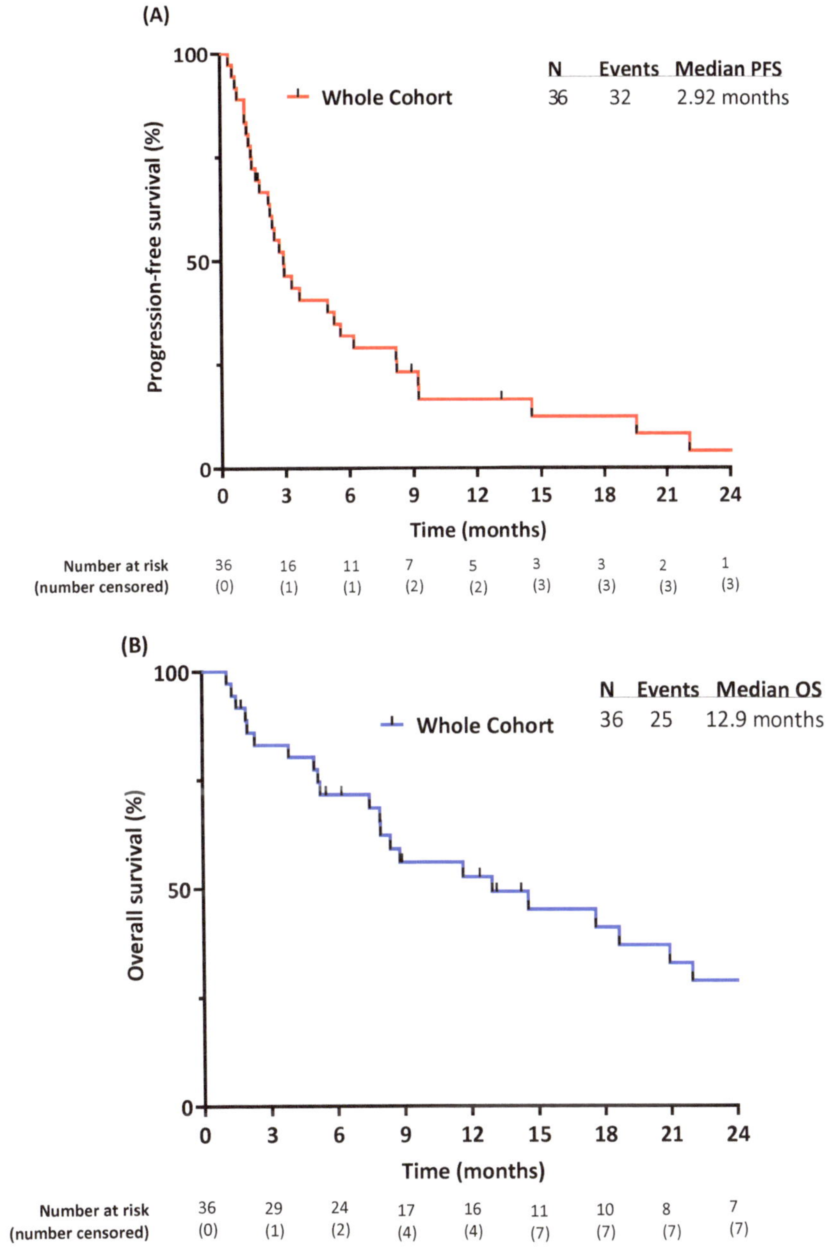

Figure 2. Outcomes of the whole cohort: (**A**) progression-free survival; (**B**) overall survival.

Table 3. Progression-free survival according to clinical characteristics.

Variable	Number of Patients	Median PFS (Months, IQR)	Univariate HR (95%CI)	*p*-Value	Cox Multivariate (a) HR (95%CI)	*p*-Value
Sex	-	-	1.33 (0.60–2.92)	0.48		
Male	24	2.7 (1.6–9)	-	-		
Female	12	3.1 (2–5.3)	-	-		
Race	-	-	0.6 (1.9–2.2)	0.4		
Caucasian	31	3.3	-	-		
African American	5	1.6	-	-		
Age	-	-	1.2 (0.5–2.6)	0.68		
0–65	25	3 (1.5–8.8)	-	-		
>65 years	11	2.5 (1.6–6.6)	-	-		
Performance Status	-					
0, 1						
2+						
Histology	-	-	0.97 (0.44–2.09)			
UPS	26	2.9 (1.68–7.64)	-	0.92		
Other unclassified pleomorphic	10	3.8 (1.64–6.21)	-			
Previous Radiotherapy	-	-	0.49 (0.24–1.01)	0.054		
Yes	22	2.2 (1.3–3.5)	-	-	**0.39 (0.18–0.86)**	**0.019**
No	14	5.4 (3–13.1)	-	-		
ICB Combination Type	-	-	**0.42 (0.23–0.92)**	**0.0207**	**0.4 (0.17–0.94)**	
Standalone	15	9.2 (2.1–11.2)	-	-		**0.036**
Combination	21	2.3 (1.4–3.32)	-	-		
Number of Previous Systemic Therapies	-	-	2.06 (0.93–4.29)	0.053		
≤2	19	3.68 (1.9–9.2)	-	-	0.65 (0.29–1.45)	0.29
>2	17	2.4 (1.3–5.3)	-	-		
Lung Metastasis	-	-	0.95 (0.45–2.0)	0.94		
Yes	26	3 (1.4–5.6)	-	-		
No	10	2.8 (1.9–8.5)		-		
Liver Metastasis	-	-	1.012 (0.34–3.0)	0.98		
Yes	4	2.5 (2–7.6)	-	-		
No	32	3 (1.5–7.2)		-		

PFS = progression-free survival; ICB = immune checkpoint blockade; HR = hazard ratio; RT = radiotherapy; UPS = undifferentiated pleomorphic sarcoma; (a) multivariate model included previous radiotherapy, number of previous systemic therapies, and type of ICB combination. Values in bold have *p*-values < 0.05.

In the multivariate analysis including RT prior to ICB treatment, the number of previous systemic therapies (>2 vs. 1 or 2), and the type of ICB treatment (standalone vs. combination), only previous RT and the type of ICB treatment were significantly and independently associated with shorter PFS (p = 0.0019 and p = 0.036, respectively, Table 3). There was no significant difference in PFS (p = 0.52) between patients receiving peri-operative RT and patients receiving RT with palliative intent in the metastatic setting (Supplementary Figure S1).

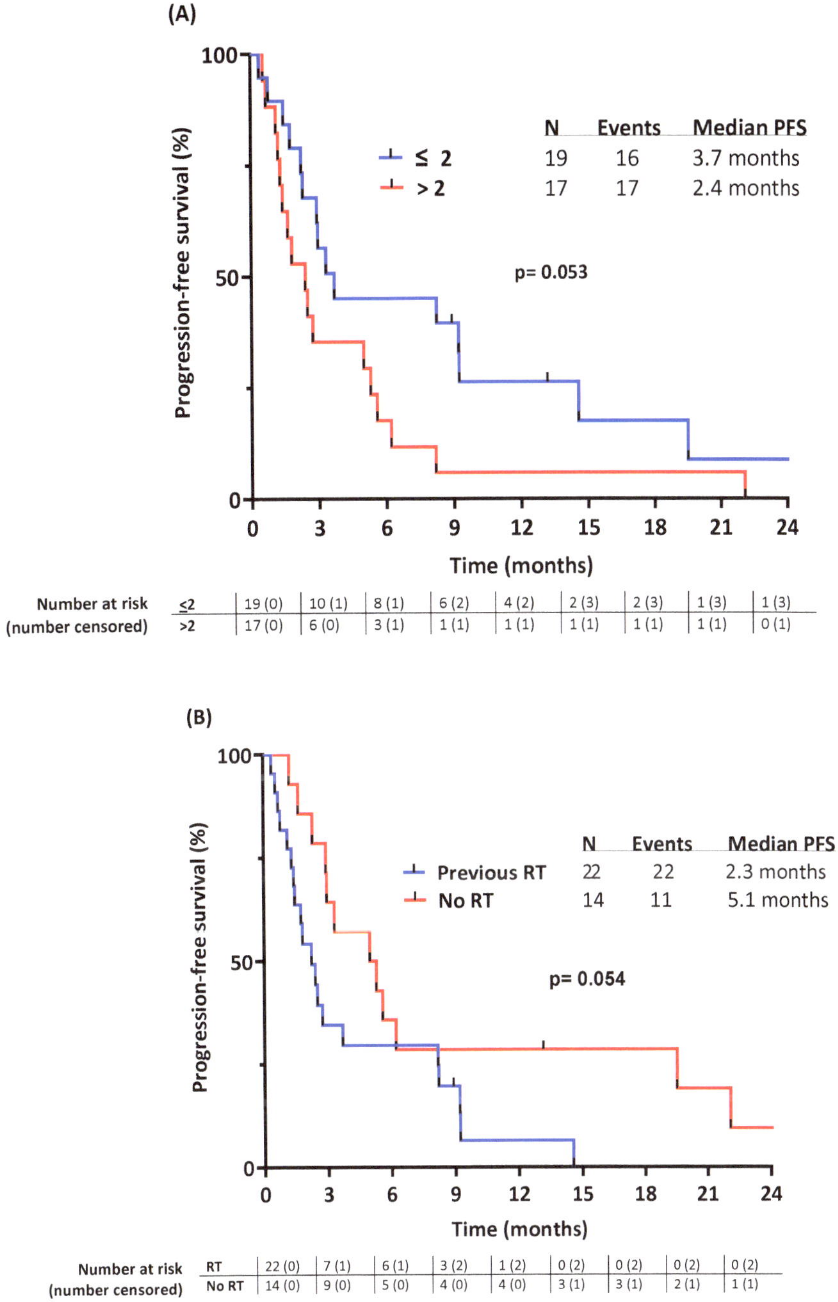

Figure 3. *Cont.*

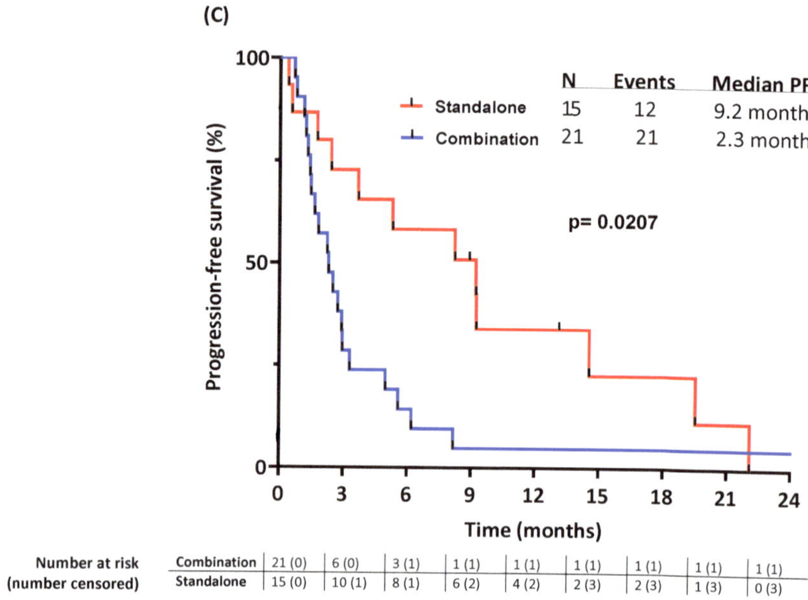

Figure 3. Progression-free survival stratified by (**A**) number of lines of systemic therapy prior to ICB, (**B**) exposure to radiotherapy (RT) prior to ICB, and (**C**) type of ICB treatment.

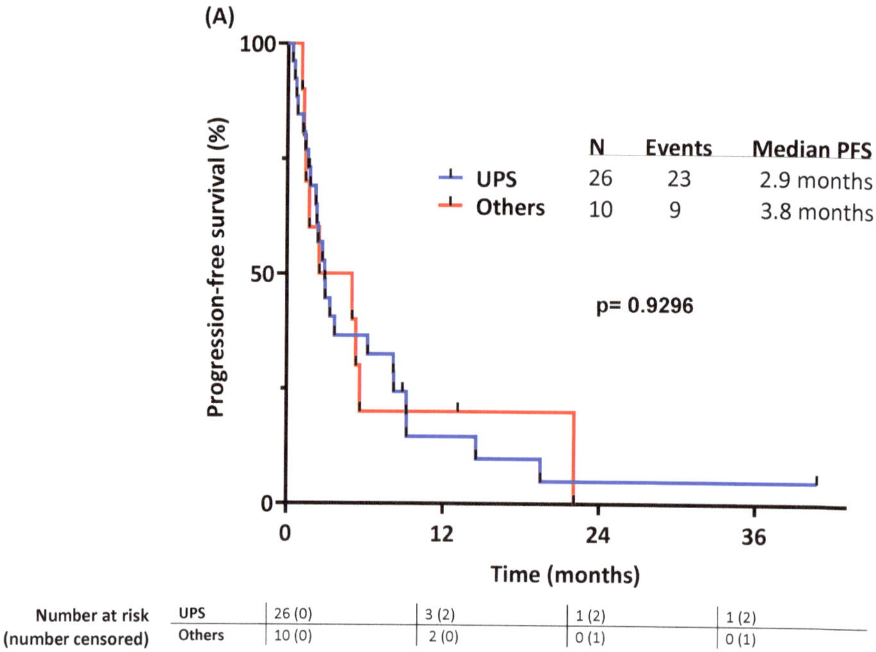

Figure 4. *Cont.*

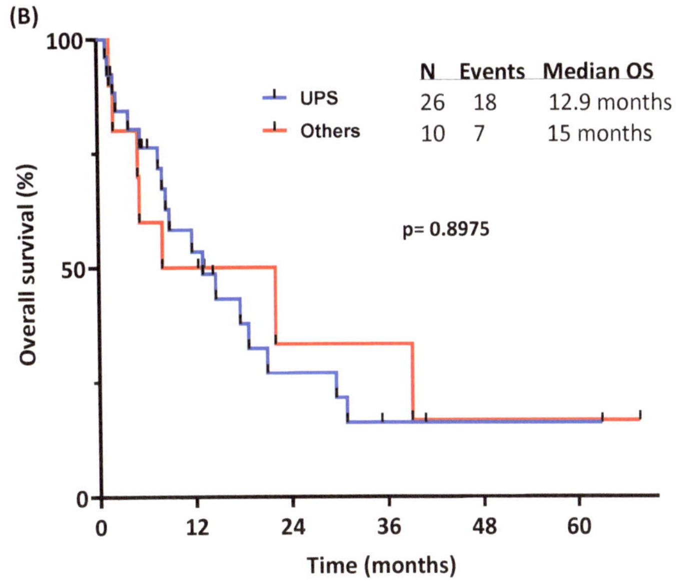

Number at risk (number censored)

UPS	26 (0)	11 (4)	5 (5)	2 (6)	1 (7)	1 (7)
Others	10 (0)	5 (0)	2 (2)	2 (2)	1 (2)	1 (2)

Figure 4. Outcomes stratified by histology: (**A**) progression-free survival; (**B**) overall survival.

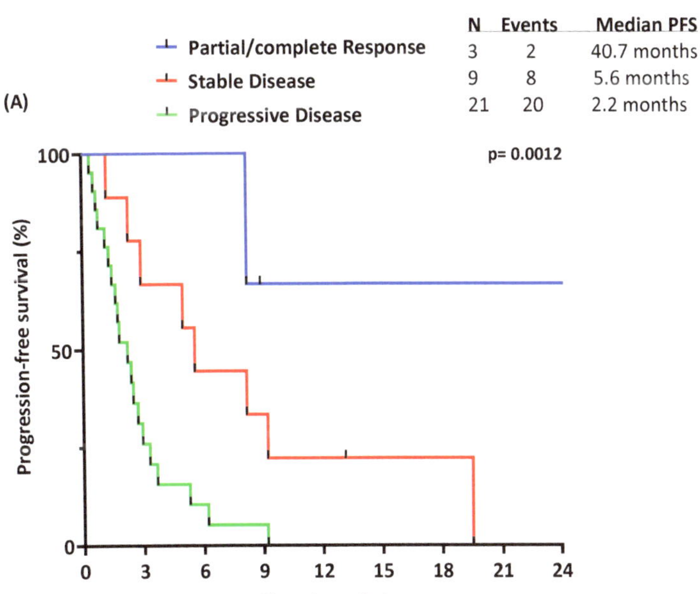

Number at risk (Number censored)

Partial/Complete	3 (0)	3 (0)	3 (0)	1 (1)	1 (1)	1 (1)	1 (1)	1 (1)	1 (1)
Stable	9 (0)	6 (0)	4 (0)	3 (0)	2 (0)	1 (1)	1 (1)	0 (1)	0 (1)
Progressive	21 (0)	5 (1)	2 (1)	2 (1)	0 (1)	0 (1)	0 (1)	0 (1)	0 (1)

Figure 5. *Cont.*

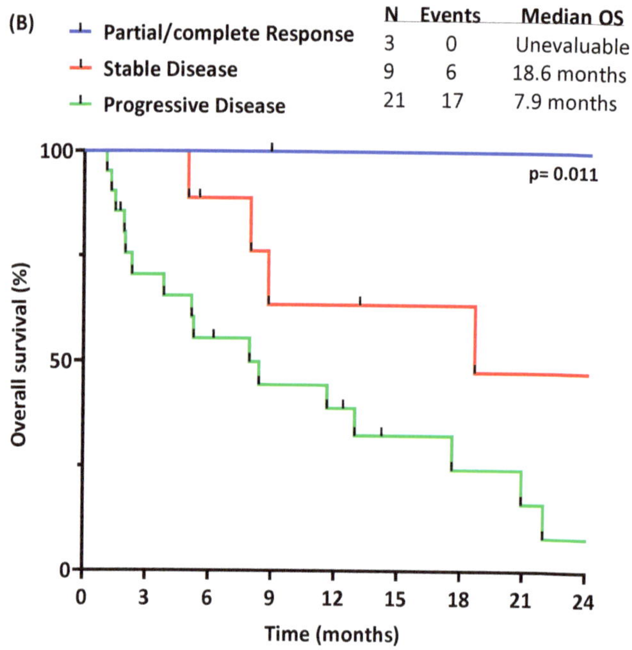

Number at risk (Number censored)

	0	3	6	9	12	15	18	21	24
Partial/Complete	3 (0)	3 (0)	3 (0)	2 (1)	2 (1)	2 (1)	2 (1)	2 (1)	2 (1)
Stable	9 (0)	9 (0)	7 (1)	5 (1)	5 (1)	4 (2)	4 (2)	3 (2)	3 (2)
Progressive	21 (0)	14 (1)	11 (1)	8 (2)	7 (2)	4 (4)	3 (3)	2 (3)	1 (3)

Figure 5. Outcomes stratified by best response to ICB: (**A**) progression-free survival; (**B**) overall survival.

3.4. Overall Survival with ICB

The median OS was 12.9 months in the whole cohort (Figure 2B). The 12-month OS and 24-month OS rates were 53% and 29%, respectively.

The median OS was 12.9 months and 15 months in the UPS group and in the other high-grade pleomorphic group, respectively.

In the univariate analyses for OS, previous RT had a statistically significant impact on OS (HR = 0.44, previous RT 7.9 months vs. no previous RT 17.5 months, $p = 0.047$, Supplementary Table S1 and Supplementary Figure S2) while age, race and ethnicity, sex, histology, presence of liver or lung metastasis, number of previous lines of systemic therapy (Supplementary Figure S3), and type of ICB combination (Supplementary Figure S4) did not. Sarcoma histology (UPSs vs. other pleomorphic sarcomas) did not have a statistically significant impact on OS ($p = 0.90$) (Figure 4B).

The best response to ICB treatment per RECIST1.1 significantly impacted OS ($p = 0.011$) (Figure 5B): the median OS was not attained and 18.6 months in patients whose tumors responded and stabilized per RECIST 1.1, respectively.

3.5. Previous Radiation Therapy

Due to the results on the impact of previous radiation therapy, we performed additional analyses to compare patients who had had previous RT and those who did not. While no significant difference was found between these two groups, patients who had previous RT tended to have larger tumors (biggest diameter 7.7 cm vs. 4 cm; Supplementary Table S2).

Additionally, we performed a subgroup analysis based on the timing of RT prior to ICB. This analysis showed no difference in PFS or OS based on the timing of RT (peri-operative vs. palliative/metastatic setting, Supplementary Figure S1).

3.6. ICB Combination

Due to the results on the impact of ICB combination therapy, we performed additional analyses to compare patients who had standalone ICB therapy and those who did not. Patients who had ICB combinations tended to have more prior surgical resections (median of 2 vs. 1 surgical resections for standalone ICB, $p = 0.013$) and more prior lines of therapy (median 3 vs. 1, $p = 0.062$), suggesting that this population was more heavily pretreated. Patients who had ICB combinations also tended to have received ICB as part of a clinical trial (90% of combination ICB vs. 40% of standalone ICB patients; Supplementary Table S3).

3.7. Toxicity

No new safety signal was identified, and the toxicity was overall manageable. There were no toxic deaths. Four patients (11%) stopped treatment due to toxicity. Seven patients (19%) experienced > grade 3 toxicities. Five patients (14%) experienced diarrhea or colitis of any grade.

4. Discussion

This report aims to describe real-world data on ICB treatment for patients with UPSs and other high-grade pleomorphic STSs. While several prospective clinical trials have identified this subtype of STS as one of the most immune-sensitive types of STS [13,14,26], few real-world histology-specific data exist.

The data reported here are overall consistent with that reported in clinical trials with a median PFS of roughly 3 months for UPS patients [13,14]; however, the ORR was slightly inferior to that previously reported, as ours is around 10% versus 20–30% in clinical trials. Real-world data are often slightly less promising than clinical trials, as patients are usually frailer and have more comorbidities and prior lines of treatment. We also found that ICB in later lines of therapy may be less effective and that RECIST response is associated with PFS with ICB treatment, which is consistent with reports across other cancer types [27–33].

High-grade pleomorphic sarcomas that do not meet all morphologic criteria for UPSs have often been treated along the same lines as UPSs and are offered ICB in clinical practice, and our report indicates that this approach is reasonable for ICB treatment, given that the 3-month PFS and 6-month PFS were 46% and 32%, respectively. Clinical trials are now including these sarcomas with UPS cohorts testing ICB treatment.

The notable difference in our report compared to clinical trials is that combination ICB seemed inferior to standalone ICB regarding PFS. This is not expected, as multiple trials have shown a benefit of combination ICB compared to single-agent PD1 [21,22,34,35]. However, a potential selection bias could explain this association since clinicians are aware that single-agent PD1 treatment may take longer to be active and can be concerned with the risk of hyperprogressive disease [36]. Thus, in cases of rapidly progressing disease or high tumor burden, clinicians are more likely to try a combination therapy. In contrast, in the case of a slow-growing disease, the added toxicity of a combination therapy is avoided by treating patients with single-agent ICB [37,38]. This selection bias is illustrated by the fact that patients who were treated with combination therapies had a higher number of previous surgical resections and lines of systemic therapies, indicating this population was more heavily pretreated and that there is likely a selection bias in this analysis.

The association between RT and ICB response observed in our study is also likely a result of selection bias. The main hypothesis to explain this is that tumors that recurred and progressed despite RT in the peri-operative setting had intrinsically bad biology and an immune-suppressive microenvironment. Likewise, RT in the metastatic setting would be offered in patients with more symptomatic disease, which may inherently be linked to bad biology. However, there may be biological ties, but the role of previous RT in resistance to ICB is controversial. While concurrent ICB and RT are effective and likely synergistic in the treatment of UPSs and other tumor types, previous RT may have a more immune-suppressive role, as several cytokines, including TGFbeta, and tissue remodeling cells, such as macrophages and neutrophils, are recruited to the tumor after RT [39]. We tried to investigate differences between patients who had previous RT compared to those who did not and did not find any significant differences in our small cohort. Ultimately, our data are hypothesis-generating in nature, and no strong association or mechanistic claim can be made from our clinical report, but the question of RT timing with respect to ICB may be an important one to address in the future. As such, concurrent RT when feasible with ICB is a very promising therapeutic strategy for patients with UPSs.

This study highlights a critical need for the identification of biomarkers of response to ICB for patients with sarcomas. As our cohort is a small, real-world, heterogeneous cohort, we were unable to identify strong predictors of response. This highlights a challenge in deriving data from real-world studies beyond clinical trials but invites further investigation into potential predictors of response.

This is a retrospective, single-center cohort with a small number of patients, and thus, there is significant bias in this analysis, and the statistical power of our findings is limited. The results discussed here are hypothesis-generating and descriptive by nature, and no causality can be inferred.

5. Conclusions

Real-world retrospective data demonstrate a median PFS of 2.9 months for patients with UPSs and other high-grade pleomorphic sarcomas treated with ICB in the metastatic setting, which is consistent with the published literature. The optimal sequencing of RT and prior lines of systemic therapy needs to be further evaluated.

Supplementary Materials: The following supporting information can be downloaded at: https://www.mdpi.com/article/10.3390/cancers16091763/s1, Table S1. Overall Survival according to clinical characteristics; Table S2. Patient and disease characteristics by exposure to RT prior to ICB; Table S3. Patient and disease characteristics by type ICB treatment; Figure S1. Progression-free survival in patients with previous RT comparing peri-op vs metastatic RT; Figure S2. Overall Survival stratified by exposure to RT prior to ICB; Figure S3. Overall Survival stratified by number of lines of systemic therapy prior to ICB; Figure S4. Overall Survival stratified by type of ICB treatment.

Author Contributions: Conceptualization, E.F.N.H.; methodology, L.F.N. and E.F.N.H.; formal analysis, L.F.N., M.Z. and E.F.N.H.; investigation, R.L., M.N., A.J.L., C.R., A.J.B., A.F., D.M., B.A.G., R.B., S.P., V.R., D.M.A., A.L., M.A.Z., A.P.C., R.R., N.S., E.Z.K. and E.F.N.H.; data curation, L.F.N., M.Z. and E.F.N.H.; writing—original draft preparation, L.F.N. and E.F.N.H.; writing—review and editing, L.F.N., M.Z., R.L., M.N., A.J.L., C.R., A.J.B., A.F., D.M., B.A.G., R.B., S.P., V.R., D.M.A., A.L., M.A.Z., A.P.C., R.R., N.S., E.Z.K. and E.F.N.H.; supervision, E.F.N.H.; project administration, L.F.N. and E.F.N.H. All authors have read and agreed to the published version of the manuscript.

Funding: This research received no external funding.

Institutional Review Board Statement: This retrospective study was IRB-approved by MD Anderson IRB committee under protocol number DR09-0245.

Informed Consent Statement: A waiver of informed consent was used for this study due to this being a retrospective chart review that involves no diagnostic or therapeutic intervention, as well as no direct patient contact. Study staff are unable to obtain consent from study subjects because, due to the international patient population of MD Anderson Cancer Center, it would be impracticable to obtain informed consent from individual patients prior to beginning the protocol. Additionally, it would bias the study if it were restricted to only those patients that we were able to contact. It would not be practicable to conduct this research without this waiver since the status of the patient is unknown, i.e., whether they are alive or deceased, and it is difficult to trace the whereabouts of the patients.

Data Availability Statement: Data will not be made publicly available. Anonymized data will be shared upon reasonable academic request and will be subject to legal data transfer agreements.

Acknowledgments: E.F.N.H. received support from the LMS SPORE Career Enhancement Program, the QuadW Foundation, the Sarcoma Foundation of America, Fondation pour la Recherche Medicale, and Fondation Nuovo-Soldati.

Conflicts of Interest: The authors declare no conflicts of interest.

References

1. Penel, N.; Coindre, J.-M.; Giraud, A.; Terrier, P.; Ranchere-Vince, D.; Collin, F.; Guellec, S.L.E.; Bazille, C.; Lae, M.; de Pinieux, G.; et al. Presentation and outcome of frequent and rare sarcoma histologic subtypes: A study of 10,262 patients with localized visceral/soft tissue sarcoma managed in reference centers. *Cancer* **2018**, *124*, 1179–1187. [CrossRef]
2. Savina, M.; Le Cesne, A.; Blay, J.-Y.; Ray-Coquard, I.; Mir, O.; Toulmonde, M.; Cousin, S.; Terrier, P.; Ranchere-Vince, D.; Meeus, P.; et al. Patterns of care and outcomes of patients with METAstatic soft tissue SARComa in a real-life setting: The METASARC observational study. *BMC Med.* **2017**, *15*, 78. [CrossRef] [PubMed]
3. Kelleher, F.; Viterbo, A. Histologic and Genetic Advances in Refining the Diagnosis of "Undifferentiated Pleomorphic Sarcoma". *Cancers* **2013**, *5*, 218–233. [CrossRef] [PubMed]
4. Fletcher, C.D. Pleomorphic malignant fibrous histiocytoma: Fact or fiction? A critical reappraisal based on 159 tumors diagnosed as pleomorphic sarcoma. *Am. J. Surg. Pathol.* **1992**, *16*, 213–228. [CrossRef]
5. Fletcher, C.D. The evolving classification of soft tissue tumours—An update based on the new 2013 WHO classification. *Histopathology* **2014**, *64*, 2–11. [CrossRef] [PubMed]
6. Benjamin, R.S.; Wiernik, P.H.; Bachur, N.R. Adriamycin chemotherapy—Efficacy, safety, and pharmacologic basis of an intermittent single high-dosage schedule. *Cancer* **1974**, *33*, 19–27. [CrossRef]
7. Gronchi, A.; Palmerini, E.; Quagliuolo, V.; Martin Broto, J.; Lopez Pousa, A.; Grignani, G.; Brunello, A.; Blay, J.Y.; Tendero, O.; Diaz Beveridge, R.; et al. Neoadjuvant Chemotherapy in High-Risk Soft Tissue Sarcomas: Final Results of a Randomized Trial From Italian (ISG), Spanish (GEIS), French (FSG), and Polish (PSG) Sarcoma Groups. *J. Clin. Oncol.* **2020**, *38*, 2178–2186. [CrossRef]
8. Seddon, B.; Strauss, S.J.; Whelan, J.; Leahy, M.; Woll, P.J.; Cowie, F.; Rothermundt, C.; Wood, Z.; Benson, C.; Ali, N.; et al. Gemcitabine and docetaxel versus doxorubicin as first-line treatment in previously untreated advanced unresectable or metastatic soft-tissue sarcomas (GeDDiS): A randomised controlled phase 3 trial. *Lancet Oncol.* **2017**, *18*, 1397–1410. [CrossRef] [PubMed]
9. Tap, W.D.; Jones, R.L.; Van Tine, B.A.; Chmielowski, B.; Elias, A.D.; Adkins, D.; Agulnik, M.; Cooney, M.M.; Livingston, M.B.; Pennock, G.; et al. Olaratumab and doxorubicin versus doxorubicin alone for treatment of soft-tissue sarcoma: An open-label phase 1b and randomised phase 2 trial. *Lancet* **2016**, *388*, 488–497. [CrossRef]
10. Tap, W.D.; Papai, Z.; Van Tine, B.A.; Attia, S.; Ganjoo, K.N.; Jones, R.L.; Schuetze, S.; Reed, D.; Chawla, S.P.; Riedel, R.F.; et al. Doxorubicin plus evofosfamide versus doxorubicin alone in locally advanced, unresectable or metastatic soft-tissue sarcoma (TH CR-406/SARC021): An international, multicentre, open-label, randomised phase 3 trial. *Lancet Oncol.* **2017**, *18*, 1089–1103. [CrossRef]
11. Maki, R.G.; Wathen, J.K.; Patel, S.R.; Priebat, D.A.; Okuno, S.H.; Samuels, B.; Fanucchi, M.; Harmon, D.C.; Schuetze, S.M.; Reinke, D.; et al. Randomized Phase II Study of Gemcitabine and Docetaxel Compared With Gemcitabine Alone in Patients With Metastatic Soft Tissue Sarcomas: Results of Sarcoma Alliance for Research Through Collaboration Study 002. *J. Clin. Oncol.* **2007**, *25*, 2755–2763. [CrossRef] [PubMed]
12. Kim, J.H.; Park, H.S.; Heo, S.J.; Kim, S.K.; Han, J.W.; Shin, K.H.; Kim, S.H.; Hur, H.; Kim, K.S.; Choi, Y.D.; et al. Differences in the Efficacies of Pazopanib and Gemcitabine/Docetaxel as Second-Line Treatments for Metastatic Soft Tissue Sarcoma. *Oncology* **2019**, *96*, 59–69. [CrossRef] [PubMed]
13. Tawbi, H.A.; Burgess, M.; Bolejack, V.; Van Tine, B.A.; Schuetze, S.M.; Hu, J.; D'Angelo, S.; Attia, S.; Riedel, R.F.; Priebat, D.A.; et al. Pembrolizumab in advanced soft-tissue sarcoma and bone sarcoma (SARC028): A multicentre, two-cohort, single-arm, open-label, phase 2 trial. *Lancet Oncol.* **2017**, *18*, 1493–1501. [CrossRef] [PubMed]
14. Burgess, M.A.; Bolejack, V.; Schuetze, S.M.; Van Tine, B.A.; Attia, S.; Riedel, R.F.; Hu, J.; Davis, L.E.; Okuno, S.H.; Priebat, D.A.; et al. Clinical activity of pembrolizumab (P) in undifferentiated pleomorphic sarcoma (UPS) and dedifferentiated/pleomorphic liposarcoma (LPS): Final results of SARC028 expansion cohorts. *J. Clin. Oncol.* **2019**, *37*, 11015. [CrossRef]
15. Keung, E.Z.; Burgess, M.; Salazar, R.; Parra, E.R.; Rodrigues-Canales, J.; Bolejack, V.; Van Tine, B.A.; Schuetze, S.M.; Attia, S.; Riedel, R.F.; et al. Correlative Analyses of the SARC028 Trial Reveal an Association Between Sarcoma-Associated Immune Infiltrate and Response to Pembrolizumab. *Clin. Cancer Res.* **2020**, *26*, 1258–1266. [CrossRef] [PubMed]
16. Petitprez, F.; de Reynies, A.; Keung, E.Z.; Chen, T.W.; Sun, C.M.; Calderaro, J.; Jeng, Y.M.; Hsiao, L.P.; Lacroix, L.; Bougouin, A.; et al. B cells are associated with survival and immunotherapy response in sarcoma. *Nature* **2020**, *577*, 556–560. [CrossRef] [PubMed]

17. Kelly, C.M.; Qin, L.X.; Whiting, K.A.; Richards, A.L.; Avutu, V.; Chan, J.E.; Chi, P.; Dickson, M.A.; Gounder, M.M.; Keohan, M.L.; et al. A Phase II Study of Epacadostat and Pembrolizumab in Patients with Advanced Sarcoma. *Clin. Cancer Res.* **2023**, *29*, 2043–2051. [CrossRef] [PubMed]
18. Liu, J.; Fan, Z.; Bai, C.; Li, S.; Xue, R.; Gao, T.; Zhang, L.; Tan, Z.; Fang, Z. Real-world experience with pembrolizumab in patients with advanced soft tissue sarcoma. *Ann. Transl. Med.* **2021**, *9*, 339. [CrossRef]
19. Pollack, S.M.; Redman, M.W.; Baker, K.K.; Wagner, M.J.; Schroeder, B.A.; Loggers, E.T.; Trieselmann, K.; Copeland, V.C.; Zhang, S.; Black, G.; et al. Assessment of Doxorubicin and Pembrolizumab in Patients With Advanced Anthracycline-Naive Sarcoma: A Phase 1/2 Nonrandomized Clinical Trial. *JAMA Oncol.* **2020**, *6*, 1778–1782. [CrossRef]
20. D'Angelo, S.P.; Melchiori, L.; Merchant, M.S.; Bernstein, D.; Glod, J.; Kaplan, R.; Grupp, S.; Tap, W.D.; Chagin, K.; Binder, G.K.; et al. Antitumor Activity Associated with Prolonged Persistence of Adoptively Transferred NY-ESO-1 (c259)T Cells in Synovial Sarcoma. *Cancer Discov.* **2018**, *8*, 944–957. [CrossRef]
21. Movva, S.; Avutu, V.; Chi, P.; Dickson, M.A.; Gounder, M.M.; Kelly, C.M.; Keohan, M.L.; Meyers, P.A.; Cohen, S.M.; Hensley, M.L.; et al. A pilot study of lenvatinib plus pembrolizumab in patients with advanced sarcoma. *J. Clin. Oncol.* **2023**, *41* (Suppl. 16), 11517.
22. Rosenbaum, E.; Qin, L.-X.; Dickson, M.A.; Keohan, M.L.; Gounder, M.M.; Chi, P.; Movva, S.; Kelly, C.M.; Avutu, V.; Chan, J.E.; et al. Interim results of a phase II trial of first line retifanlimab (R) plus gemcitabine and docetaxel (GD) in patients (pts) with advanced soft tissue sarcoma (STS). *J. Clin. Oncol.* **2023**, *41* (Suppl. 16), 11518. [CrossRef]
23. Van Tine, B.A.; Eulo, V.; Toeniskoetter, J.; Ruff, T.; Luo, J.; Kemp, L.; Moreno Tellez, C.; Weiss, M.C.; Hirbe, A.C.; Meyer, C.F.; et al. Randomized phase II trial of cabozantinib combined with PD-1 and CTLA-4 inhibition versus cabozantinib in metastatic soft tissue sarcoma. *J. Clin. Oncol.* **2023**, *41* (Suppl. 17), LBA11504. [CrossRef]
24. Roland, C.L.; Nassif Haddad, E.F.; Keung, E.Z.; Wang, W.L.; Lazar, A.J.; Lin, H.; Chelvanambi, M.; Parra, E.R.; Wani, K.; Guadagnolo, B.A.; et al. A randomized, non-comparative phase 2 study of neoadjuvant immune-checkpoint blockade in retroperitoneal dedifferentiated liposarcoma and extremity/truncal undifferentiated pleomorphic sarcoma. *Nat. Cancer* **2024**, *5*, 625–641. [CrossRef] [PubMed]
25. Eisenhauer, E.A.; Therasse, P.; Bogaerts, J.; Schwartz, L.H.; Sargent, D.; Ford, R.; Dancey, J.; Arbuck, S.; Gwyther, S.; Mooney, M.; et al. New response evaluation criteria in solid tumours: Revised RECIST guideline (version 1.1). *Eur. J. Cancer* **2009**, *45*, 228–247. [CrossRef] [PubMed]
26. Keung, E.Z.-Y.; Nassif, E.F.; Lin, H.Y.; Lazar, A.J.; Torres, K.E.; Wang, W.-L.; Guadagnolo, B.A.; Bishop, A.J.; Hunt, K.; Feig, B.W.; et al. Randomized phase II study of neoadjuvant checkpoint blockade for surgically resectable undifferentiated pleomorphic sarcoma (UPS) and dedifferentiated liposarcoma (DDLPS): Survival results after 2 years of follow-up and intratumoral B-cell receptor (BCR) correlates. *J. Clin. Oncol.* **2022**, *40* (Suppl. 17), LBA11501.
27. Dall'Olio, F.G.; Marabelle, A.; Caramella, C.; Garcia, C.; Aldea, M.; Chaput, N.; Robert, C.; Besse, B. Tumour burden and efficacy of immune-checkpoint inhibitors. *Nat. Rev. Clin. Oncol.* **2022**, *19*, 75–90. [CrossRef] [PubMed]
28. Chalabi, M.; Fanchi, L.F.; Dijkstra, K.K.; Van den Berg, J.G.; Aalbers, A.G.; Sikorska, K.; Lopez-Yurda, M.; Grootscholten, C.; Beets, G.L.; Snaebjornsson, P.; et al. Neoadjuvant immunotherapy leads to pathological responses in MMR-proficient and MMR-deficient early-stage colon cancers. *Nat. Med.* **2020**, *26*, 566–576. [CrossRef] [PubMed]
29. Cascone, T.; William, W.N.; Jr Weissferdt, A.; Leung, C.H.; Lin, H.Y.; Pataer, A.; Godoy, M.C.B.; Carter, B.W.; Federico, L.; Reuben, A.; et al. Neoadjuvant nivolumab or nivolumab plus ipilimumab in operable non-small cell lung cancer: The phase 2 randomized NEOSTAR trial. *Nat. Med.* **2021**, *27*, 504–514. [CrossRef] [PubMed]
30. Amaria, R.N.; Reddy, S.M.; Tawbi, H.A.; Davies, M.A.; Ross, M.I.; Glitza, I.C.; Cormier, J.N.; Lewis, C.; Hwu, W.J.; Hanna, E.; et al. Neoadjuvant immune checkpoint blockade in high-risk resectable melanoma. *Nat. Med.* **2018**, *24*, 1649–1654. [CrossRef]
31. Gross, N.D.; Miller, D.M.; Khushalani, N.I.; Divi, V.; Ruiz, E.S.; Lipson, E.J.; Meier, F.; Su, Y.B.; Swiecicki, P.L.; Atlas, J.; et al. Neoadjuvant Cemiplimab for Stage II to IV Cutaneous Squamous-Cell Carcinoma. *N. Engl. J. Med.* **2022**, *387*, 1557–1568. [CrossRef]
32. Klemen, N.D.; Hwang, S.; Bradic, M.; Rosenbaum, E.; Dickson, M.A.; Gounder, M.M.; Kelly, C.M.; Keohan, M.L.; Movva, S.; Thornton, K.A.; et al. Long-term Follow-up and Patterns of Response, Progression, and Hyperprogression in Patients after PD-1 Blockade in Advanced Sarcoma. *Clin. Cancer Res.* **2021**, *28*, 939–947. [CrossRef] [PubMed]
33. Nishino, M.; Ramaiya, N.H.; Hatabu, H.; Hodi, F.S. Monitoring immune-checkpoint blockade: Response evaluation and biomarker development. *Nat. Rev. Clin. Oncol.* **2017**, *14*, 655–668. [CrossRef] [PubMed]
34. Somaiah, N.; Conley, A.P.; Parra, E.R.; Lin, H.; Amini, B.; Solis Soto, L.; Salazar, R.; Barreto, C.; Chen, H.; Gite, S.; et al. Durvalumab plus tremelimumab in advanced or metastatic soft tissue and bone sarcomas: A single-centre phase 2 trial. *Lancet Oncol.* **2022**, *23*, 1156–1166. [CrossRef] [PubMed]
35. D'Angelo, S.P.; Mahoney, M.R.; Van Tine, B.A.; Atkins, J.; Milhem, M.M.; Jahagirdar, B.N.; Antonescu, C.R.; Horvath, E.; Tap, W.D.; Schwartz, G.K.; et al. Nivolumab with or without ipilimumab treatment for metastatic sarcoma (Alliance A091401): Two open-label, non-comparative, randomised, phase 2 trials. *Lancet Oncol.* **2018**, *19*, 416–426. [CrossRef] [PubMed]
36. Champiat, S.; Ferrara, R.; Massard, C.; Besse, B.; Marabelle, A.; Soria, J.C.; Ferte, C. Hyperprogressive disease: Recognizing a novel pattern to improve patient management. *Nat. Rev. Clin. Oncol.* **2018**, *15*, 748–762. [CrossRef]
37. Morad, G.; Helmink, B.A.; Sharma, P.; Wargo, J.A. Hallmarks of response, resistance, and toxicity to immune checkpoint blockade. *Cell* **2021**, *184*, 5309–5337. [CrossRef] [PubMed]

38. Sharma, P.; Allison, J.P. The future of immune checkpoint therapy. *Science* **2015**, *348*, 56–61. [CrossRef]
39. Rodriguez-Ruiz, M.E.; Vitale, I.; Harrington, K.J.; Melero, I.; Galluzzi, L. Immunological impact of cell death signaling driven by radiation on the tumor microenvironment. *Nat. Immunol.* **2020**, *21*, 120–134. [CrossRef]

cancers

MDPI

Article

Immunohistochemical Investigation into Protein Expression Patterns of FOXO4, IRF8 and LEF1 in Canine Osteosarcoma

Simone de Brot [1,2], Jack Cobb [1], Aziza A. Alibhai [1], Jorja Jackson-Oxley [1], Maria Haque [1], Rodhan Patke [1], Anna E. Harris [1], Corinne L. Woodcock [1], Jennifer Lothion-Roy [1], Dhruvika Varun [1], Rachel Thompson [1], Claudia Gomes [1], Valentina Kubale [3], Mark D. Dunning [1,4], Jennie N. Jeyapalan [1,5], Nigel P. Mongan [1,4,6,*] and Catrin S. Rutland [1,5,*]

1 School of Veterinary Medicine and Science, Faculty of Medicine and Health Sciences, University of Nottingham, Nottingham NG7 2RD, UK; simone.debrot@unibe.ch (S.d.B.); mzyjc18@exmail.nottingham.ac.uk (J.C.); svaaa3@exmail.nottingham.ac.uk (A.A.A.); svxjj1@exmail.nottingham.ac.uk (J.J.-O.); stxmh36@exmail.nottingham.ac.uk (M.H.); stxrp16@exmail.nottingham.ac.uk (R.P.); svzaeh@exmail.nottingham.ac.uk (A.E.H.); svzclw@exmail.nottingham.ac.uk (C.L.W.); svzjhl@exmail.nottingham.ac.uk (J.L.-R.); stxdv8@exmail.nottingham.ac.uk (D.V.); styrt5@exmail.nottingham.ac.uk (R.T.); svxcg4@exmail.nottingham.ac.uk (C.G.); mark.dunning@nottingham.ac.uk (M.D.D.); plzjnj@exmail.nottingham.ac.uk (J.N.J.)
2 Comparative Pathology Platform of the University of Bern (COMPATH), Institute of Animal Pathology, University of Bern, 3012 Bern, Switzerland
3 Institute of Preclinical Sciences, Veterinary Faculty, University of Ljubljana, 1000 Ljubljana, Slovenia; valentina.kubaledvojmoc@vf.uni-lj.si
4 Willows Veterinary Centre and Referral Service, Solihull B90 4NH, UK
5 Faculty of Medicine and Health Science, Biodiscovery Institute, University of Nottingham, Nottingham NG7 2RD, UK
6 Department of Pharmacology, Weill Cornell Medicine, New York, NY 10075, USA
* Correspondence: nigel.mongan@nottingham.ac.uk (N.P.M.); catrin.rutland@nottingham.ac.uk (C.S.R.)

Citation: Brot, S.d.; Cobb, J.; Alibhai, A.A.; Jackson-Oxley, J.; Haque, M.; Patke, R.; Harris, A.E.; Woodcock, C.L.; Lothion-Roy, J.; Varun, D.; et al. Immunohistochemical Investigation into Protein Expression Patterns of FOXO4, IRF8 and LEF1 in Canine Osteosarcoma. *Cancers* **2024**, *16*, 1945. https://doi.org/10.3390/cancers16101945

Academic Editor: Shinji Miwa

Received: 30 April 2024
Revised: 16 May 2024
Accepted: 18 May 2024
Published: 20 May 2024

Simple Summary: There have been limited advances in the diagnosis, prognosis, and treatment of both canine and human osteosarcoma (OSA), the most common type of primary bone cancer. OSA has an aggressive nature, with incidence rates ranging from 13.9 to 27.2 cases per 100,000 dogs, yet there have been limited advances in patient outcomes in recent decades. Recent developments have identified similarities between human and canine OSA; therefore, researching naturally occurring canine bone cancer may help inform research into OSA in people. The present research investigated three proteins, FOXO4, IRF8, and LEF1, to visualise their expression in OSA tissue. This research helps us understand where the proteins are being expressed in the tumours, which genetic pathways are changing, and may help us identify potentially informative diagnostic, prognostic, and treatment avenues for this cancer in dogs and people.

Abstract: Osteosarcoma (OSA) is the most common type of primary bone malignancy in people and dogs. Our previous molecular comparisons of canine OSA against healthy bone resulted in the identification of differentially expressed protein-expressing genes (forkhead box protein O4 (*FOXO4*), interferon regulatory factor 8 (*IRF8*), and lymphoid enhancer binding factor 1 (*LEF1*)). Immunohistochemistry (IHC) and H-scoring provided semi-quantitative assessment of nuclear and cytoplasmic staining alongside qualitative data to contextualise staining ($n = 26$ patients). FOXO4 was expressed predominantly in the cytoplasm with significantly lower nuclear H-scores. IRF8 H-scores ranged from 0 to 3 throughout the cohort in the nucleus and cytoplasm. LEF1 was expressed in all patients with significantly lower cytoplasmic staining compared to nuclear. No sex or anatomical location differences were observed. While reduced levels of FOXO4 might indicate malignancy, the weak or absent protein expression limits its primary use as diagnostic tumour marker. IRF8 and LEF1 have more potential for prognostic and diagnostic uses and facilitate further understanding of their roles within their respective molecular pathways, including Wnt/beta-catenin/LEF1 signalling and differential regulation of tumour suppressor genes. Deeper understanding of the mechanisms involved in OSA are essential contributions towards the development of novel diagnostic, prognostic, and treatment options in human and veterinary medicine contexts.

Keywords: cancer identification; forkhead box protein O4; interferon regulatory factor 8; lymphoid enhancer binding factor 1; osteosarcoma; pathology

1. Introduction

Osteosarcoma (OSA) is a neoplasia of mesenchymal origin, which tends to derive from the medullary cavity of metaphyseal bones and subsequently expands to the cortical bone; this pathological process is named central OSA [1,2]. Rarely, OSA can originate from the periosteal surface and is thought to be overall less aggressive compared to central OSA [2]. Canine OSA is considered the most common bone tumour identified, with a documented prevalence of approximately 85% of all primary malignancies arising in the skeleton of this species and 3–4% of all malignant tumours in dogs [3–6]. Reported OSA incidence is greater in canines than in any other species, with an estimated rate of 13.9–27.2 cases per 100,000 dogs [7–10]. OSA incidence in humans is much lower, at 0.89–1.2 per 100,000 [8,10,11]. This high incidence in dogs not only emphasises the veterinary challenge posed by OSA, but also enhances the efficacy of the canine model as the low incidence rate in humans is a big factor in the lack of understanding observed to date. The low incidence in people also contributes towards the lack of diagnosis and treatment options, and the relatively poor outcomes following OSA diagnosis.

Linking OSA incidence with specific risk factors can allude to the involvement of certain biological pathways. Canine OSA predominantly affects middle-aged, naturally larger breeds including Rottweilers, Great Danes, Saint Bernards, Doberman Pinschers, and Irish Wolfhounds [7,11–13]. In addition, increased body weight (even after controlling for breed), height, and age are risk factors for OSA in dogs [12]. These link with human OSA risk factors, as taller and heavier individuals are more prone to OSA formation [14–16]. This is further reinforced by higher incidence in males across both species, who on average naturally grow to be slightly larger [10,12,16,17]. These risk factors also implicate bone growth as potentially causative in OSA formation. The growth risk factor may also correspond to the respective risk of certain age groups. In people, OSA formation follows a bimodal trend with the primary peak being in adolescence, contributing to over 50% of cases. There is then another smaller peak in seniors [10,15,16]. A similar bimodal trend is observed in canines, with a peak in dogs aged less than 3 years old, and then 80% of cases presenting in dogs aged 7 years (middle-aged) and older [9,18]. The association with OSA formation and high growth levels in puberty implicate growth and developmental factors in OSA aetiology.

Clinical presentation of canine OSA is characterised by progressive lameness, hard bony swelling, or even pathological fracture of the affected bone [5,19]. This neoplasia is very aggressive and invasive in dogs, causing local skeletal destruction and is also highly metastatic, predominantly to the lungs, with a lower frequency of spread to distant bones, regional lymph nodes, and other soft tissues [3,5]. The accepted treatment at present is a combination of radiotherapy, chemotherapy (both adjuvant and/or neo-adjuvant) and surgery [10,17,20]. For people with OSA with no concomitant metastases at diagnosis, the 5-year event-free survival stands at around 70%; however, approximately 20% of patients will exhibit metastases upon diagnosis and their 5-year event-free survival drops to 27%. In canine OSA, the 1-year survival after treatment is just 45%, whereas the median time from diagnosis to euthanasia has been reported to be as low as 111 days (range, 28 to 447 days) in one study [21]. For those dogs that survive past 1 year, over 50% will develop metastases and present with a median survival time of 243 days [7,17,22]; patient outcomes are currently limited due to the highly metastatic nature of OSA and multi-drug resistance limiting the effects of chemotherapy. Investigations into the canine model could yield new treatments specifically targeting OSA molecular pathways.

Our previous research showed that *FOXO4*, *IRF8*, and *LEF1* were differentially expressed (via RNA sequencing) in canine osteosarcoma compared to patient-matched non-tumour tissue [11,23]. FOXO4 (also known as AFX1) belongs to the forkhead box class O (FOXO) family, a group of transcription factors involved in numerous cellular processes, including development, proliferation, survival, apoptosis, metabolism, and homeostasis [24–26]. Post-translational changes in the FOXO group can alter their nuclear import/export, modify DNA binding affinity, and change the transcriptional activity of target genes [24]. Growth factors such as insulin and insulin-like growth factor 1 (IGF-1) can regulate the activity of FOXO4 by repressing it through the phosphoinositide—3 kinase (PI3K)/Akt signalling pathway [24,25]. Upon activation of the PI3K-AKT pathway, AKT phosphorylates FOXO proteins, impeding their localisation to the nucleus and transcriptional activity [27,28]. Without growth factors present, the FOXO transcription factors are localised in the nucleus and upregulate key genes, causing cell cycle arrest and cell death [29]. Additionally, through an interaction with p53 that represses p53-mediated apoptosis, FOXO4 has been shown to have a key role in senescent cell viability [30]. The FOXO4 protein, along with FOXO1 and FOXO3, is also important for bone development. The loss of these proteins in osteoclast progenitors can result in an increase in proliferation, bone resorption, and osteoclast formation [31]. In human OSA, a study by Chen and colleagues found that the oncogenic miRNA, miR-664, promoted cell proliferation by supressing *FOXO4* expression, suggesting that FOXO4 has a role as a tumour suppressor in osteosarcoma [32]. Consistent with this finding, FOXO4 has been reported to have reduced expression in cancer compared to non-malignant tissue and have a tumour suppressor role in several other cancer types, including colorectal, gastric, and head and neck squamous cell cancer [33–35]. Findings by Paik et al. [36] revealed that FOXO1, 3, and 4 are largely functionally redundant in their tumour suppressor function. Oncogenic splice variants of FOXO4 have also been reported [37].

The transcription factor, interferon regulatory factor 8 (IRF8), originally named interferon consensus sequence binding protein (ICSBP), is a member of the IRF protein family [38]. IRF8 is constitutively expressed, is IFNγ inducible, and plays key roles in multiple biological processes, including modulation of the immune response and other physiological processes (reviewed in [39]). Considering that IRF8 is expressed in hematopoietic cells, recent studies have shown that the formation of mammalian dendritic cells (DCs) requires the transcription factor IRF8 [40]. Specifically, type 1 dendritic cells (DC1s) and the tumour-associated macrophages (TAMs) require and express IRF8 [41]. This is required for the TAM ability to present cancer cell antigens, indicating that IRF8 may play a role in promoting tumour growth [41]. Like many other transcription factors, IRF8 can be dysregulated in cancer, and therefore, this study aimed to determine the protein expression of IRF8 in canine OSA.

The *LEF1* gene encodes the Lymphoid enhancer-binding factor 1 protein, which belongs to the TCF/LEF family of transcription factors known to act via the *Wnt* signalling pathway [42,43]. LEF1 is principally involved in the process of T cell, B cell, and natural killer cell development [44–46]; however, it also plays a role in the regulation of skin development, the hair cycle, and development of the mammary gland [47–49]. As an effector of the *Wnt* signalling pathway, LEF1 is also associated with regulating the cell cycle, epithelial-to-mesenchymal transition, and with tumour development and progression [50–53]. Increased LEF1 expression has been associated with carcinogenesis in many different cancer types, including melanoma, pancreatic, colorectal, and breast, as well as several myeloid and blood cancers [54–60].

Our previous research [11] showed that *FOXO4*, *IRF8*, and *LEF1* were differentially expressed (via RNA sequencing) in canine OSA compared to patient-matched non-tumour tissue. *FOXO4* in OSA tissue had a 1.42 Log2 fold decrease, but not significant at $p = 0.056$, compared to patient-matched non-tumour bone; *IRF8* showed a 2.33 Log2 fold decrease,

p = 0.01, and *LEF1* exhibited a 2.2 Log2 fold increase, p = 0.04. Therefore, to determine the prognostic or diagnostic potential of FOXO4, IRF8, and LEF1 in canine OSA, we performed IHC on OSA specimens to determine protein expression. The present research ascertained the H-scores in the nucleus and cytoplasm (and total H-score), alongside descriptive qualitative analysis, of these proteins in OSA specimens. The present study additionally investigated tumour location (appendicular vs. axial) and sex in relation to protein expression.

2. Materials and Methods

2.1. Specimen Preparation

All animal tissue work in this study was approved by the ethics committee at the University of Nottingham School of Veterinary Medicine and Science. The ethics complied with national (Home Office) and international ethics procedures (permission numbers—1823 160714, 959 130925, UG 20331). The samples were from patients (Figure 1) under veterinary practice care for OSA, not related to research. A board-certified veterinary pathologist histologically confirmed the diagnosis of OSA.

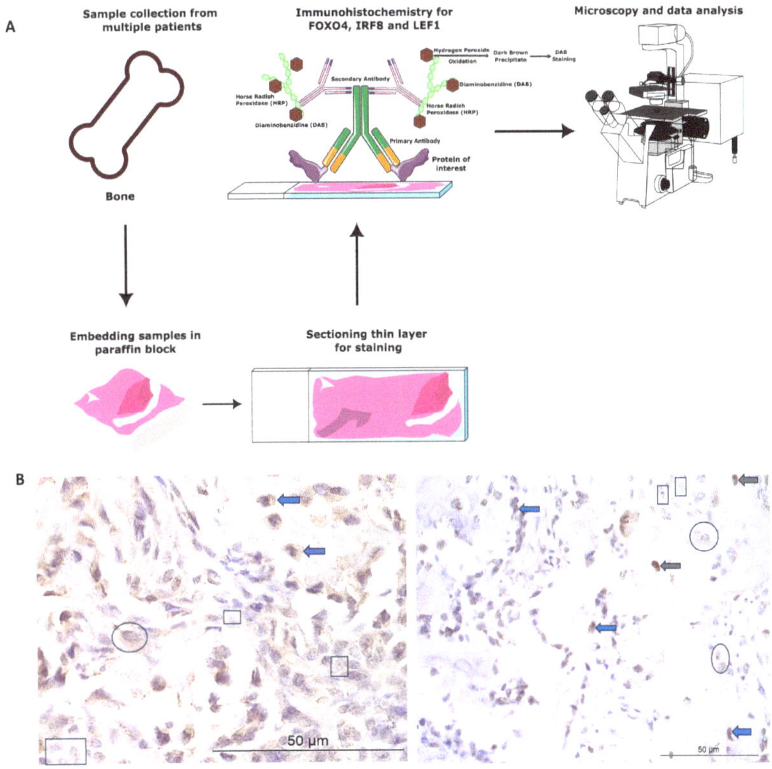

Figure 1. Canine osteosarcoma protein expression materials and methods. (**A**) Overview of methods. (**B**) Examples of cytoplasmic and nuclear H-scores 0, 1, 2, and 3, on IRF8 and LEF1 immunohistochemistry photomicrographs. H-score 0 (rectangles), 1+ (circles), 2+ (blue arrows), 3+ (grey arrows).

Canine diagnostic OSA tissues from Rottweilers (n = 26) were obtained from Bridge Pathology, UK and the tissue had been formalin-fixed, paraffin-embedded. There were 13 females, 12 males, and 1 not specified. The males were between 4 and 10 years old, and the females ages ranged between 5 and 12 years old. OSA samples were from a variety of bones, including four from the head and two mammary/thoracic wall (n = 6 axial), and n = 20 from the appendicular skeleton (humerus, ulna, stifle, including the femur and tibia).

2.2. Immunohistochemistry and Microscopy

Paraffin-embedded samples were sectioned at 7 μm. Positive protein expression of FOXO4, IRF8, and LEF1 was visualised using a Leica Novolink Polymer Detection Kit (Leica, Wetzlar, Germany) according to manufacturer's protocols. Primary antibodies were diluted in foetal calf serum; these included anti-FOXO4/AFX1 polyclonal unconjugated human antibody raised in rabbit (1:100 dilution, LS-C112273; LSBio, Cambridge, UK), anti-IRF8 antibody (1:500 dilution; ab28696, Abcam, Cambridge, UK), and anti-LEF1 rabbit polyclonal (1:100 dilution, GTX129186, GeneTex, Irvine, CA, USA). Negative controls were incubated in foetal calf serum only, without the primary antibody. Positive controls consisted of canine blood vessels, skeletal muscle, and nasal epithelium, as the proteins were known to be expressed in these cells and tissue types [34,61,62]. Cytoplasmic and nuclear staining was assessed following microscopy (Leica, Wetzlar, Germany), and $n = 5$ photomicrographs/specimen at 40× magnification were taken using systematic random sampling for H-scoring analysis (for methods overview, see Figure 1A).

2.3. H-Scoring and Statistical Analysis

H-scoring was used to semi-quantitatively analyse the IHC staining, as it is considered as one of the "gold standards" for IHC evaluation [63–65]. Staining intensity for each cell was designated into scores of 0, 1+, 2+, or 3+ (none, weak, moderate, strong staining signal) for each target protein (examples shown in Figure 1B). The percentage of positive staining for each score for each cell (nuclear and cytoplasmic independently) was scored to the nearest 5% for a fixed field of $n = 5$ photomicrographs per sample ($n = 26$ OSA samples) for each antibody. H-scores were calculated using the following formula: H-score = [1 × (% cells 1+) + 2 × (% cells 2+) + 3 × (% cells 3+)]. Cytoplasmic, nuclear, and total H-scores were calculated for each specimen (0–300) for each marker of interest. One double-blinded researcher undertook H-scores and established a scoring definition. Thereafter, an additional researcher scored a random 10% of the samples, to ensure concordance (intraclass correlation coefficient (ICC) > 90% for all proteins) and interpretation consistency. The mean, standard error of the mean, minimum, maximum, and range of H-scores were tabulated and plotted for FOXO4, LEF1, and IRF8 to demonstrate score distributions and staining intensities. Additionally, representative staining classifications were demonstrated as benchmarks. The H-score low/moderate/high classifications were calculated based on the ranges for each individual antibody: FOXO4 = low $\leq$ 34, moderate 35–69, high $\geq$ 70; IRF8 low $\leq$ 83, moderate 84–166, high $\geq$ 167; and LEF1 = low $\leq$ 62, moderate 63–124, high $\geq$ 125. Statistical analysis between cytoplasmic and nuclear H-scores, male vs. female H-scores, and OSA location (appendicular vs. axial) were conducted using paired *t*-test (SPSS v26). Fisher's exact test 2 × 3 Contingency Table was used to compare the number of specimens with low, moderate, and high H-score staining categories in both the cytoplasm and nucleus.

Qualitative data were also recorded to describe general immunohistochemical staining patterns. Specifically, the tissue structures and cell types with positive immunostaining were indicated, the general staining distribution was identified for each sample (diffuse, multifocal, focal), and in addition, the overall predominant cytoplasmic and nuclear staining intensity was described following H-scoring (absent, low, moderate, high) and the main staining location was identified (cytoplasmic or nuclear).

3. Results

The IHC staining of the three proteins is summarised in Tables 1–3. FOXO4 staining showed H-score variations between the different patients, and 8/26 specimens exhibited no nuclear or cytoplasmic staining, 15/25 expressed cytoplasmic staining only, and 3/26 exhibited both nuclear and cytoplasmic staining. Cytoplasmic H-scores for FOXO4 were low in the majority of patients (20/26; 77%), with moderate (3/26; 11.5%) and high (3/26; 11.5%) scores in the remaining dogs (Table 1). Nuclear FOXO4 scores were either absent (23/26; 88.5%) or low (3/26; 11.5%; Table 1) in all samples. Hence, a total 20 of the 26 patients (77%) exhibited both low cytoplasmic and low nuclear average scores, 3/26 (11.5%) exhibited low nuclear and moderate cytoplasmic scores, and 3/26 (11.5%) showed low nuclear and high cytoplasmic scores (Table 2). Overall, the staining was diffuse, and the nuclear H-scores were significantly lower than the cytoplasmic staining scores ($p > 0.002$, Table 3, Figure 2).

Table 1. Immunohistochemistry staining overview for H-scores.

Cytoplasmic	**FOXO4**	**IRF8**	**LEF1**
Absent	-	2 (8%)	8 (31%)
Low	20 (77%)	-	4 (15.5%)
Moderate	3 (11.5%)	15 (58%)	10 (38%)
High	3 (11.5%)	9 (34%)	4 (15.5%)
Nuclear			
Absent	23 (88.5%)	-	-
Low	3 (11.5%)	17 (65.5%)	26 (100%)
Moderate	-	6 (23%)	-
High	-	3 (11.5%)	-

Table 2. Subcellular staining H-scores (nuclear and cytoplasmic).

		Cytoplasmic Score			
	FOXO4	**Absent**	**Low**	**Moderate**	**High**
Nuclear score	**Absent**	-	-	-	-
	Low	-	20 (77%)	3 (11.5%)	3 (11.5%)
	Moderate	-	-	-	-
	High	-	-	-	-
	IRF8	**Absent**	**Low**	**Moderate**	**High**
	Absent	-	-	-	-
	Low	-	1 (4%)	-	1 (4%)
	Moderate	-	11 (42%)	3 (11.5%)	1 (4%)
	High	-	5 (19%)	3 (11.5%)	1 (4%)
	LEF1	**Absent**	**Low**	**Moderate**	**High**
	Absent	-	-	-	-
	Low	8 (31%)	4 (15.5%)	10 (38%)	4 (15.5%)
	Moderate	-	-	-	-
	High	-	-	-	-

Table 3. H-scores for FOXO4, IRF8, and LEF1.

		H-Score		
Protein (n = 26)	**Cellular Location**	**Mean ± SEM**	***p*-Value (Cytoplasmic vs. Nuclear)**	**Range (Min-Max)**
FOXO4	Cytoplasmic Nuclear	23.17 ± 5.53 1.38 ± 0.96	0.002	103 (0–103) 27 (0–27)
IRF8	Cytoplasmic Nuclear	63.65 ± 12.15 146.81 ± 9.16	0.0001	230 (0–230) 48 (20–68)
LEF1	Cytoplasmic Nuclear	70.66 ± 8.64 6.13 ± 1.77	0.0001	185.5 (7–192.5) 38.5 (0–38.5)

n = 26 immunostained canine OSA specimens showing inter case variation. $p < 0.05$ = significant difference.

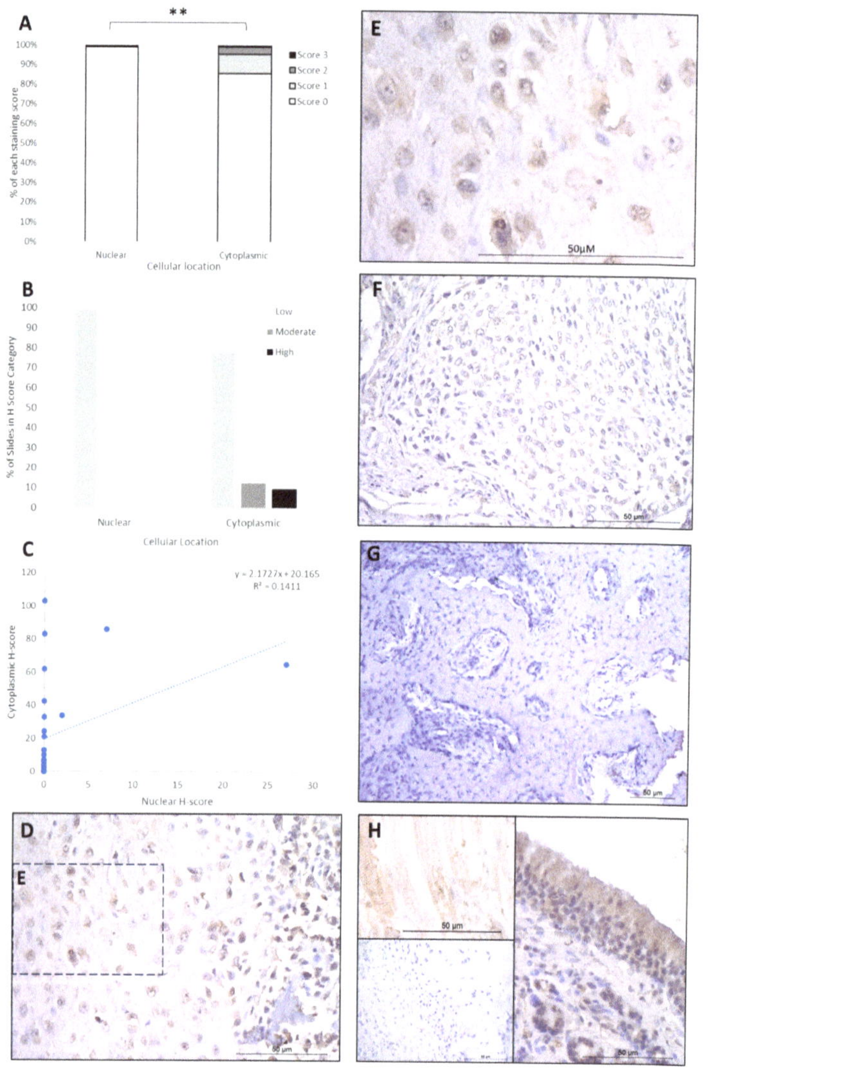

Figure 2. FOXO4 cytoplasmic and nuclear H-scores. (**A**) H-score (scores 0, 1, 2, and 3) distributions. (**B**) H-score low/moderate/high classifications across the cases (p = 0.02). (**C**) Nuclear and cytoplasmic H-score distributions and correlation. Overall, the nuclear H-scores were significantly lower than cytoplasmic (** p = 0.002), n = 26. (**D–G**) Immunohistochemical staining photomicrographs of canine osteosarcoma FOXO4 expression, 40× magnification. (**H**) Right-hand side: positive control nasal mucosa lined by well-differentiated pseudostratified tall columnar ciliated epithelium, inset upper left: muscle positive control, insert lower left: negative control, 40× magnification. All scale bars represent 50 µm.

IRF8 staining showed H-score variations between the different patients, and 2/26 patients exhibited nuclear staining only, whilst the remaining 24/26 had both nuclear and cytoplasmic staining. The cytoplasmic score was moderate in the majority of patients (15/26; 58%), with high (9/26; 34%) and low (2/26; 8%) scores in the remaining cases (Table 1). Nuclear scores showed 17/26 patients (65.5%) with low H scores, 6/26 (23%) at moderate, and 3/26 (12%) at high (Table 1). When assessing subcellular score combinations, low cytoplasmic scores were combined with low, moderate, and high nuclear scores in 1/26

(4%), 11/26 (42%), and 5/26 (19%) of patients, respectively (Table 2). Moderate cytoplasmic score was combined with moderate and high nuclear scores in 3/26 (11.5%) and 3/26 (11.5%), respectively (Table 2). High cytoplasmic score was combined with low, moderate, and high nuclear scores in one case each. Overall, the staining was diffuse, and the nuclear H-scores were significantly higher than the cytoplasmic staining scores ($p > 0.0001$, Table 3, Figure 3), and there was little correlation between nuclear and cytoplasmic H-scores within individual samples (Figure 3).

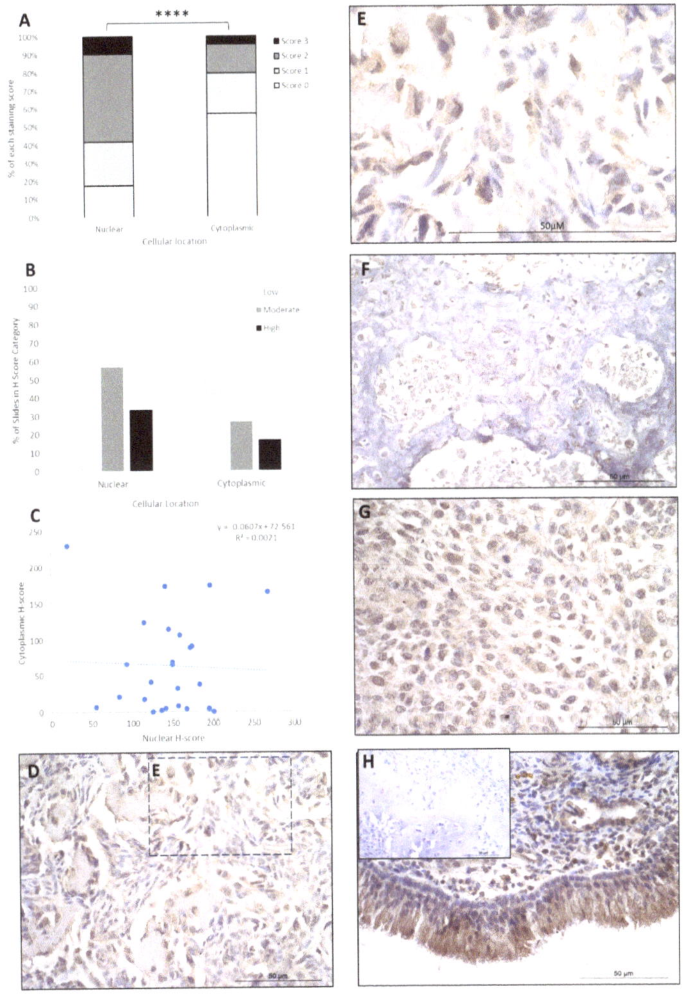

Figure 3. IRF8 cytoplasmic and nuclear H-scores. (**A**) H-score (scores 0, 1, 2, and 3) distributions. (**B**) H-score low/moderate/high classifications across the cases ($p = 0.0001$). (**C**) Nuclear and cytoplasmic H-score distributions and correlation. Overall, the nuclear scores were significantly higher than cytoplasmic (**** $p = 0.0001$), $n = 26$. (**D–G**) Immunohistochemical staining photomicrographs of canine osteosarcoma IRF8 expression, 40× magnification. (**H**) Positive control nasal mucosa lined by well-differentiated pseudostratified tall columnar ciliated epithelium, inset upper left: negative control, 40× magnification. All scale bars represent 50 μm.

LEF1 staining showed H-score variations between the different patients, and 8/26 expressed nuclear staining only, while the remaining 18/26 had both nuclear and cytoplasmic staining. Cytoplasmic scores were low, moderate, and high in 12/26 (46.6%), 10/26

(38%), and 4/26 (15.5%) patients, respectively. Nuclear scores were low in 26/26 (100%) of the patients, with no cases expressing moderate or high H-scores. Low nuclear score was combined with low, moderate, and high cytoplasmic scores in 12/26 (46.5%), 10/26 (38%), and 4/26 (15.5%) of cases, respectively (Table 2). Overall, the staining was diffuse, and nuclear H-scores for LEF1 were significantly lower than the cytoplasmic staining scores ($p > 0.0001$, Table 3, Figure 4), and there was a small positive correlation between nuclear and cytoplasmic H-scores within individual samples (Figure 4).

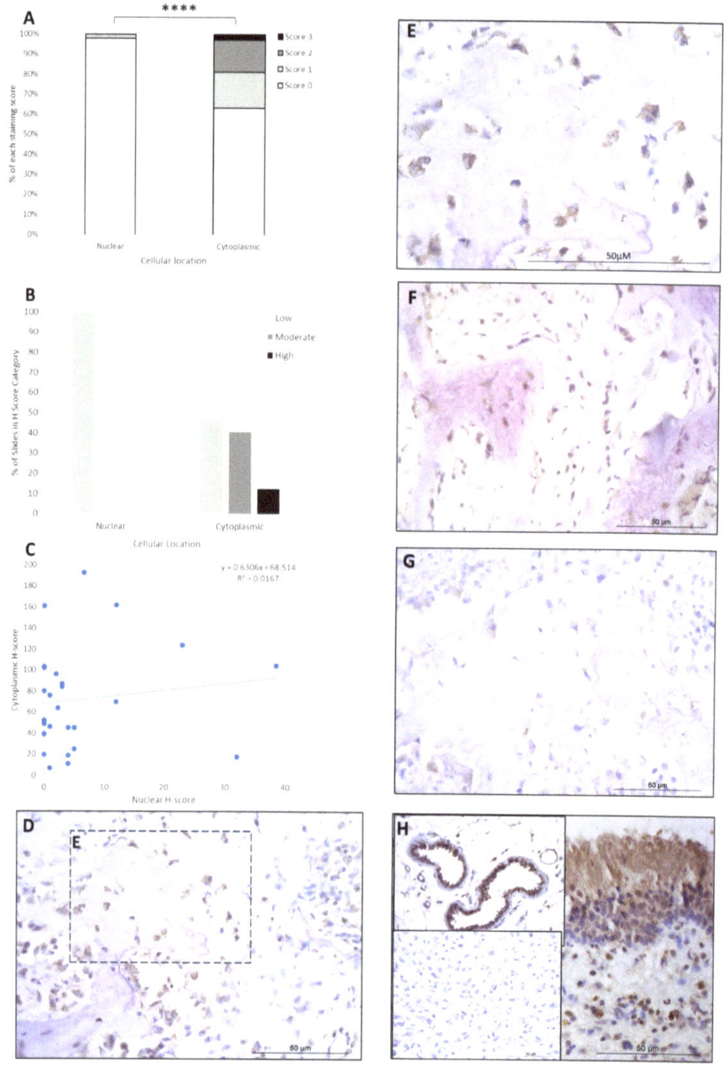

Figure 4. LEF1 cytoplasmic and nuclear H-scores. (**A**) H-score (scores 0, 1, 2, and 3) distributions. (**B**) H-score low/moderate/high classifications across the cases ($p = 0.0001$). (**C**) Nuclear and cytoplasmic H-score distributions and correlation. Overall, the nuclear scores were significantly lower than cytoplasmic (**** $p = 0.0001$), $n = 26$. (**D–G**) Immunohistochemical staining photomicrographs of canine osteosarcoma LEF1 expression, 40× magnification. (**H**) Right-hand side: positive control nasal mucosa lined by well-differentiated pseudostratified tall columnar ciliated epithelium, inset upper left: endothelial cells of vasculature positive control, inset lower left: negative control, 40× magnification. All scale bars represent 50 μm.

Comparisons between the sexes and anatomical locations were also analysed for each protein. Nuclear, cytoplasmic, and combined H-scores for FOXO4, IRF8, and LEF1 in males and females (n = 12 and 13, respectively) showed no statistically significant differences (t-tests, $p > 0.05$; Figure 5A). Differing anatomical location of the bone tumours—either appendicular (n = 20) or axial (head + thorax; n = 6)—showed no significant differences for any of the proteins (t-test, $p > 0.05$; Figure 5B).

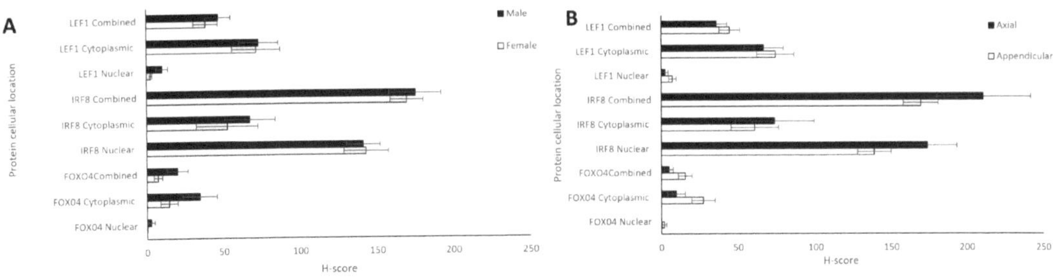

Figure 5. Nuclear, cytoplasmic, and combined H-scores for FOXO4, IRF8, and LEF1 by sex and anatomical location. (**A**) Males and females (n = 12 and 13, respectively), and (**B**) differing bone locations—appendicular and axial (n = 20 and 6, respectively). No statistically significant differences in sex or bone location were observed for nuclear, cytoplasmic, or total H-scores (t-test, $p > 0.05$).

4. Discussion

The influences of genetic factors in relation to the aetiology and progression of OSA have been widely recognised, and p53 is the most frequently investigated gene in canine OSA [3]. Furthermore, there is strong documented evidence of chromosomal rearrangements, gene mutations, alterations in gene expression, alterations in microRNA expression, and DNA methylation pattern changes between human and canine OSA [66]. Another important genetic cause is the RB tumour suppressor gene, which has been associated to the development of canine OSA. In a pilot study, the expression of MET proto-oncogene was identified in the majority of the histopathological samples of seven large breed dogs with spontaneous skeletal OSA [3,67]. Importantly, canine OSA has been compared to human OSA due to similar genetic, biologic, and clinical pathological features, and it has a 14 times higher incidence rate compared to human OSA; hence, it has been used as translational medicine to understand human OSA [6,68]. A study ascribed an interesting role of miR-1 and miR-133b as biomarkers for canine OSA's treatment and validated the high molecular homology between human and canine OSA [6]. Despite these advances, very little is known about the genetics of OSA and protein expression, the present study aimed to elucidate the expression of three proteins, FOXO4, LEF1, and IRF8, in OSA.

Previously, we have shown a 1.42 Log2 fold decrease in *FOXO4* transcripts in OSA tissue compared to patient-matched non-tumour bone [11]. In the present study, the majority of OSA specimens exhibited no nuclear protein staining of FOXO4. As little cytoplasmic expression was also observed, FOXO4 holds limited value for prognostic or diagnostic use, although expression in non-malignant bone is yet to be determined. Firstly, it is of interest that our research shows that it is expressed in bone. The Human Protein Atlas data by Santos and coauthors found FOXO4 transcript expression across numerous tissues; however, protein was detected by IHC in only the testis, placenta, heart, skeletal muscle, and smooth muscle [69], notably not bone. This discrepancy between mRNA and protein expression was not consistent with our analysis of canine OSA tissue, where the protein was expressed. There was little previously known about FOXO4 in relation to canine OSA; however, a study investigating the role of FOXO4 in human colorectal cancer found that it had a role as a tumour suppressor, as FOXO4 was downregulated in colorectal cancers when compared to the control [33]. Overexpression of FOXO4 was found to have reduced migration and in vivo metastasis of the colorectal cancer cells by regulating the colorectal cancer tumour suppressor gene adenomatous polyposis coli 2 (APC2) in the APC2/β-catenin axis; therefore, this inhibitory effect could be reversed by APC2 knockdown [33]. Another study supports this claim that FOXO4 is a tumour suppressor, as it found that its expression was decreased in human gastric cancer tissue and gastric cancer cell lines. The upregulation of this protein inhibited tumour growth and progression, whereas downregulation of this protein promoted tumour growth and progression [34]. Therefore, although FOXO4 expression was not expected in bone, its expression and its downregulation in other tumours (reviewed in [70,71]), including OSA, as shown in the present research, support its role relating to tumour suppression, and the subsequent growth and progression where it is downregulated.

We previously identified *IRF8* to be downregulated in canine OSA as compared to matched non-malignant specimens. In agreement with previous studies analysing cellular localisation of the protein in other cell types [72,73], we found positive staining in both the nuclear and cytoplasmic compartment of OSA cells, with cytoplasmic staining present at a lower level of expression. Currently, little is known about the role of IRF8 in human or canine OSA. Muhitch and coauthors [74] observed that high expression of IRF8 in combination with low levels of TAMs has a significantly better survival outcome in comparison to low levels in both TAMs and IRF8 expression metastatic renal cell carcinoma tumour. It has been shown that IRF8 promoted epithelial–mesenchymal transition (EMT)-like phenomena, cell motility, and invasion in a human OSA cell line, suggesting that it may play a role in metastasis [75]. Another group found that PD-L1 was induced by IRF8 and that in human OSA cells, PD-L1 and IRF8 were involved in growth and tumorigenicity, and that PD-L1 knockdown combined with doxorubicin treatment resulted in inhibition of cell growth [76]. Furthermore, a study identified that *IRF8* was among one of the many genes deleted in >25% of cases, according to an analysis of 28 human OSA samples [77]. Our current study presents the expression of IFR8 in canine OSA. Taken together, these studies identify a role for IRF8 in OSA, with potential as a prognostic marker, and further studies on the role and clinical relevance of IRF8 in both human and canine OSA are warranted.

Our present research shows LEF1 expression in both the cytoplasm and nucleus, and our previous work showed that *LEF1* exhibited a 2.2 Log2 fold increase in OSA samples compared to patient-matched non-tumour tissue. *LEF1* expression has also been demonstrated to be upregulated in OSA cells and patient samples compared with non-malignant osteoblasts and tissue in people [78–81]. *LEF1* has been associated with metastasis in OSA. Overexpression of *LEF1* was observed in highly metastatic OSA cell lines compared with OSA cells with low metastatic potential and, moreover, knock-out of *LEF1* resulted in significantly reduced extravasation of OSA cells to the lungs [82]. *LEF1* overexpression has also been found to abrogate the inhibitory effect of miR-34c on metastasis and chemoresistance in OSA cells [83]. It is notable that all studies investigating the role of LEF1 in OSA appear to have been conducted only in people and mice. To the best of the authors' knowledge, the current study is the first to report on the potential role of LEF1 in canine OSA. LEF1 expression has also been shown to be downregulated in numerous cancers through promoter hypermethylation and also higher levels of *IRF8* [61,84,85]. LEF1 was initially thought to be an effective therapeutic target [86]. In 2010, two small molecule inhibitors of Wnt/beta-catenin/LEF1 signalling (CGP049090 and PKF115-584) significantly inhibited the proliferation of CLL cells in vivo [60]. A large number of other small molecule inhibitors targeting the Wnt/beta-catenin/LEF1 pathway have since been discovered, and work to improve their utility and specificity as anti-cancer treatments is ongoing [87]. Unfortunately, very few of the compounds that have shown promise in vitro and in vivo have progressed to clinical trials, and among the ones that have, many have resulted in unsatisfactory outcomes due to inhibition of the wide-ranging essential functions of this pathway in normal physiological processes [88]. The authors are not aware of any inhibitors that have advanced beyond the very early phases of clinical trials to date. LEF1 has also been purported to have utility as a prognostic biomarker, since its high expression has been significantly associated with disease progression and poorer prognosis in chronic lymphocytic leukaemia (CLL) [89], acute lymphoblastic leukaemia (ALL) [90,91], small B-cell lymphomas [86], solid pseudopapillary neoplasms and pancreatic neuroendocrine tumours [92], oesophageal squamous cell carcinoma [93], nasopharyngeal carcinoma [94], deep penetrating nevi [62], and with metastasis in colorectal cancer [95]. Proteins such as GLUT1, MMP3, and NRF2 have shown promise as canine OSA biomarkers and are involved in Wnt activation [10]. The Wnt/β-catenin/LEF1 signaling pathway has also been shown to be involved in human osteosarcoma cells and tissues via RT-qPCR, where it was indicated that LEF1 translation via degradation of DKK3 was mediated through miR-214-3p, and that cantharidin could be a prospective candidate for osteosarcoma by targeting the pathways involved [96].

Primary OSA occurs more frequently on the appendicular skeleton in around 75% of the cases; 24% on the axial skeleton; and also, very rarely, approximately 1%, in extraskeletal tissues, for instance, mammary tissue, subcutaneous tissue, spleen, bowel, liver, kidney, testicle, vagina, eye, gastric ligament, synovium, meninges, and adrenal gland [3,4]. Interestingly, this does differ in relation to dog size, with one study showing that 5% of the diagnosed large and giant breed dogs with OSA presented with axial tumours compared to 59% in small breed dogs (less of 15 kg) [97]. Appendicular canine OSA is more commonly in the metaphysis of long bones, especially of the forelimbs, with higher frequency rates affecting locations such as the proximal humerus, the distal radius, and the distal tibia in the hind limb [18,98,99]. One crucial risk factor is related to the body size, since the tumour tends to occur in major weight-bearing bones adjacent to late closing physes [13,100]. Obesity has also been postulated to promote osteoblast proliferation in the limbs, which can contribute to remodelling in response to increased stress on weight-bearing limbs [100]. In light of this, and given the fact that appendicular OSA is the most frequent presentation in large and giant breed dogs with rapid early bone growth, it is reasonable to argue that the combination of these factors can help elucidate the complex aetiopathophysiology of OSA in this species [19]. No overall differences were observed between the axial and appendicular samples within this study, but given the body of evidence relating to axial and appendicular OSA, future studies should note potential differences and consider whether the anatomical location impacts the tumours, their environment, and prognosis and treatment factors. A limitation of the present study is a relatively smaller number of samples in the axial bones, and ideally a larger number of anatomical locations should also be investigated in the future.

The males and females in this study showed no significant differences between protein expression for any of the markers investigated; these data add interesting evidence to the sex susceptibility discussions which are ongoing about OSA. There is contradictory evidence concerning sex predisposition in canine OSA [101]. Historically, males have been thought to be slightly more frequently affected than females, with a reported ratio of 1.1–1.5:1 [3]. In contrast, another study showed that females were more prone to be affected with OSA, with a ratio of 2.1:1, but this was not consistent with respect to the location of OSA [97]. Ru et al. concluded in their study there was no sex susceptibility, but neutered males and females were noted to have twice the risk for OSA compared to intact dogs for both of the sexes [12]. A retrospective case series with 744 dogs diagnosed with appendicular OSA revealed that the male-to-female ratio was 0.95:1.0, and 80.9% of the population with OSA were neutered [99]. Despite these findings, research has not found any strong evidence that sex or neuter status is a risk factor for the development of OSA in dogs. Additionally, in some of these older reports, males have been overrepresented and/or there was bias towards male dogs or neutered animals [19,98–100]. Nevertheless, it has been thought that endogenous sex hormones have a significant impact on OSA, and some reports go towards the consideration of a protective influence in intact dogs [99].

Understanding whether proteins can assist with diagnosis, prognosis, or treatment development is important. Several negative prognostic factors for canine OSA have been described in the literature, and these include histological grade, distant metastasis at diagnosis, age at diagnosis, large primary tumour size, high body weight, high serum alkaline phosphate (ALP) activity, proximal humeral location, prolonged duration of clinical signs before surgery, lymph node metastasis, and delayed initiation of chemotherapy following surgery [98,102,103]. Schmidt et al. confirmed in their study that tumour location and ALP activity levels are prognostic factors for both mortality and metastasis; age was only a prognostic factor for mortality [103]. Understanding the expression of these proteins in canine, and indeed human OSA, could prove beneficial for diagnosis, prognosis, and treatment development. Further studies elucidating their roles, mechanisms of action, further protein expression level studies (e.g., Western blots), and drug discovery avenues of research are recommended.

5. Conclusions

Unlike human medicine, diagnosis and prognosis are not presently facilitated by the use of IHC for canine OSA; however, the present work enhances the knowledge required to understand protein expression in these tissues in different OSA samples. The diagnosis of OSA can be made through a combination of signalment, clinical presentation, and radiographic findings such as lytic, proliferative, or mixed bone lesions [18,101]. Nevertheless, histopathological samples are warranted for a final diagnosis and for tumour classification based on the formation of osteoid matrix with osteoblastic, fibroblastic, chondroblastic, telangiectic, and combined subtypes [98]. The aetiology of canine OSA has not been completely established but is considered to be complex, involving physical, genetic, and molecular factors. The investigations undertaken in the present research facilitate further understanding of the roles played by these proteins within their respective molecular pathways, including altered Wnt/beta-catenin/LEF1 signalling and via differential regulation of tumour suppressor genes and proliferation, and the effects of promoter hypermethylation. This is especially important given the opportunities that advances in molecular methods for investigating canine cancer offer for diagnosis, prognosis, and treatment development [23]. Given the staining observed and their involvement in various signalling pathways, IRF8 and LEF1 are promising biomarker candidates for prognostic and diagnostic purposes and may have mechanisms which can be targeted for the development of therapeutics. A deeper understanding of the mechanisms involved in OSA represents essential contributions towards the development of novel diagnostic, prognostic, and treatment options in human and veterinary medicine contexts.

Author Contributions: Conceptualization, C.S.R., N.P.M. and J.N.J.; methodology, C.S.R., A.A.A., J.C., N.P.M. and S.d.B.; validation, C.S.R., J.C. and A.A.A.; formal analysis, all authors; investigation, all authors; resources, C.S.R., J.N.J. and N.P.M.; data curation, C.S.R. and N.P.M.; writing—original draft preparation, all authors; writing—review and editing, all authors; visualization, C.S.R., A.A.A., J.C. and R.P.; supervision, N.P.M., C.S.R., S.d.B., J.N.J. and M.D.D.; project administration, C.S.R., N.P.M. and J.N.J.; funding acquisition, C.S.R., J.N.J. and N.P.M. All authors have read and agreed to the published version of the manuscript.

Funding: This research was funded by the BBSRC Doctoral training program and the School of Veterinary Medicine and Science, University of Nottingham (BB/M008770/1: A.E.H., C.L.W., J.L.-R., supervised by C.S.R., N.P.M., and BB/T008369/1: R.P., D.V., M.H., supervised by C.S.R., J.J., N.P.M.). We also gratefully acknowledge the Cancer Sciences Undergraduate Bursary awarded by the University of Nottingham (to J.C. supervised by C.S.R., J.J. and N.P.M.). The authors would like to thank the Irish Wolfhound Charitable Trust for their valued charitable donation which helps us to store biobank material.

Institutional Review Board Statement: All animal tissue work in this study was approved by the Ethics committee at the University of Nottingham School of Veterinary Medicine and Science and complied with national (Home Office) and international ethics procedures (permission numbers—1823 160714 (02/07/2016), 959 130925 (02/02/2012), UG 20331 (26/08/2020)). Patients were under veterinary practice care under circumstances unrelated to research.

Informed Consent Statement: Not applicable.

Data Availability Statement: Data are available upon request from the corresponding authors.

Acknowledgments: The authors would like to thank Mohamed Saeed, Emily Stephens, Elysia Pughe, and Harrison Downing for conducting some optimisation staining prior to the study under the supervision of Catrin Rutland and Aziza Alibhai.

Conflicts of Interest: The authors declare no conflicts of interest.

References

1. Schmidt, A.F.; Groenwold, R.H.H.; Amsellem, P.; Bacon, N.; Klungel, O.H.; Hoes, A.W.; de Boer, A.; Kow, K.; Maritato, K.; Kirpensteijn, J.; et al. Which dogs with appendicular osteosarcoma benefit most from chemotherapy after surgery? Results from an individual patient data meta-analysis. *Prev. Vet. Med.* **2016**, *125*, 116–125. [CrossRef] [PubMed]
2. Cook, M.R.; Lorbach, J.; Husbands, B.D.; Kisseberth, W.C.; Samuels, S.; Silveira, C.; Wustefeld-Janssens, B.G.; Wouda, R.; Keepman, S.; Oblak, M.L.; et al. A retrospective analysis of 11 dogs with surface osteosarcoma. *Vet. Comp. Oncol.* **2022**, *20*, 82–90. [CrossRef]
3. Greene, S. Withrow & MacEwen's Small Animal Clinical Oncology, 6th edition. *Am. J. Vet. Res.* **2020**, *81*, 391.
4. Morello, E.; Martano, M.; Buracco, P. Biology, diagnosis and treatment of canine appendicular osteosarcoma: Similarities and differences with human osteosarcoma. *Vet. J.* **2011**, *189*, 268–277. [CrossRef]
5. Selvarajah, G.T.; Kirpensteijn, J. Prognostic and predictive biomarkers of canine osteosarcoma. *Vet. J.* **2010**, *185*, 28–35. [CrossRef]
6. Leonardo, L.; Laura, P.; Serena, B.M. miR-1 and miR-133b expression in canine osteosarcoma. *Res. Vet. Sci.* **2018**, *117*, 133–137. [CrossRef]
7. Egenvall, A.; Nodtvedt, A.; von Euler, H. Bone tumors in a population of 400 000 insured Swedish dogs up to 10 y of age: Incidence and survival. *Can. J. Vet. Res.* **2007**, *71*, 292–299. [PubMed]
8. Fenger, J.M.; London, C.A.; Kisseberth, W.C. Canine osteosarcoma: A naturally occurring disease to inform pediatric oncology. *ILAR J.* **2014**, *55*, 69–85. [CrossRef]
9. Makielski, K.M.; Mills, L.J.; Sarver, A.L.; Henson, M.S.; Spector, L.G.; Naik, S.; Modiano, J.F. Risk Factors for Development of Canine and Human Osteosarcoma: A Comparative Review. *Vet. Sci.* **2019**, *6*, 48. [CrossRef]
10. Rutland, C.S.; Cockcroft, J.M.; Lothion-Roy, J.; Harris, A.E.; Jeyapalan, J.N.; Simpson, S.; Alibhai, A.; Bailey, C.; Ballard-Reisch, A.C.; Rizvanov, A.A.; et al. Immunohistochemical Characterisation of GLUT1, MMP3 and NRF2 in Osteosarcoma. *Front. Vet. Sci.* **2021**, *8*, 704598. [CrossRef]
11. Simpson, S.; Dunning, M.; de Brot, S.; Alibhai, A.; Bailey, C.; Woodcock, C.L.; Mestas, M.; Akhtar, S.; Jeyapalan, J.N.; Lothion-Roy, J.; et al. Molecular Characterisation of Canine Osteosarcoma in High Risk Breeds. *Cancers* **2020**, *12*, 2405. [CrossRef] [PubMed]
12. Ru, G.; Terracini, B.; Glickman, L.T. Host related risk factors for canine osteosarcoma. *Vet. J.* **1998**, *156*, 31–39. [CrossRef] [PubMed]
13. Rosenberger, J.A.; Pablo, N.V.; Crawford, P.C. Prevalence of and intrinsic risk factors for appendicular osteosarcoma in dogs: 179 cases (1996-2005). *J. Am. Vet. Med. Assoc.* **2007**, *231*, 1076–1080. [CrossRef] [PubMed]
14. Mirabello, L.; Pfeiffer, R.; Murphy, G.; Daw, N.C.; Patino-Garcia, A.; Troisi, R.J.; Hoover, R.N.; Douglass, C.; Schuz, J.; Craft, A.W.; et al. Height at diagnosis and birth-weight as risk factors for osteosarcoma. *Cancer Causes Control* **2011**, *22*, 899–908. [CrossRef] [PubMed]
15. Ottaviani, G.; Jaffe, N. The epidemiology of osteosarcoma. *Cancer Treat. Res.* **2009**, *152*, 3–13. [CrossRef]
16. Sadykova, L.R.; Ntekim, A.I.; Muyangwa-Semenova, M.; Rutland, C.S.; Jeyapalan, J.N.; Blatt, N.; Rizvanov, A.A. Epidemiology and Risk Factors of Osteosarcoma. *Cancer Investig.* **2020**, *38*, 259–269. [CrossRef]
17. Simpson, S.; Dunning, M.D.; de Brot, S.; Grau-Roma, L.; Mongan, N.P.; Rutland, C.S. Comparative review of human and canine osteosarcoma: Morphology, epidemiology, prognosis, treatment and genetics. *Acta Vet. Scand.* **2017**, *59*, 71. [CrossRef] [PubMed]
18. Edmunds, G.L.; Smalley, M.J.; Beck, S.; Errington, R.J.; Gould, S.; Winter, H.; Brodbelt, D.C.; O'Neill, D.G. Dog breeds and body conformations with predisposition to osteosarcoma in the UK: A case-control study. *Canine Med. Genet.* **2021**, *8*, 2. [CrossRef]
19. Selvarajah, G.T.; Kirpensteijn, J.; van Wolferen, M.E.; Rao, N.A.; Fieten, H.; Mol, J.A. Gene expression profiling of canine osteosarcoma reveals genes associated with short and long survival times. *Mol. Cancer* **2009**, *8*, 72. [CrossRef]
20. Szewczyk, M.; Lechowski, R.; Zabielska, K. What do we know about canine osteosarcoma treatment? Review. *Vet. Res. Commun.* **2015**, *39*, 61–67. [CrossRef]
21. Rubin, J.A.; Suran, J.N.; Brown, D.C.; Agnello, K.A. Factors associated with pathological fractures in dogs with appendicular primary bone neoplasia: 84 cases (2007-2013). *J. Am. Vet. Med. Assoc.* **2015**, *247*, 917–923. [CrossRef] [PubMed]
22. Paoloni, M.; Davis, S.; Lana, S.; Withrow, S.; Sangiorgi, L.; Picci, P.; Hewitt, S.; Triche, T.; Meltzer, P.; Khanna, C. Canine tumor cross-species genomics uncovers targets linked to osteosarcoma progression. *BMC Genom.* **2009**, *10*, 625. [CrossRef]
23. Kehl, A.; Aupperle-Lellbach, H.; de Brot, S.; van der Weyden, L. Review of Molecular Technologies for Investigating Canine Cancer. *Animals* **2024**, *14*, 769. [CrossRef] [PubMed]
24. Tsuchiya, K.; Ogawa, Y. Forkhead box class O family member proteins: The biology and pathophysiological roles in diabetes. *J. Diabetes Investig.* **2017**, *8*, 726–731. [CrossRef] [PubMed]
25. Liu, W.; Li, Y.; Luo, B. Current perspective on the regulation of FOXO4 and its role in disease progression. *Cell. Mol. Life Sci.* **2020**, *77*, 651–663. [CrossRef] [PubMed]

26. Rached, M.T.; Kode, A.; Xu, L.; Yoshikawa, Y.; Paik, J.H.; Depinho, R.A.; Kousteni, S. FoxO1 is a positive regulator of bone formation by favoring protein synthesis and resistance to oxidative stress in osteoblasts. *Cell Metab.* **2010**, *11*, 147–160. [CrossRef] [PubMed]
27. Kops, G.J.; de Ruiter, N.D.; De Vries-Smits, A.M.; Powell, D.R.; Bos, J.L.; Burgering, B.M. Direct control of the Forkhead transcription factor AFX by protein kinase B. *Nature* **1999**, *398*, 630–634. [CrossRef] [PubMed]
28. Brunet, A.; Bonni, A.; Zigmond, M.J.; Lin, M.Z.; Juo, P.; Hu, L.S.; Anderson, M.J.; Arden, K.C.; Blenis, J.; Greenberg, M.E. Akt promotes cell survival by phosphorylating and inhibiting a Forkhead transcription factor. *Cell* **1999**, *96*, 857–868. [CrossRef] [PubMed]
29. Greer, E.L.; Brunet, A. FOXO transcription factors at the interface between longevity and tumor suppression. *Oncogene* **2005**, *24*, 7410–7425. [CrossRef]
30. Baar, M.P.; Brandt, R.M.C.; Putavet, D.A.; Klein, J.D.D.; Derks, K.W.J.; Bourgeois, B.R.M.; Stryeck, S.; Rijksen, Y.; van Willigenburg, H.; Feijtel, D.A.; et al. Targeted Apoptosis of Senescent Cells Restores Tissue Homeostasis in Response to Chemotoxicity and Aging. *Cell* **2017**, *169*, 132–147.e116. [CrossRef]
31. Sergi, C.; Shen, F.; Liu, S.M. Insulin/IGF-1R, SIRT1, and FOXOs Pathways-An Intriguing Interaction Platform for Bone and Osteosarcoma. *Front. Endocrinol.* **2019**, *10*, 93. [CrossRef] [PubMed]
32. Chen, B.; Bao, Y.; Chen, X.; Yi, J.; Liu, S.; Fang, Z.; Zheng, S.; Chen, J. Mir-664 promotes osteosarcoma cells proliferation via downregulating of FOXO4. *Biomed. Pharmacother.* **2015**, *75*, 1–7. [CrossRef] [PubMed]
33. Sun, Y.; Wang, L.; Xu, X.; Han, P.; Wu, J.; Tian, X.; Li, M. FOXO4 Inhibits the Migration and Metastasis of Colorectal Cancer by Regulating the APC2/beta-Catenin Axis. *Front. Cell Dev. Biol.* **2021**, *9*, 659731. [CrossRef] [PubMed]
34. Su, L.; Liu, X.; Chai, N.; Lv, L.; Wang, R.; Li, X.; Nie, Y.; Shi, Y.; Fan, D. The transcription factor FOXO4 is down-regulated and inhibits tumor proliferation and metastasis in gastric cancer. *BMC Cancer* **2014**, *14*, 378. [CrossRef]
35. Lu, Y.; Shen, Y.; Li, L.; Zhang, M.; Wang, M.; Ge, L.; Yang, J.; Tang, X. Clinicopathological Significance of FOXO4 Expression and Correlation with Prx1 in Head and Neck Squamous Cell Carcinoma. *Anal. Cell. Pathol.* **2021**, *2021*, 5510753. [CrossRef]
36. Paik, J.-H.; Kollipara, R.; Chu, G.; Ji, H.; Xiao, Y.; Ding, Z.; Miao, L.; Tothova, Z.; Horner, J.W.; Carrasco, D.R.; et al. FoxOs Are Lineage-Restricted Redundant Tumor Suppressors and Regulate Endothelial Cell Homeostasis. *Cell* **2007**, *128*, 309–323. [CrossRef]
37. Lee, E.J.; Kim, J.M.; Lee, M.K.; Jameson, J.L. Splice variants of the forkhead box protein AFX exhibit dominant negative activity and inhibit AFXalpha-mediated tumor cell apoptosis. *PLoS ONE* **2008**, *3*, e2743. [CrossRef]
38. Tamura, T.; Yanai, H.; Savitsky, D.; Taniguchi, T. The IRF family transcription factors in immunity and oncogenesis. *Annu. Rev. Immunol.* **2008**, *26*, 535–584. [CrossRef]
39. Moorman, H.R.; Reategui, Y.; Poschel, D.B.; Liu, K. IRF8: Mechanism of Action and Health Implications. *Cells* **2022**, *11*, 2630. [CrossRef]
40. Cytlak, U.; Resteu, A.; Pagan, S.; Green, K.; Milne, P.; Maisuria, S.; McDonald, D.; Hulme, G.; Filby, A.; Carpenter, B.; et al. Differential IRF8 Transcription Factor Requirement Defines Two Pathways of Dendritic Cell Development in Humans. *Immunity* **2020**, *53*, 353–370.e358. [CrossRef]
41. Nixon, B.G.; Kuo, F.; Ji, L.; Liu, M.; Capistrano, K.; Do, M.; Franklin, R.A.; Wu, X.; Kansler, E.R.; Srivastava, R.M.; et al. Tumor-associated macrophages expressing the transcription factor IRF8 promote T cell exhaustion in cancer. *Immunity* **2022**, *55*, 2044–2058.e2045. [CrossRef] [PubMed]
42. Durmowicz, M.C.; Cui, C.Y.; Schlessinger, D. The EDA gene is a target of, but does not regulate Wnt signaling. *Gene* **2002**, *285*, 203–211. [CrossRef]
43. Eastman, Q.; Grosschedl, R. Regulation of LEF-1/TCF transcription factors by Wnt and other signals. *Curr. Opin. Cell Biol.* **1999**, *11*, 233–240. [CrossRef] [PubMed]
44. Steinke, F.C.; Xue, H.H. From inception to output, Tcf1 and Lef1 safeguard development of T cells and innate immune cells. *Immunol. Res.* **2014**, *59*, 45–55. [CrossRef]
45. Reya, T.; O'Riordan, M.; Okamura, R.; Devaney, E.; Willert, K.; Nusse, R.; Grosschedl, R. Wnt signaling regulates B lymphocyte proliferation through a LEF-1 dependent mechanism. *Immunity* **2000**, *13*, 15–24. [CrossRef] [PubMed]
46. Held, W.; Clevers, H.; Grosschedl, R. Redundant functions of TCF-1 and LEF-1 during T and NK cell development, but unique role of TCF-1 for Ly49 NK cell receptor acquisition. *Eur. J. Immunol.* **2003**, *33*, 1393–1398. [CrossRef]
47. Phan, Q.M.; Fine, G.M.; Salz, L.; Herrera, G.G.; Wildman, B.; Driskell, I.M.; Driskell, R.R. Lef1 expression in fibroblasts maintains developmental potential in adult skin to regenerate wounds. *eLife* **2020**, *9*, e60066. [CrossRef] [PubMed]
48. Sun, H.; He, Z.; Xi, Q.; Zhao, F.; Hu, J.; Wang, J.; Liu, X.; Zhao, Z.; Li, M.; Luo, Y.; et al. Lef1 and Dlx3 May Facilitate the Maturation of Secondary Hair Follicles in the Skin of Gansu Alpine Merino. *Genes* **2022**, *13*, 1326. [CrossRef] [PubMed]
49. Lima, B.M.; Azevedo, A.L.K.; Giner, I.S.; Gomig, T.H.B.; Ribeiro, E.; Cavalli, I.J. Biomarker potential of the LEF1/TCF family members in breast cancer: Bioinformatic investigation on expression and clinical significance. *Genet. Mol. Biol.* **2023**, *46*, e20220346. [CrossRef]
50. Cordray, P.; Satterwhite, D.J. TGF-beta induces novel Lef-1 splice variants through a Smad-independent signaling pathway. *Dev. Dyn.* **2005**, *232*, 969–978. [CrossRef]
51. Chen, W.Y.; Liu, S.Y.; Chang, Y.S.; Yin, J.J.; Yeh, H.L.; Mouhieddine, T.H.; Hadadeh, O.; Abou-Kheir, W.; Liu, Y.N. MicroRNA-34a regulates WNT/TCF7 signaling and inhibits bone metastasis in Ras-activated prostate cancer. *Oncotarget* **2015**, *6*, 441–457. [CrossRef] [PubMed]

52. Blazquez, R.; Rietkotter, E.; Wenske, B.; Wlochowitz, D.; Sparrer, D.; Vollmer, E.; Muller, G.; Seegerer, J.; Sun, X.; Dettmer, K.; et al. LEF1 supports metastatic brain colonization by regulating glutathione metabolism and increasing ROS resistance in breast cancer. *Int. J. Cancer* **2020**, *146*, 3170–3183. [CrossRef]
53. Chen, C.L.; Tsai, Y.S.; Huang, Y.H.; Liang, Y.J.; Sun, Y.Y.; Su, C.W.; Chau, G.Y.; Yeh, Y.C.; Chang, Y.S.; Hu, J.T.; et al. Lymphoid Enhancer Factor 1 Contributes to Hepatocellular Carcinoma Progression Through Transcriptional Regulation of Epithelial-Mesenchymal Transition Regulators and Stemness Genes. *Hepatol. Commun.* **2018**, *2*, 1392–1407. [CrossRef] [PubMed]
54. Keller, J.J.; Moon, R.T.; Chien, A.J. Wnt and related signaling pathways in melanomagenesis. *Cancers* **2010**, *2*, 1000–1012. [CrossRef] [PubMed]
55. White, B.D.; Chien, A.J.; Dawson, D.W. Dysregulation of Wnt/beta-catenin signaling in gastrointestinal cancers. *Gastroenterology* **2012**, *142*, 219–232. [CrossRef] [PubMed]
56. Wang, W.J.; Yao, Y.; Jiang, L.L.; Hu, T.H.; Ma, J.Q.; Liao, Z.J.; Yao, J.T.; Li, D.F.; Wang, S.H.; Nan, K.J. Knockdown of lymphoid enhancer factor 1 inhibits colon cancer progression in vitro and in vivo. *PLoS ONE* **2013**, *8*, e76596. [CrossRef] [PubMed]
57. Wu, W.; Zhu, H.; Fu, Y.; Shen, W.; Miao, K.; Hong, M.; Xu, W.; Fan, L.; Young, K.H.; Liu, P.; et al. High LEF1 expression predicts adverse prognosis in chronic lymphocytic leukemia and may be targeted by ethacrynic acid. *Oncotarget* **2016**, *7*, 21631–21643. [CrossRef]
58. Schmeel, L.C.; Schmeel, F.C.; Kim, Y.; Endo, T.; Lu, D.; Schmidt-Wolf, I.G. Targeting the Wnt/beta-catenin pathway in multiple myeloma. *Anticancer Res.* **2013**, *33*, 4719–4726. [PubMed]
59. Nguyen, A.; Rosner, A.; Milovanovic, T.; Hope, C.; Planutis, K.; Saha, B.; Chaiwun, B.; Lin, F.; Imam, S.A.; Marsh, J.L.; et al. Wnt pathway component LEF1 mediates tumor cell invasion and is expressed in human and murine breast cancers lacking ErbB2 (her-2/neu) overexpression. *Int. J. Oncol.* **2005**, *27*, 949–956. [CrossRef]
60. Gandhirajan, R.K.; Staib, P.A.; Minke, K.; Gehrke, I.; Plickert, G.; Schlosser, A.; Schmitt, E.K.; Hallek, M.; Kreuzer, K.A. Small molecule inhibitors of Wnt/beta-catenin/lef-1 signaling induces apoptosis in chronic lymphocytic leukemia cells in vitro and in vivo. *Neoplasia* **2010**, *12*, 326–335. [CrossRef]
61. Gatti, G.; Betts, C.; Rocha, D.; Nicola, M.; Grupe, V.; Ditada, C.; Nuñez, N.G.; Roselli, E.; Araya, P.; Dutto, J.; et al. High IRF8 expression correlates with CD8 T cell infiltration and is a predictive biomarker of therapy response in ER-negative breast cancer. *Breast Cancer Res.* **2021**, *23*, 40. [CrossRef] [PubMed]
62. Raghavan, S.S.; Saleem, A.; Wang, J.Y.; Rieger, K.E.; Brown, R.A.; Novoa, R.A. Diagnostic Utility of LEF1 Immunohistochemistry in Differentiating Deep Penetrating Nevi From Histologic Mimics. *Am. J. Surg. Pathol.* **2020**, *44*, 1413–1418. [CrossRef]
63. Fedchenko, N.; Reifenrath, J. Different approaches for interpretation and reporting of immunohistochemistry analysis results in the bone tissue—A review. *Diagn. Pathol.* **2014**, *9*, 221. [CrossRef]
64. McCarty, K.S., Jr.; Miller, L.S.; Cox, E.B.; Konrath, J.; McCarty, K.S., Sr. Estrogen receptor analyses. Correlation of biochemical and immunohistochemical methods using monoclonal antireceptor antibodies. *Arch. Pathol. Lab. Med.* **1985**, *109*, 716–721. [PubMed]
65. Detre, S.; Saclani Jotti, G.; Dowsett, M. A "quickscore" method for immunohistochemical semiquantitation: Validation for oestrogen receptor in breast carcinomas. *J. Clin. Pathol.* **1995**, *48*, 876–878. [CrossRef]
66. Bryan, J.N. Updates in Osteosarcoma. *Vet. Clin. N. Am. Small Anim. Pract.* **2024**, *54*, 523–539. [CrossRef]
67. Ferracini, R.; Angelini, P.; Cagliero, E.; Linari, A.; Martano, M.; Wunder, J.; Buracco, P. MET oncogene aberrant expression in canine osteosarcoma. *J. Orthop. Res.* **2000**, *18*, 253–256. [CrossRef] [PubMed]
68. Yang, Y.T.; Engleberg, A.I.; Yuzbasiyan-Gurkan, V. Establishment and Characterization of Cell Lines from Canine Metastatic Osteosarcoma. *Cells* **2024**, *13*, 25. [CrossRef] [PubMed]
69. Santos, B.F.; Grenho, I.; Martel, P.J.; Ferreira, B.I.; Link, W. FOXO family isoforms. *Cell Death Dis.* **2023**, *14*, 702. [CrossRef]
70. Yadav, R.K.; Chauhan, A.S.; Zhuang, L.; Gan, B. FoxO transcription factors in cancer metabolism. *Semin. Cancer Biol.* **2018**, *50*, 65–76. [CrossRef]
71. Dansen, T.B.; Burgering, B.M. Unravelling the tumor-suppressive functions of FOXO proteins. *Trends Cell Biol.* **2008**, *18*, 421–429. [CrossRef] [PubMed]
72. Minderman, H.; Maguire, O.; O'Loughlin, K.L.; Muhitch, J.; Wallace, P.K.; Abrams, S.I. Total cellular protein presence of the transcription factor IRF8 does not necessarily correlate with its nuclear presence. *Methods* **2017**, *112*, 84–90. [CrossRef] [PubMed]
73. Liss, F.; Frech, M.; Wang, Y.; Giel, G.; Fischer, S.; Simon, C.; Weber, L.M.; Nist, A.; Stiewe, T.; Neubauer, A.; et al. IRF8 Is an AML-Specific Susceptibility Factor That Regulates Signaling Pathways and Proliferation of AML Cells. *Cancers* **2021**, *13*, 764. [CrossRef] [PubMed]
74. Muhitch, J.B.; Hoffend, N.C.; Azabdaftari, G.; Miller, A.; Bshara, W.; Morrison, C.D.; Schwaab, T.; Abrams, S.I. Tumor-associated macrophage expression of interferon regulatory Factor-8 (IRF8) is a predictor of progression and patient survival in renal cell carcinoma. *J. Immunother. Cancer* **2019**, *7*, 155. [CrossRef] [PubMed]
75. Sung, J.Y.; Park, S.Y.; Kim, J.H.; Kang, H.G.; Yoon, J.H.; Na, Y.S.; Kim, Y.N.; Park, B.K. Interferon consensus sequence-binding protein (ICSBP) promotes epithelial-to-mesenchymal transition (EMT)-like phenomena, cell-motility, and invasion via TGF-β signaling in U2OS cells. *Cell Death Dis.* **2014**, *5*, e1224. [CrossRef] [PubMed]
76. Sung, J.Y.; Kim, J.H.; Kang, H.G.; Park, J.W.; Park, S.Y.; Park, B.K.; Kim, Y.N. ICSBP-induced PD-L1 enhances osteosarcoma cell growth. *Front. Oncol.* **2022**, *12*, 918216. [CrossRef] [PubMed]

77. Pires, S.F.; Barros, J.S.; Costa, S.S.D.; Carmo, G.B.D.; Scliar, M.O.; Lengert, A.V.H.; Boldrini, É.; Silva, S.; Vidal, D.O.; Maschietto, M.; et al. Analysis of the Mutational Landscape of Osteosarcomas Identifies Genes Related to Metastasis and Prognosis and Disrupted Biological Pathways of Immune Response and Bone Development. *Int. J. Mol. Sci.* **2023**, *24*, 10463. [CrossRef] [PubMed]
78. Ma, Y.; Ren, Y.; Han, E.Q.; Li, H.; Chen, D.; Jacobs, J.J.; Gitelis, S.; O'Keefe, R.J.; Konttinen, Y.T.; Yin, G.; et al. Inhibition of the Wnt-beta-catenin and Notch signaling pathways sensitizes osteosarcoma cells to chemotherapy. *Biochem. Biophys. Res. Commun.* **2013**, *431*, 274–279. [CrossRef] [PubMed]
79. Lu, X.; Qiao, L.; Liu, Y. Long noncoding RNA LEF1-AS1 binds with HNRNPL to boost the proliferation, migration, and invasion in osteosarcoma by enhancing the mRNA stability of LEF1. *J. Cell. Biochem.* **2020**, *121*, 4064–4073. [CrossRef]
80. Han, G.; Guo, Q.; Li, N.; Bi, W.; Xu, M.; Jia, J. Screening and Analysis of Biomarkers in the miRNA-mRNA Regulatory Network of Osteosarcoma. *J. Healthc. Eng.* **2022**, *2022*, 8055052. [CrossRef]
81. Xu, G.; Zhang, H.; Shi, Y.; Yang, F. Circular RNA circDOCK1 contributes to osteosarcoma progression by acting as a ceRNA for miR-936 to regulate LEF1. *J. Bone Oncol.* **2022**, *36*, 100453. [CrossRef] [PubMed]
82. Pongsuchart, M.; Kuchimaru, T.; Yonezawa, S.; Tran, D.T.P.; Kha, N.T.; Hoang, N.T.H.; Kadonosono, T.; Kizaka-Kondoh, S. Novel lymphoid enhancer-binding factor 1-cytoglobin axis promotes extravasation of osteosarcoma cells into the lungs. *Cancer Sci.* **2018**, *109*, 2746–2756. [CrossRef] [PubMed]
83. Xu, M.; Jin, H.; Xu, C.X.; Bi, W.Z.; Wang, Y. MiR-34c inhibits osteosarcoma metastasis and chemoresistance. *Med. Oncol.* **2014**, *31*, 972. [CrossRef] [PubMed]
84. Suzuki, M.; Ikeda, K.; Shiraishi, K.; Eguchi, A.; Mori, T.; Yoshimoto, K.; Shibata, H.; Ito, T.; Baba, Y.; Baba, H. Aberrant methylation and silencing of IRF8 expression in non-small cell lung cancer. *Oncol. Lett.* **2014**, *8*, 1025–1030. [CrossRef] [PubMed]
85. Luo, X.; Xiong, X.; Shao, Q.; Xiang, T.; Li, L.; Yin, X.; Li, X.; Tao, Q.; Ren, G. The tumor suppressor interferon regulatory factor 8 inhibits β-catenin signaling in breast cancers, but is frequently silenced by promoter methylation. *Oncotarget* **2017**, *8*, 48875–48888. [CrossRef] [PubMed]
86. Rangan, A.; Reinig, E.; McPhail, E.D.; Rech, K.L. Immunohistochemistry for LEF1 and SOX11 adds diagnostic specificity in small B-cell lymphomas. *Hum. Pathol.* **2022**, *121*, 29–35. [CrossRef] [PubMed]
87. Yan, M.; Li, G.; An, J. Discovery of small molecule inhibitors of the Wnt/beta-catenin signaling pathway by targeting beta-catenin/Tcf4 interactions. *Exp. Biol. Med.* **2017**, *242*, 1185–1197. [CrossRef]
88. Yu, F.; Yu, C.; Li, F.; Zuo, Y.; Wang, Y.; Yao, L.; Wu, C.; Wang, C.; Ye, L. Wnt/beta-catenin signaling in cancers and targeted therapies. *Signal Transduct. Target. Ther.* **2021**, *6*, 307. [CrossRef]
89. Erdfelder, F.; Hertweck, M.; Filipovich, A.; Uhrmacher, S.; Kreuzer, K.A. High lymphoid enhancer-binding factor-1 expression is associated with disease progression and poor prognosis in chronic lymphocytic leukemia. *Hematol. Rep.* **2010**, *2*, e3. [CrossRef]
90. Erbilgin, Y.; Hatirnaz Ng, O.; Can, I.; Firtina, S.; Kucukcankurt, F.; Karaman, S.; Karakas, Z.; Celkan, T.T.; Zengin, E.; Aylan Gelen, S.; et al. Prognostic evidence of LEF1 isoforms in childhood acute lymphoblastic leukemia. *Int. J. Lab. Hematol.* **2021**, *43*, 1093–1103. [CrossRef]
91. Kuhnl, A.; Gokbuget, N.; Kaiser, M.; Schlee, C.; Stroux, A.; Burmeister, T.; Mochmann, L.H.; Hoelzer, D.; Hofmann, W.K.; Thiel, E.; et al. Overexpression of LEF1 predicts unfavorable outcome in adult patients with B-precursor acute lymphoblastic leukemia. *Blood* **2011**, *118*, 6362–6367. [CrossRef]
92. Lu, H.; Allende, D.; Liu, X.; Zhang, Y. Lymphoid Enhancer Binding Factor 1 (LEF1) and Paired Box Gene 8 (PAX8): A Limited Immunohistochemistry Panel to Distinguish Solid Pseudopapillary Neoplasms and Pancreatic Neuroendocrine Tumors. *Appl. Immunohistochem. Mol. Morphol.* **2020**, *28*, 776–780. [CrossRef] [PubMed]
93. Zhao, Y.; Li, C.; Huang, L.; Niu, S.; Lu, Q.; Gong, D.; Huang, S.; Yuan, Y.; Chen, H. Prognostic value of association of OCT4 with LEF1 expression in esophageal squamous cell carcinoma and their impact on epithelial-mesenchymal transition, invasion, and migration. *Cancer Med.* **2018**, *7*, 3977–3987. [CrossRef] [PubMed]
94. Zhan, Y.; Feng, J.; Lu, J.; Xu, L.; Wang, W.; Fan, S. Expression of LEF1 and TCF1 (TCF7) proteins associates with clinical progression of nasopharyngeal carcinoma. *J. Clin. Pathol.* **2019**, *72*, 425–430. [CrossRef] [PubMed]
95. Eskandari, E.; Mahjoubi, F.; Motalebzadeh, J. An integrated study on TFs and miRNAs in colorectal cancer metastasis and evaluation of three co-regulated candidate genes as prognostic markers. *Gene* **2018**, *679*, 150–159. [CrossRef] [PubMed]
96. Hu, S.; Chang, J.; Ruan, H.; Zhi, W.; Wang, X.; Zhao, F.; Ma, X.; Sun, X.; Liang, Q.; Xu, H.; et al. Cantharidin inhibits osteosarcoma proliferation and metastasis by directly targeting miR-214-3p/DKK3 axis to inactivate β-catenin nuclear translocation and LEF1 translation. *Int. J. Biol. Sci.* **2021**, *17*, 2504–2522. [CrossRef] [PubMed]
97. Heyman, S.J.; Diefenderfer, D.L.; Goldschmidt, M.H.; Newton, C.D. Canine axial skeletal osteosarcoma. A retrospective study of 116 cases (1986 to 1989). *Vet. Surg.* **1992**, *21*, 304–310. [CrossRef] [PubMed]
98. Boerman, I.; Selvarajah, G.T.; Nielen, M.; Kirpensteijn, J. Prognostic factors in canine appendicular osteosarcoma—A meta-analysis. *BMC Vet. Res.* **2012**, *8*, 56. [CrossRef] [PubMed]
99. Tuohy, J.L.; Shaevitz, M.H.; Garrett, L.D.; Ruple, A.; Selmic, L.E. Demographic characteristics, site and phylogenetic distribution of dogs with appendicular osteosarcoma: 744 dogs (2000–2015). *PLoS ONE* **2019**, *14*, e0223243. [CrossRef]
100. Williams, K.; Parker, S.; MacDonald-Dickinson, V. Risk factors for appendicular osteosarcoma occurrence in large and giant breed dogs in western Canada. *Can. Vet. J.* **2023**, *64*, 167–173. [PubMed]
101. Phillips, J.C.; Stephenson, B.; Hauck, M.; Dillberger, J. Heritability and segregation analysis of osteosarcoma in the Scottish deerhound. *Genomics* **2007**, *90*, 354–363. [CrossRef] [PubMed]

102. Saam, D.E.; Liptak, J.M.; Stalker, M.J.; Chun, R. Predictors of outcome in dogs treated with adjuvant carboplatin for appendicular osteosarcoma: 65 cases (1996–2006). *J. Am. Vet. Med. Assoc.* **2011**, *238*, 195–206. [CrossRef] [PubMed]
103. Schmidt, A.F.; Nielen, M.; Klungel, O.H.; Hoes, A.W.; de Boer, A.; Groenwold, R.H.; Kirpensteijn, J.; Investigators, V.S.S.O. Prognostic factors of early metastasis and mortality in dogs with appendicular osteosarcoma after receiving surgery: An individual patient data meta-analysis. *Prev. Vet. Med.* **2013**, *112*, 414–422. [CrossRef] [PubMed]

cancers

MDPI

Article

Quantitative Bone SPECT/CT of Central Cartilaginous Bone Tumors: Relationship between SUVmax and Radiodensity in Hounsfield Unit

Hyukjin Yoon [1], Seul Ki Lee [2,*], Jee-Young Kim [2] and Min Wook Joo [3]

1 Division of Nuclear Medicine, Department of Radiology, St. Vincent's Hospital, College of Medicine, The Catholic University of Korea, Seoul 06591, Republic of Korea
2 Department of Radiology, St. Vincent's Hospital, College of Medicine, The Catholic University of Korea, Seoul 06591, Republic of Korea
3 Department of Orthopaedic Surgery, St. Vincent's Hospital, College of Medicine, The Catholic University of Korea, Seoul 06591, Republic of Korea
* Correspondence: beneffy@catholic.ac.kr

Simple Summary: Radionuclide bone imaging, which reflects osteoblastic activity, is used in evaluating cartilaginous bone tumors; higher SUVmax is more indicative of an ACT rather than an enchondroma in SPECT/CT. However, SUVmax can be influenced by several factors, including radiodensity. Therefore, this study was designed to correlate radiodensity measurements with SUVmax of central cartilaginous bone tumors, including enchondroma, and low-to-intermediate grade chondrosarcomas. Our findings revealed a significant negative correlation between SUVmax and radiodensity measurements in HUmax, HUmean, and HU_{SD}. The subgroup analysis showed significantly higher SUVmax and lower HU_{SD} in the malignant group (grade 1 and 2 chondrosarcoma) than in the benign group (enchondroma). It was observed that higher SUVmax and lower HU_{SD} were associated with a higher probability of having a low-to-intermediate grade chondrosarcoma with aggressive features and a less calcified tumor matrix.

Abstract: (1) Background: it is challenging to determine the accurate grades of cartilaginous bone tumors. Using bone single photon emission computed tomography (SPECT)/computed tomography (CT), maximum standardized uptake value (SUVmax) was found to be significantly associated with different grades of cartilaginous bone tumor. The inquiry focused on the effect of the tumor matrix on SUVmax. (2) Methods: a total of 65 patients from 2017 to 2022 with central cartilaginous bone tumors, including enchondromas and low-to-intermediate grade chondrosarcomas, who had undergone bone SPECT/CT were retrospectively enrolled. The SUVmax was recorded and any aggressive CT findings of cartilaginous bone tumor and Hounsfield units (HU) of the chondroid matrix as mean, minimum, maximum, and standard deviation (SD) were reviewed on CT scans. Pearson's correlation analysis was performed to determine the relationship between CT features and SUVmax. Subgroup analysis was also performed between the benign group (enchondroma) and the malignant group (grade 1 and 2 chondrosarcoma) for comparison of HU values and SUVmax. (3) Results: a significant negative correlation between SUVmax and HU measurements, including HUmax, HUmean, and HU_{SD}, was found. The subgroup analysis showed significantly higher SUVmax in the malignant group, with more frequent CT aggressive features, and significantly lower HU_{SD} in the malignant group than in the benign group. (4) Conclusions: it was observed that higher SUVmax and lower HU_{SD} were associated with a higher probability of having a low-to-intermediate chondrosarcoma with aggressive features and a less calcified tumor matrix.

Keywords: cartilaginous bone tumor; bone SPECT/CT; SUVmax; hounsfield units; chondroid matrix mineralization; correlation

Citation: Yoon, H.; Lee, S.K.; Kim, J.-Y.; Joo, M.W. Quantitative Bone SPECT/CT of Central Cartilaginous Bone Tumors: Relationship between SUVmax and Radiodensity in Hounsfield Unit. *Cancers* **2024**, *16*, 1968. https://doi.org/10.3390/cancers16111968

Academic Editor: Catrin Sian Rutland

Received: 30 April 2024
Revised: 17 May 2024
Accepted: 19 May 2024
Published: 22 May 2024

1. Introduction

Cartilaginous bone tumors are among the most common bone tumors [1]. Enchondroma represents the most prevalent benign tumor and chondrosarcoma represents the most common malignant tumor [2]. Distinguishing between the grades of these tumors, particularly between enchondroma and atypical cartilaginous tumor (ACT), is often challenging due to their similar radiological and histologic features [2–6]. While most enchondromas do not necessitate treatment unless symptomatic or causing complications, ACTs require curettage due to their locally aggressive nature [7]. Thus, efforts have been made to improve radiological differentiation between these tumors using imaging modalities like computed tomography (CT) or magnetic resonance imaging (MRI) [8].

Radionuclide bone imaging, which reflects osteoblastic activity, is also used in evaluating cartilaginous bone tumors. Higher-grade tumors typically exhibit an increased uptake on scintigraphy, reflecting cortical destruction and permeation due to cartilaginous tumor growth [9]. Single photon emission computed tomography (SPECT) provides three-dimensional information on radiotracer uptake, while the combination of SPECT and CT (SPECT/CT) enables accurate localization of the uptake [10]. Recent advancements have allowed for quantitative analysis of radiotracer distribution, with maximum standardized uptake value (SUVmax) being widely used in clinical practice due to its simplicity [10,11]. A previous study by Choi et al. suggests that a higher SUVmax is more indicative of an ACT rather than an enchondroma, with a cutoff value of 15.6 [10].

However, SUVmax can also be high in enchondromas, raising questions about other factors influencing it besides cortical destruction and permeation. While previous studies have focused on imaging features determining tumor grading, such as endosteal scalloping, cortical expansion, and disruption [8,12–14], the significance of chondroid matrix mineralization within the tumor has not been thoroughly explored. Cartilaginous bone tumors often exhibit chondroid matrix mineralization [2], which can be analyzed by measuring the radiodensity using the Hounsfield units (HU) scale on CT, indicating strength and distribution [15]. Given a report suggesting that bone mineral density (BMD) influences bone SPECT/CT radiotracer uptake [16], we planned to investigate the relationship between radiodensity on CT and SUVmax of a central cartilaginous bone tumor on SPECT.

Therefore, our aim is to investigate the relationship between chondroid matrix mineralization and bone radionuclide uptake in central cartilaginous bone tumors. Combined SPECT/CT images provide both radiodensity of HU information from CT and SUVmax from SPECT simultaneously, making it an excellent tool for analysis. Also, we performed subgroup analysis for comparison of radiodensity in HU measurements and SUVmax between benign and malignant groups.

2. Materials and Methods

2.1. Patient Selection

This retrospective study was approved by the institutional review board at our institution and informed consent was waived. From July 2017 to December 2022, 108 patients with suspected cartilaginous bone tumors underwent bone SPECT/CT. Moreover, 37 patients with suspected cartilaginous bone tumors in the hand or foot were excluded from the analysis in order to avoid a selection bias due to the different radiological and histopathologic appearances that may falsely suggest aggressiveness (such as pathologic fracture), particularly in small bones, even if the tumors were benign and a disproportionate amount of enchondromas was present in these regions. Five patients with secondary chondrosarcomas arising from osteochondroma or enchondromatosis and one patient with dedifferentiated chondrosarcoma were also excluded. Accordingly, 65 patients (22 males, 43 females; mean age 52.7 ± 14.7 years; range 18–83 years) with the diagnosis of an enchondroma, ACT/chondrosarcoma grade 1 (CS1, low grade), and chondrosarcoma grade 2 (CS2, intermediate grade) were included in the analysis. All diagnoses were made by pathological findings (n = 27) via surgery or biopsy with preoperative imaging studies and by clinical and radiological findings (n = 38) such as X-ray, CT, or MRI without pathological confirmation due to a high suspicion of benign conditions. In none of these patients was the diagnosis changed during follow-ups of at least two years. Clinical and radiological information such as age, gender, and tumor location were obtained from medical records. Patients had central cartilaginous bone tumors in the proximal humerus (n = 20; 18 enchondromas and 2 ACTs), distal femur (n = 28; 19 enchondromas, 8 ACTs, and 1 CS2), proximal femur (n = 7; 4 enchondromas, 2 ACTs, and 1 CS2), proximal fibula (n = 4; 2 enchondromas and 2 ACTs), distal radius (n = 1, ACT), scapula (n = 2; 1 enchondroma and 1 CS2), and pelvic bone (n = 3; 1 enchondroma, 1 CS1, and 1 CS2).

2.2. Bone SPECT/CT Acquisition

All bone SPECT/CT scans were conducted using an NMCT/670 SPECT/CT scanner (GE Healthcare, Waukesha, WI, USA). First, 800–1100 MBq of Tc-99m hydroxymethylene diphosphonate (HDP) was injected. SPECT/CT images of the tumor site were obtained 4 h after the radiotracer injection. CT acquisition was done with the following parameters: peak energy at 140 keV with 10% window and step-and-shot mode acquisition (25 s per step and 30 steps per detector) with 6° angular increments. For SPECT image reconstruction, an iterative ordered subset expectation maximization algorithm was employed (four iterations; 10 subsets), with CT-based attenuation correction, scatter correction, and resolution recovery carried out on a Xeleris imaging workstation (version 4.0, GE Healthcare, Waukesha, WI, USA). The reconstructed images had a matrix size of 128 × 128 with a section thickness of 4.42 mm. The minimal source-to-collimator distance for the parallel-hole collimation of Tc-99m was set to 4 mm. The camera sensitivity of the scanner was calibrated as 68.06 count/second/Mbq, using a dedicated point source provided by GE healthcare.

The patient information and acquisition parameters were obtained at the time of injection. Patient height and body weight was measured prior to injection. The pre-injection and post-injection activity of the syringe was measured before and after injection, respectively. The time of each measurement was also recorded. The injected radioactivity was automatically calculated with a decay correction on the Xeleris workstation as follows:

$$\text{Injected radioactivity} = \text{post-injection activity} - \text{pre-injection activity}$$

2.3. Image Analysis

2.3.1. SPECT Image Evaluation

All images were evaluated by experienced nuclear medicine physician blinded to histological results. All SPECT/CT images were evaluated on a dedicated workstation (Xeleris 4.0, GE Healthcare, Waukesha, WI, USA) that displayed CT, SPECT, and fused SPECT/CT images. For quantitative analysis, the volumes of interest (VOIs) were generated by automatic segmentation function on the dedicated workstation by clicking the seed point on the tumor center. The generated VOI was manually inspected and corrected if needed. Quantitative parameters were obtained from VOIs using the Q.Metrix toolkit installed on the dedicated workstation. SUVmax in a given VOI was calculated as follows:

$$\text{SUVmax} = (\text{maximum radioactivity/voxel volume})/(\text{injected radioactivity/bodyweight})\ (\text{g/mL})$$

2.3.2. CT Image Evaluation

First, radiodensity measurements were performed using a picture archiving and communication system (PACS) workstation (Zetta PACS, TaeYoung Soft, Anyang-si, Republic of Korea). The independent evaluation of images was performed by two musculoskeletal radiologists. Both readers were blinded regarding clinical information including surgery and histopathological results. A region of interest (ROI) marker was placed around the lesion with the use of the freehand ROI tool, which produced the maximum, minimum, mean, and standard deviation (SD) values of the lesions' radiodensity in Hounsfield units (HU; HUmax, HUmin, HUmean, and HU_{SD}). The CT slice on which the lesion had the largest cross-sectional area was selected. Among these, the axial slice with an abundant mineralized matrix and increased radionuclide uptake as seen on fused SPECT/CT images were selected. The freehand ROI was drawn to contain the lesion only within the intramedullary canal, ensuring that the bony cortex was not included within the ROI. When the lesion margin was not well visualized, the fused SPECT/CT image or MRI were referenced (Figure 1).

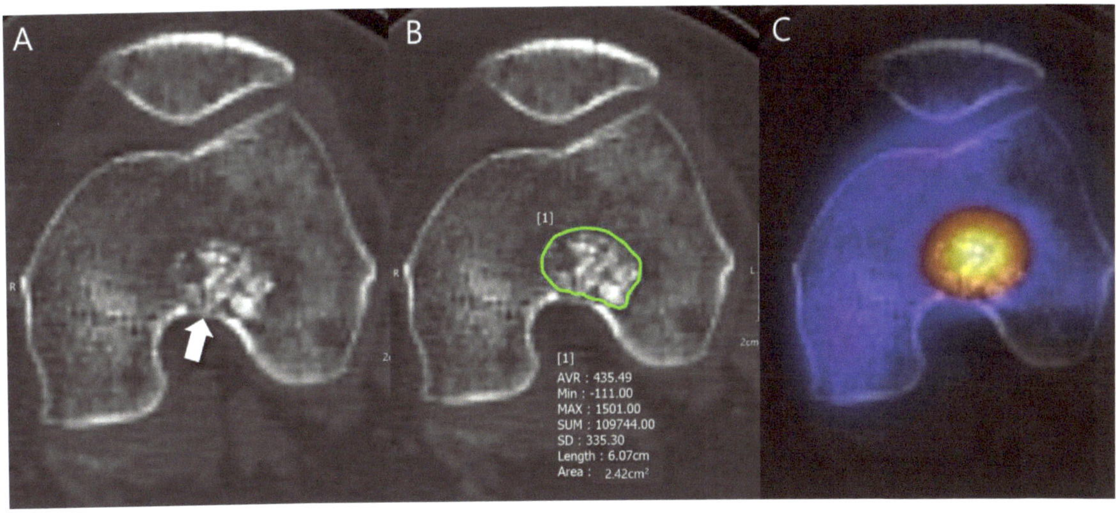

Figure 1. Radiodensity measurement in a patient with ACT in the distal femur. (**A**) Axial CT image shows a lobulated mass containing chondroid matrix mineralization and focal deep endosteal scalloping $\geq 2/3$ of the normal cortical thickness (arrow). (**B**) Mean (±SD) attenuation of this lesion was measured using the freehand ROI tool (green line) and HU values were found to be 435.49 ± 335.30 [−111 to 1501]. (**C**) Fused SPECT/CT image shows the radioactive uptake with SUVmax of this lesion which was calculated to be 23.71.

Any aggressive CT features for the grading of central cartilaginous bone tumors including (i) deep endosteal scalloping $\geq 2/3$ of the normal cortical thickness (Figure 1), (ii) extensive endosteal scalloping $\geq 2/3$ of the lesion length (Figure 2), (iii) expansile cortical remodeling (Figure 3), and (iv) cortical destruction with or without soft tissue extension (Figure 4) were also evaluated. The CT images were evaluated in conjunction with the plain radiographs and/or MRI. After finishing the independent review, a consensus review of the CT was performed. The two radiologists reviewed the CT images together to reach a final consensus on discrepant interpretations from the independent reading.

2.4. Statistical Analyses

To reveal the correlative relationships between SUVmax and radiodensity measurements in HU, the Pearson's correlation coefficient was calculated. Strength of correlation was interpreted as follows: Spearman's rho (denoted as r)—0.0 to 0.1 no correlation, 0.1 to 0.3 poor correlation, 0.3 to 0.5 fair correlation, 0.5 to 0.7 moderate correlation, 0.7 to 1 very strong correlation, and 1 perfect correlation. Assessment of the relative importance of regressors in the multiple linear regression analysis was performed. A student's t-test was performed to compare SUVmax and radiodensity measurements in HU, and a chi-square test was conducted to compare aggressive CT features of central cartilaginous bone tumors between the benign group (enchondromas) and the malignant group (low-to-intermediate grade chondrosarcomas). Intraclass correlation coefficients (ICCs) were used to evaluate interobserver reliability for the SUVmax and radiodensity measurements in HU. The ICC was calculated using a two-way random model by absolute agreement. The degree of agreement was interpreted as follows: ICC < 0.40 poor, 0.4–0.59 fair, 0.60–0.74 good, 0.75–1.00 excellent. Interobserver variability for the aggressive CT features of central cartilaginous bone tumors was assessed using kappa statistics. A kappa value lower than 0.40 indicated poor agreement, 0.40 to 0.59 moderate agreement, 0.60 to 0.79 good agreement, and 0.80 or greater excellent agreement. For all statistical comparisons, the significance level was set to $p < 0.05$. Statistical analysis was conducted using a software package (SPSS v. 20.0, Chicago, IL, USA).

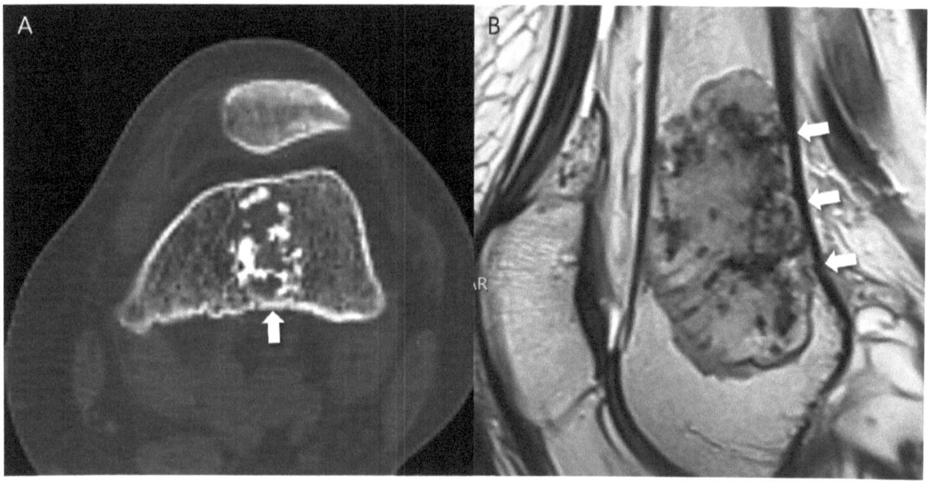

Figure 2. Extensive endosteal scalloping $\geq 2/3$ of the lesion length in a patient with ACT in the distal femur. (**A**) An axial CT image shows a lobulated mass containing chondroid matrix mineralization and focal endosteal scalloping (arrow). (**B**) A sagittal T2-weighted image shows a lobulated mass with heterogeneously increased signal intensity and extensive endosteal scalloping $\geq 2/3$ of the lesion length (arrows).

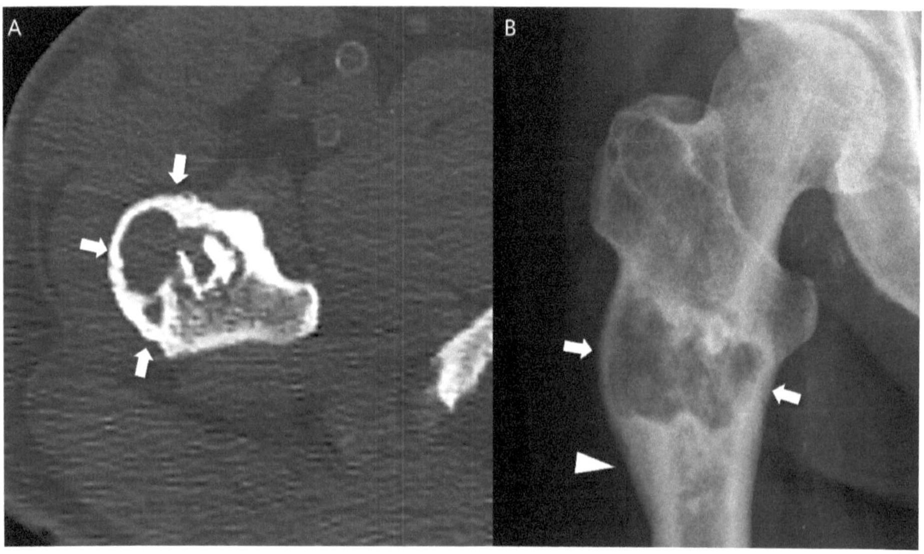

Figure 3. Expansile cortical remodeling in a patient with CS2 in the proximal femur. (**A**) An axial CT image shows a lobulated mass containing chondroid matrix mineralization and expansile cortical remodeling (arrows). (**B**) A plain radiograph shows a lobulated mass with expansile cortical remodeling (arrows) with cortical thickening (arrowhead).

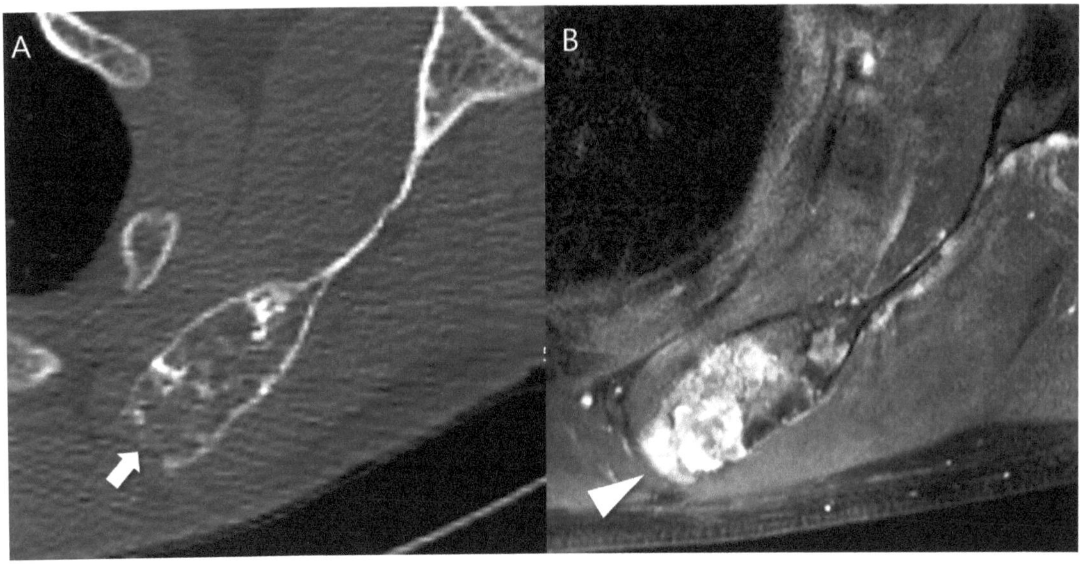

Figure 4. Cortical destruction with small extraosseous soft tissue extension in a patient with CS2 in the scapula. (**A**) An axial CT image shows a lobulated mass containing chondroid matrix mineralization and focal cortical destruction (arrow). (**B**) An axial T1-weighted fat-suppressed enhanced MRI shows a lobulated mass with extraosseous soft tissue extension (arrowhead).

3. Results

3.1. Relationship of SUVmax with Radiodensity in HU Measurements

Pearson's correlation analysis showed that SUVmax demonstrated a fair negative correlation with HUmax ($r = -0.45$, $p < 0.001$), a fair positive correlation with HUmin ($r = 0.32$, $p = 0.010$), a fair negative correlation with HUmean ($r = -0.31$, $p = 0.012$), and a moderate negative correlation with HU_{SD} ($r = -0.52$, $p < 0.001$). Figure 5 shows these relationships with the linear fit trend lines and r^2 values (goodness-of-fit of linear regression). The multiple linear regression analysis demonstrated that the HU_{SD} ($r = -0.52$, $r^2 = 0.256$, $p < 0.001$) was significantly and independently associated with SUVmax.

The interobserver reliability of the radiodensity between both readers were 'excellent' for HUmax (ICC of 0.861, $p < 0.001$), HUmean (ICC of 0.933, $p < 0.001$), and HU_{SD} (ICC of 0.944, $p < 0.001$), and 'fair' for HUmin (ICC of 0.710, $p = 0.001$).

3.2. Association of SUVmax and CT Features between Benign and Malignant Groups

Since most of the relationships between radiodensity in HU measurements and SUVmax showed a negative correlation, further subgroup analysis was performed to determine the association of each parameter between the benign (enchondroma) and malignant (ACT/CS1 + CS2) groups.

First of all, most of the aggressive CT features of central cartilaginous bone tumors were significantly more frequent in the malignant group than in the benign group (deep endosteal scalloping, 60.0% vs. 11.1%, $p < 0.001$; extensive endosteal scalloping, 80.0% vs. 26.7%, $p < 0.001$; expansile cortical remodeling, 35.0% vs. 2.2%, $p = 0.001$) (Table 1). Interobserver agreement between the two readers was 'good' for deep endosteal scalloping (κ, 0.710) and extensive endosteal scalloping (κ, 0.757), and 'excellent' for expansile cortical remodeling (κ, 0.867) and cortical destruction (κ, 0.936).

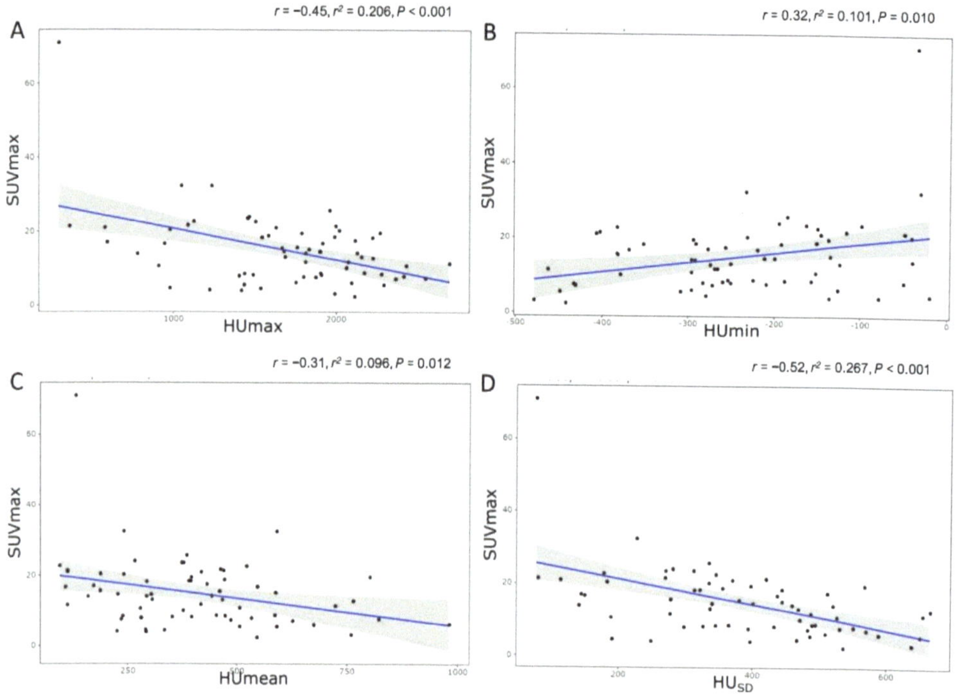

Figure 5. The relationship between SUVmax and radiodensity. (**A**) HUmax, (**B**) HUmin, (**C**) HUmean, and (**D**) HU_{SD}. r = Pearson's correlation; r^2 = goodness-of-fit of linear regression.

Table 1. Comparison of aggressive CT features between benign and malignant groups.

	Benign, Enchondroma	**Malignant, ACT/CS1 + CS2**	***p* Value**
	(n = 45)	**(n = 20)**	
Deep endosteal scalloping			
<1/3 of normal cortical thickness	40 (88.9%)	8 (40.0%)	<0.001
≥2/3 of normal cortical thickness	5 (11.1%)	12 (60.0%)	
Extensive endosteal scalloping			
<1/3 of the lesion length	33 (73.3%)	4 (20.0%)	<0.001
≥2/3 of the lesion length	12 (26.7%)	16 (80.0%)	
Expansile cortical remodeling			
Absent	44 (97.8%)	13 (65.0%)	0.001
Present	1 (2.2%)	7 (35.0%)	
Cortical destruction			
Absent	45 (100.0%)	18 (90.0%)	0.169
Present	0 (0.0%)	2 (10.0%)	

SUVmax was also significantly higher in the malignant group than in the benign group (22.3 ± 13.2 vs. 11.8 ± 5.9, p = 0.003) (Table 2). Among the radiodensity measurements in HU, only HU_{SD} was significantly lower in the malignant group than in the benign group (322.5 ± 149.1 vs. 405.1 ± 140.3, p = 0.036) (Table 2). In addition, HUmax (1522.6 ± 623.4 vs. 1748.8 ± 480.9, p = 0.116) and HUmean (381.5 ± 193.3 vs. 412.8 ± 197.9, p = 0.556) were lower in the malignant group than in the benign group, although without statistically significant differences, and HUmin (−223.0 ± 137.2 vs. −244.2 ± 113.1, p = 0.515) was higher in the malignant group than in the benign group, again without statistically significant difference (Table 2). A representative case is shown in Figure 6.

Table 2. Comparison of SUVmax and radiodensity in HU measurements between benign and malignant groups.

	Benign, Enchondroma	**Malignant, ACT/CS1 + CS2**	***p* Value**
	(n = 45)	**(n = 20)**	
SUVmax	11.8 ± 5.9	22.3 ± 13.2	0.003
HUmax	1748.8 ± 480.9	1522.6 ± 623.4	0.116
HUmin	−244.2 ± 113.1	−223.0 ± 137.2	0.515
HUmean	412.8 ± 197.9	381.5 ± 193.3	0.556
HU_{SD}	405.1 ± 140.3	322.5 ± 149.1	0.036

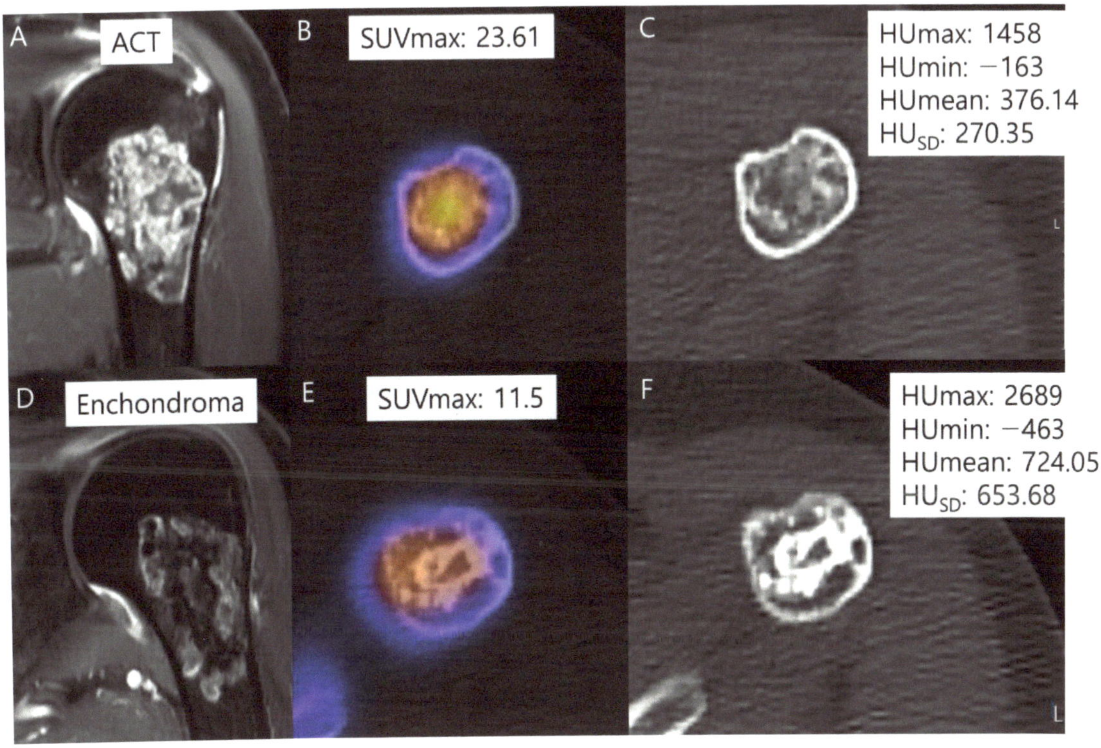

Figure 6. Comparison of SUVmax and radiodensity in HU measurements between ACT (**A**–**C**) and enchondroma (**D**–**F**). (**A**) A coronal T1-weighted fat-suppressed enhanced MRI shows a lobulated mass with septal enhancement with pathology of ACT. (**B**) A fused SPECT/CT image shows the radioactive uptake with SUVmax of this lesion which was calculated to be 23.61. (**C**) The radiodensity in HU measurement of this lesion was found to be 376.14 ± 270.35 [−163 to 1458]. (**D**) A coronal T1-weighted fat-suppressed enhanced MRI shows a lobulated mass with faint septal enhancement with pathology of enchondroma. (**E**) A fused SPECT/CT image shows the radioactive uptake with SUVmax of this lesion which was calculated to be 11.5. (**F**) The radiodensity in HU measurement of this lesion was found to be 724.05 ± 653.68 [−463 to 2689].

4. Discussion

Our findings revealed a significant negative correlation between SUVmax and radiodensity measurements in HU, including HUmax, HUmean, and HU_{SD}. This contradicts the assumption that the strength and distribution of chondroid matrix mineralization directly influences SUVmax, suggesting an inverse relationship instead. To further explore this paradoxical association, subgroup analyses were conducted. The data were divided into benign (enchondroma) and malignant (low-to-intermediate grade chondrosarcoma) groups, and comparisons were made in relation to SUVmax and radiodensity measurements in HU between the two groups. The subgroup analysis showed significantly higher SUVmax in the low-to-intermediate grade malignant group, which exhibited more frequent CT aggressive features compared to the benign group. It also revealed significantly lower HU_{SD} in the low-to-intermediate grade malignant group compared to the benign group. Although statistical significance was not established, a trend of lower radiodensity of HUmax and HUmean was observed in the low-to-intermediate malignant group relative to the benign group.

Bone scintigraphy uptake in cartilaginous tumors typically increase with a higher-grade [17]. This is thought to reflect the cortical destruction and bone permeation, as important histological features of high-grade tumors include infiltration and encasement of the existing trabecular bone [18]. Increased bone radiotracer uptake can be objectively described by SUV in SPECT/CT images. Previous research by Choi et al. [10] demonstrated that ACT exhibits a higher SUVmax compared to enchondroma. This suggests that the higher SUVmax in ACT may reflect a greater degree of reaction in the surrounding bone, such as cortical extension or permeation, as observed in ACT, leading to increased radionuclide uptake [17]. However, SUVmax can be influenced by several factors, including BMD which is known to be positively correlated with SUVs [19]. Therefore, there was curiosity regarding whether chondroid matrix mineralization might influence the SUVmax of tumors. This curiosity led us to initiate a study correlating radiodensity in HU measurements with SUVmax of central cartilaginous bone tumors.

Based on existing literature, a positive correlation between SUVmax and the radiodensity of the tumor matrix was anticipated [19]; however, our actual results contradicted our expectations. Our findings revealed a significant negative correlation between SUVmax and radiodensity measurements in HUmax, HUmean, and HU_{SD}. This result led us to the discovery that the strength and distribution of chondroid matrix mineralization do not directly influence SUVmax; instead, they exhibit an inverse relationship. Consequently, further subgroup analysis was prompted.

Subgroup analysis was conducted, dividing the patients into two groups: benign (enchondroma) and low-to-intermediate grade malignant (ACT/CS1 + CS2) groups. SUVmax and radiodensity measurements in HU were subsequently compared between the two groups. The subgroup analysis revealed significantly higher SUVmax in the malignant group, which exhibited more frequent CT aggressive features compared to the benign group. This finding is consistent with previous studies aiming to differentiate benign from malignant cartilaginous bone tumors using bone SPECT/CT [10,20]. In bone SPECT/CT, SUVmax serves as an indicator of osteoblastic activity, reflecting cortical destruction and permeation in cartilaginous bone tumors. Our results confirm that low-to-intermediate grade chondrosarcomas are associated with higher SUVmax and more frequent CT aggressive features, such as deep and extensive endosteal scalloping and expansile cortical remodeling [21–23].

Subgroup analysis also revealed a significantly lower HU_{SD} in the low-to-intermediate grade malignant group compared to the benign group. Although statistical significance was not established, a trend of lower radiodensity was observed in the low-to-intermediate grade malignant group (HUmax and HUmean). Balta et al. conducted a study analyzing cartilaginous bone tumors using HU measurements and reported that ACT exhibited lower HU values than enchondroma. However, their study analyzed only the maximum and minimum values of HU without the SD of HU [15]. Although our study showed statistical significance only in HU_{SD}, and not in HUmax, HUmin, and HUmean, between the two groups, there appears to be some degree of similarity in the results between the study by Balta and ours. In other words, a higher SD of HU in a tumor matrix with dense calcification indicates a greater likelihood of a stable enchondroma, while a lower SD of HU in a less calcified tumor matrix suggests a greater likelihood of an active aggressive low-to-intermediate grade chondrosarcoma [15].

Relevant previous studies have demonstrated that HU measurements in CT examinations, including the proximal femur and lumbar vertebrae, can predict BMD and strength [24,25] as well as bone neoplasms [26–29]. Thus, it can be concluded that HU measurements on CT scans can serve as a tool to differentiate between enchondroma and ACT, reflecting the characteristics of the tumor matrix. This aspect will be further analyzed in the future with a larger sample size and additional tools such as texture analysis [30–33]. It is also notable that, in the combined modality of SPECT/CT, both radiotracer and radiodensity information can be obtained simultaneously without additional examinations. Performing multiple examinations during disease evaluation increases the medical cost and decreases patient satisfaction due to the need for repeated visits to the hospital. By reducing the total number of examinations, SPECT/CT may improve patient convenience and reduce the total medical cost.

There are several limitations in this study. Firstly, it is retrospective in nature and conducted at a single center. Larger studies are required to confirm the results in the future. Secondly, there is a lack of histological confirmation for some portions of the enchondromas. Instead, we consider the diagnosis of enchondroma based on radiological imaging and follow-up results to be clinically valid. Third, the study did not encompass grade 3 chondrosarcoma (high-grade malignancy), thus limiting the capacity to definitely establish a conclusive relationship among chondrosarcoma malignant features, SUVmax, and HU values. Therefore, our results derive from the incorporation of low-to-intermediate grade chondrosarcomas only, with plans to broaden the analysis by including more patients across multiple centers in the future. Lastly, the heterogeneous tumor locations, such as in the trunk, pose a limitation. Future studies will include a large number of patients with tumors located only in the extremities or only in the trunk.

5. Conclusions

The current study revealed a negative correlation between SUVmax and radiodensity in HU measurements of the tumor matrix in central cartilaginous bone tumors. It was observed that higher SUVmax and lower HU_{SD} were associated with a higher probability of having a malignant cartilaginous bone tumor (low-to-intermediate grade) with aggressive feature and a less calcified tumor matrix. This study highlights the potential usefulness of SPECT/CT scans to tumor diagnosis and characterization for central cartilaginous bone tumors. Further research is needed to validate these findings in larger patient cohorts and to explore their clinical applications.

Author Contributions: Conceptualization, S.K.L.; methodology, H.Y. and S.K.L.; software, J.-Y.K.; validation, H.Y., S.K.L. and J.-Y.K.; formal analysis, S.K.L.; investigation, H.Y. and S.K.L.; data curation, H.Y. and S.K.L.; writing—original draft preparation, H.Y. and S.K.L.; writing—review and editing, H.Y., S.K.L. and M.W.J.; visualization, S.K.L.; supervision, J.-Y.K. and M.W.J.; funding acquisition, S.K.L. All authors have read and agreed to the published version of the manuscript.

Funding: This research was funded by The Catholic Medical Center Research Foundation in the program year of 2024.

Institutional Review Board Statement: This retrospective study was approved by the institutional review board at The Catholic University of Korea St. Vincent's Hospital (VC24RISI0075).

Informed Consent Statement: The informed consent was waived because this retrospective study was a minimal risk study and no personally identifiable information was collected.

Data Availability Statement: Dataset available on request from the authors.

Acknowledgments: The authors wish to acknowledge the financial support of the Catholic Medical Center Research Foundation made in the program year of 2024.

Conflicts of Interest: The authors declare no conflicts of interest.

Abbreviations

ACT	Atypical cartilaginous tumor
BMD	Bone mineral density
CS1	Chondrosarcoma grade 1
CS2	Chondrosarcoma grade 2
CT	Computed tomography
HDP	Hydroxymethylene diphosphonate
HU	Hounsfield units
ICC	Intraclass correlation coefficients
MRI	Magnetic resonance imaging
SD	Standard deviation
SPECT	Single photon emission computed tomography
SUVmax	Maximum standardized uptake value
PACS	Picture archiving and communication system
ROI	Region of interest
VOI	Volume of interest

References

1. WHO Classification of Tumours Editorial Board. *WHO Classification of Tumours: Soft Tissue and Bone Tumours*; International Agency for Research on Cancer: Lyon, France, 2020.
2. Kim, J.H.; Lee, S.K. Classification of Chondrosarcoma: From Characteristic to Challenging Imaging Findings. *Cancers* **2023**, *15*, 1703. [CrossRef]
3. Skeletal Lesions Interobserver Correlation among Expert Diagnosticians (SLICED) Study Group. Reliability of histopathologic and radiologic grading of cartilaginous neoplasms in long bones. *J. Bone Jt. Surg. Am.* **2007**, *89*, 2113–2123. [CrossRef] [PubMed]
4. Scheitza, P.; Uhl, M.; Hauschild, O.; Zwingmann, J.; Bannasch, H.; Kayser, C.; Südkamp, N.P.; Herget, G.W. Interobserver Variability in the Differential Diagnosis of Benign Bone Tumors and Tumor-like Lesions. *Rofo* **2016**, *188*, 479–487. [CrossRef] [PubMed]
5. Donthineni, R.; Ofluoglu, O. Solitary enchondromas of long bones: Pattern of referral and outcome. *Acta Orthop. Traumatol. Turc.* **2010**, *44*, 397–402. [CrossRef] [PubMed]
6. Ferrer-Santacreu, E.M.; Ortiz-Cruz, E.J.; González-López, J.M.; Pérez Fernández, E. Enchondroma versus low-grade chondrosarcoma in appendicular skeleton: Clinical and radiological criteria. *J. Oncol.* **2012**, *2012*, 437958. [CrossRef] [PubMed]
7. Dierselhuis, E.F.; Gerbers, J.G.; Ploegmakers, J.J.W.; Stevens, M.; Suurmeijer, A.J.H.; Jutte, P.C. Local Treatment with Adjuvant Therapy for Central Atypical Cartilaginous Tumors in the Long Bones: Analysis of Outcome and Complications in One Hundred and Eight Patients with a Minimum Follow-up of Two Years. *JBJS* **2016**, *98*, 303–313. [CrossRef] [PubMed]
8. Gassert, F.G.; Breden, S.; Neumann, J.; Gassert, F.T.; Bollwein, C.; Knebel, C.; Lenze, U.; von Eisenhart-Rothe, R.; Mogler, C.; Makowski, M.R.; et al. Differentiating Enchondromas and Atypical Cartilaginous Tumors in Long Bones with Computed Tomography and Magnetic Resonance Imaging. *Diagnostics* **2022**, *12*, 2186. [CrossRef]
9. Hudson, T.M.; Chew, F.S.; Manaster, B.J. Radionuclide bone scanning of medullary chondrosarcoma. *AJR Am. J. Roentgenol.* **1982**, *139*, 1071–1076. [CrossRef]
10. Choi, W.H.; Han, E.J.; Chang, K.B.; Joo, M.W. Quantitative SPECT/CT for differentiating between enchondroma and grade I chondrosarcoma. *Sci. Rep.* **2020**, *10*, 10587. [CrossRef]
11. Bailey, D.L.; Willowson, K.P. Quantitative SPECT/CT: SPECT joins PET as a quantitative imaging modality. *Eur. J. Nucl. Med. Mol. Imaging* **2014**, *41*, 17–25. [CrossRef]
12. Lee, S.; Yoon, M.A. Assessment of central cartilaginous tumor of the appendicular bone: Inter-observer and intermodality agreement and comparison of diagnostic performance of CT and MRI. *Acta Radiol* **2022**, *63*, 376–386. [CrossRef] [PubMed]
13. Miwa, S.; Yamamoto, N.; Hayashi, K.; Takeuchi, A.; Igarashi, K.; Tada, K.; Yonezawa, H.; Morinaga, S.; Araki, Y.; Asano, Y.; et al. A Radiological Scoring System for Differentiation between Enchondroma and Chondrosarcoma. *Cancers* **2021**, *13*, 3558. [CrossRef] [PubMed]
14. Douis, H.; Parry, M.; Vaiyapuri, S.; Davies, A.M. What are the differentiating clinical and MRI-features of enchondromas from low-grade chondrosarcomas? *Eur. Radiol.* **2018**, *28*, 398–409. [CrossRef] [PubMed]
15. Balta, O.; Altinayak, H.; Zengin, E.C.; Eren, M.B.; Demir, O.; Aytekin, K. Incidental Enchondromas of the Lower Extremity Long Bones and Atypical Chondroid Tumors Differentiation Based on Hounsfield Units. *Eurasian J. Med. Investig.* **2022**, *6*, 245–258. [CrossRef]
16. Huang, K.; Feng, Y.; Liu, D.; Liang, W.; Li, L. Quantification evaluation of 99mTc-MDP concentration in the lumbar spine with SPECT/CT: Compare with bone mineral density. *Ann. Nucl. Med.* **2020**, *34*, 136–143. [CrossRef]
17. Murphey, M.D.; Flemming, D.J.; Boyea, S.R.; Bojescul, J.A.; Sweet, D.E.; Temple, H.T. Enchondroma versus chondrosarcoma in the appendicular skeleton: Differentiating features. *Radiographics* **1998**, *18*, 1213–1237; quiz 1244–1215. [CrossRef] [PubMed]

18. Suster, D.; Hung, Y.P.; Nielsen, G.P. Differential Diagnosis of Cartilaginous Lesions of Bone. *Arch. Pathol. Lab. Med.* **2020**, *144*, 71–82. [CrossRef]
19. Cachovan, M.; Vija, A.H.; Hornegger, J.; Kuwert, T. Quantification of 99m Tc-DPD concentration in the lumbar spine with SPECT/CT. *EJNMMI Res.* **2013**, *3*, 45. [CrossRef]
20. Kitajima, K.; Futani, H.; Tsuchitani, T.; Takahashi, Y.; Tachibana, T.; Yamakado, K. Quantitative bone SPECT/CT applications for cartilaginous bone neoplasms. *Hell. J. Nucl. Med.* **2020**, *23*, 133–137.
21. Murphey, M.D.; Walker, E.A.; Wilson, A.J.; Kransdorf, M.J.; Temple, H.T.; Gannon, F.H. From the archives of the AFIP: Imaging of primary chondrosarcoma: Radiologic-pathologic correlation. *Radiographics* **2003**, *23*, 1245–1278. [CrossRef]
22. Engel, H.; Herget, G.W.; Füllgraf, H.; Sutter, R.; Benndorf, M.; Bamberg, F.; Jungmann, P.M. Chondrogenic bone tumors: The importance of imaging characteristics. *Rofo* **2021**, *193*, 262–275. [CrossRef] [PubMed]
23. Mayes, G.B.; Wallace, S.; Bernardino, M.E. Computed tomography of chondrosarcoma. *J. Comput. Tomogr.* **1981**, *5*, 345–348. [CrossRef]
24. Christensen, D.L.; Nappo, K.E.; Wolfe, J.A.; Wade, S.M.; Brooks, D.I.; Potter, B.K.; Forsberg, J.A.; Tintle, S.M. Proximal femur hounsfield units on CT colonoscopy correlate with dual-energy X-ray absorptiometry. *Clin. Orthop. Relat. Res.* **2019**, *477*, 850–860. [CrossRef] [PubMed]
25. Schreiber, J.J.; Anderson, P.A.; Rosas, H.G.; Buchholz, A.L.; Au, A.G. Hounsfield units for assessing bone mineral density and strength: A tool for osteoporosis management. *JBJS* **2011**, *93*, 1057–1063. [CrossRef] [PubMed]
26. Hong, J.H.; Jung, J.-Y.; Jo, A.; Nam, Y.; Pak, S.; Lee, S.-Y.; Park, H.; Lee, S.E.; Kim, S. Development and validation of a radiomics model for differentiating bone islands and osteoblastic bone metastases at abdominal CT. *Radiology* **2021**, *299*, 626–632. [CrossRef] [PubMed]
27. Ulano, A.; Bredella, M.A.; Burke, P.; Chebib, I.; Simeone, F.J.; Huang, A.J.; Torriani, M.; Chang, C.Y. Distinguishing untreated osteoblastic metastases from enostoses using CT attenuation measurements. *Am. J. Roentgenol.* **2016**, *207*, 362–368. [CrossRef] [PubMed]
28. Sala, F.; Dapoto, A.; Morzenti, C.; Firetto, M.C.; Valle, C.; Tomasoni, A.; Sironi, S. Bone islands incidentally detected on computed tomography: Frequency of enostosis and differentiation from untreated osteoblastic metastases based on CT attenuation value. *Br. J. Radiol.* **2019**, *92*, 20190249. [CrossRef] [PubMed]
29. Mhuircheartaigh, J.N.; McMahon, C.; Lin, Y.-C.; Wu, J. Diagnostic yield of percutaneous biopsy for sclerotic bone lesions: Influence of mean Hounsfield units. *Clin. Imaging* **2017**, *46*, 53–56. [CrossRef]
30. Yoon, H.; Choi, W.H.; Joo, M.W.; Ha, S.; Chung, Y.-A. SPECT/CT Radiomics for Differentiating between Enchondroma and Grade I Chondrosarcoma. *Tomography* **2023**, *9*, 1868–1875. [CrossRef]
31. Gitto, S.; Cuocolo, R.; Annovazzi, A.; Anelli, V.; Acquasanta, M.; Cincotta, A.; Albano, D.; Chianca, V.; Ferraresi, V.; Messina, C. CT radiomics-based machine learning classification of atypical cartilaginous tumours and appendicular chondrosarcomas. *EBioMedicine* **2021**, *68*, 103407. [CrossRef]
32. Deng, X.-Y.; Chen, H.-Y.; Yu, J.-N.; Zhu, X.-L.; Chen, J.-Y.; Shao, G.-L.; Yu, R.-S. Diagnostic value of CT-and MRI-based texture analysis and imaging findings for grading cartilaginous tumors in long bones. *Front. Oncol.* **2021**, *11*, 700204. [CrossRef] [PubMed]
33. Gitto, S.; Cuocolo, R.; Emili, I.; Tofanelli, L.; Chianca, V.; Albano, D.; Messina, C.; Imbriaco, M.; Sconfienza, L.M. Effects of Interobserver variability on 2D and 3D CT-and MRI-based texture feature reproducibility of cartilaginous bone tumors. *J. Digit. Imaging* **2021**, *34*, 820–832. [CrossRef] [PubMed]

cancers

MDPI

Article

Conventional Spinal Chordomas: Investigation of SMARCB1/INI1 Protein Expression, Genetic Alterations in *SMARCB1* Gene, and Clinicopathological Features in 89 Patients

Margherita Maioli [1], Stefania Cocchi [1,*], Marco Gambarotti [1,*], Stefania Benini [1], Giovanna Magagnoli [1], Gabriella Gamberi [1], Cristiana Griffoni [2], Alessandro Gasbarrini [2], Riccardo Ghermandi [2], Luigi Emanuele Noli [2], Chiara Alcherigi [2], Cristina Ferrari [3], Giuseppe Bianchi [4], Sofia Asioli [5,6], Elettra Pignotti [1] and Alberto Righi [1]

1 Department of Pathology, IRCCS Istituto Ortopedico Rizzoli, 40136 Bologna, Italy
2 Department of Spine Surgery, IRCCS Istituto Ortopedico Rizzoli, 40136 Bologna, Italy
3 Experimental Oncology Laboratory, IRCCS Istituto Ortopedico Rizzoli, 40136 Bologna, Italy
4 Department of Orthopedic Oncology, IRCCS Istituto Ortopedico Rizzoli, 40136 Bologna, Italy
5 Department of Biomedical and Neuromotor Sciences (DIBINEM), Alma Mater Studiorum—University of Bologna, 40126 Bologna, Italy
6 IRCCS Istituto delle Scienze Neurologiche di Bologna, 40139 Bologna, Italy
* Correspondence: stefania.cocchi@ior.it (S.C.); marco.gambarotti@ior.it (M.G.); Tel.: +39-051-6366665 (S.C. & M.G.)

Simple Summary: Alterations in the SMARCB1/INI1 expression pattern have been detected in many tumors, including chordomas. We studied a large group of patients with conventional spinal chordomas, and the aims were to assess the differences in the immunohistochemical expression of SMARCB1/INI1 and the underlying alterations in the *SMARCB1* gene and to investigate the correlation between clinicopathological features and patient survival. Partial SMARCB1/INI1 loss was identified in several patients, and this pattern correlated with mobile spine location and inadequate surgical margins. Moreover, mobile spine tumor location and inadequate surgical margins negatively impacted disease-free survival. The complete loss of SMARCB1/INI1 is currently ongoing as a target for molecular therapy; therefore, the partial loss of SMARCB1/INI1 in tumors could also have therapeutic implications.

Abstract: The partial loss of SMARCB1/INI1 expression has recently been reported in skull base conventional chordomas, with possible therapeutic implications. We retrospectively analyzed 89 patients with conventional spinal chordomas to investigate the differences in the immunohistochemical expression of SMARCB1/INI1 and the underlying genetic alterations in the *SMARCB1* gene. Moreover, we assessed the correlation of clinicopathological features (age, gender, tumor size, tumor location, surgical margins, Ki67 labelling index, SMARCB1/INI1 pattern, previous surgery, previous treatment, type of surgery, and the Charlson Comorbidity Index) with patient survival. Our cohort included 51 males and 38 females, with a median age at diagnosis of 61 years. The median tumor size at presentation was 5.9 cm. The 5-year overall survival (OS) and 5-year disease-free survival (DFS) rates were 90.8% and 54.9%, respectively. Partial SMARCB1/INI1 loss was identified in 37 (41.6%) patients with conventional spinal chordomas (27 mosaic and 10 clonal). The most frequent genetic alteration detected was the monoallelic deletion of a portion of the long arm of chromosome 22, which includes the *SMARCB1* gene. Partial loss of SMARCB1/INI1 was correlated with cervical–thoracic–lumbar tumor location ($p = 0.033$) and inadequate surgical margins ($p = 0.007$), possibly due to the high degree of tumor invasiveness in this site. Among all the considered clinicopathological features related to patient survival, only tumor location in the sacrococcygeal region and adequate surgical margins positively impacted DFS. In conclusion, partial SMARCB1/INI1 loss, mostly due to 22q deletion, was detected in a significant number of patients with conventional spinal chordomas and was correlated with mobile spine location and inadequate surgical margins.

Citation: Maioli, M.; Cocchi, S.; Gambarotti, M.; Benini, S.; Magagnoli, G.; Gamberi, G.; Griffoni, C.; Gasbarrini, A.; Ghermandi, R.; Noli, L.E.; et al. Conventional Spinal Chordomas: Investigation of SMARCB1/INI1 Protein Expression, Genetic Alterations in *SMARCB1* Gene, and Clinicopathological Features in 89 Patients. *Cancers* **2024**, *16*, 2808. https://doi.org/10.3390/cancers16162808

Academic Editor: Catrin Sian Rutland

Received: 19 June 2024
Revised: 5 August 2024
Accepted: 6 August 2024
Published: 9 August 2024

Keywords: conventional chordoma; SMARCB1/INI1; *SMARCB1* gene; FISH analysis

1. Introduction

Chordomas are rare malignant neoplasms that develop from embryonic remnants of the notochord. They exhibit distinct histotypes (conventional, poorly differentiated, and dedifferentiated) with different clinical behavior [1]. Conventional chordoma accounts for approximately 95% of cases [1,2]. Chordomas are locally destructive tumors characterized by very slow growth, with possible local recurrence and metastases. The 5- and 10-year OS rates are estimated to be 68.4% and 39.2%, respectively, and the 5- and 10-year DFS rates are 80.9% and 60.1%, respectively [3]. The diagnostic hallmark of chordomas is the nuclear expression of the brachyury protein [1,4]. Complete loss of the SMARCB1/INI1 nuclear protein has also been reported as a peculiar feature of poorly differentiated chordoma [3,5,6]. Recently, the partial loss of SMARCB1/INI1 protein expression has been detected in conventional chordomas localized in the skull base [7]. SMARCB1/INI1 is a tumor suppressor encoded by the *SMARCB1* gene (SWI/SNF-related, matrix-associated, actin-dependent regulator of chromatin, subfamily B, member 1), which is located on the long arm of chromosome 22 (22q11.23). This protein is part of the multisubunit 'SWItch/Sucrose Non-Fermentable ATP-dependent chromatin remodelling complex' (SWI/SNF), which regulates different cellular mechanisms, including gene expression and cell proliferation and differentiation [8,9]. Abnormal expression of SMARCB1/INI1 has been detected extensively in different tumor types, and three distinct expression patterns have been identified: complete loss, partial loss, and reduced expression [10,11]. However, the type of abnormal expression pattern and the type of mutation in the *SMARCB1* gene do not always match; in some cases, no DNA or RNA changes are detected [10]. Among tumors with focal expression of SMARCB1/INI1, different types of genetic alterations have been described, the most frequent being the monoallelic deletion of a portion of the long arm of chromosome 22, which includes the *SMARCB1* gene [7,10]. However, several studies have revealed that SMARCB1/INI1-deficient tumors, despite being very different from each other in location and type, generally share an aggressive clinical course with high local recurrence rates and a prognosis that is often poor [11–14].

From a treatment perspective, chordoma appears to be resistant to common chemotherapy, and clinical studies are currently ongoing to treat some of these forms with new targeted molecules, including tyrosine kinase inhibitors, CDK4 inhibitors, and immunotherapy based on monoclonal antibodies [2,3,15]. Specifically, the complete loss of SMARCB1/INI1 expression is considered a marker for the evaluation of the effectiveness of Enhancer of Zeste Homolog 2 (EZH2) inhibitors (Tazemetostat) [15,16]. The most frequent cytogenetic abnormalities observed in conventional chordomas are monosomy of chromosome 1 and copy number gains of chromosomes 2, 6, and 7 [1,3]. Loss of chromosome 22 and/or genetic alterations in the *SMARCB1* gene seem to be rare [17–19].

This study aimed to compare SMARCB1/INI1 protein expression patterns in spinal conventional chordomas with genetic alterations detectable in the *SMARCB1* gene by FISH, clinicopathological features, OS, and DFS.

2. Materials and Methods

A retrospective study of 89 patients with conventional spinal chordoma diagnosed at the Anatomy and Pathological Histology Unit of the Rizzoli Orthopedic Institute from 2010 to 2019 was carried out. In order to perform morphological, immunohistochemical, and molecular analyses, a formalin-fixed paraffin-embedded (FFPE) tumor tissue sample of adequate size and quality was used, after selection by pathologists (MG and AR). The diagnosis of all the original tumor slides was confirmed independently by two pathologists (MG and AR) via the immunohistochemical expression of brachyury and pan-cytokeratin AE1/AE3. The clinicopathological parameters investigated were: age, gender, tumor size, tumor location, surgical margins, Ki67 labelling index, SMARCB1/INI1 pattern, previous surgery, previous treatment, type of surgery, and comorbidities. The surgical margins were classified according to the Enneking classification [20] and to the Weinstein–Boriani–Biagini (WBB) system [21]. The comorbidities were evaluated by the Charlson Comorbidity Index (CCI) [22]. Ethical committee approval was obtained from the Comitato Etico di Area Vasta Emilia Centro on 27/04/2023 (protocol # CE AVEC: 312/2023/Oss/IOR). As a comparison group, 4 patients with poorly differentiated chordoma were included in the analysis.

Immunohistochemical staining was performed using an automated immunostainer following the manufacturer's guidelines (Ventana BenchMark-Ventana Medical Systems, Tucson, AZ, USA), with a mouse monoclonal anti-INI-1 antibody at a concentration of 0.4 μg/mL (MRQ-27; Cell Marque, Rocklin, CA, USA) and a rabbit monoclonal primary anti-Ki-67 antibody at a concentration of 0.2 μg/mL (clone 30-9, Ventana). The immunohistochemical evaluation was executed independently by two pathologists to determine the percentage of proliferating cells (Ki67 labelling index) and to select only samples with partial SMARCB1/INI1 expression and a minimum 10% cut-off of neoplastic nuclei. Regarding SMARCB1/INI1, both patients with mosaic expression (defined by the presence of negative nuclei mixed with positive nuclei) and patients with clonal expression (characterized by the presence of a completely negative high-magnification field alongside a fully positive high-magnification field) were considered eligible; homogeneous nuclear staining in the background of inflammatory cells, stromal fibroblasts, normal epithelial cells, and/or vascular endothelial cells were used as an internal control.

FISH for the *SMARCB1* gene was performed using a commercial SPEC SMARCB1/22q12 Dual color CE/IVD Probe (ZytoVision, Bremerhaven, Germany). The analysis was performed on conventional chordomas with focal SMARCB1/INI1 expression and four poorly differentiated chordomas. The probe included a 545 kb sequence mapped to the 22q11.23 region (ZyGreen fluorochrome label) harboring the *SMARCB1* gene and a 335 kb sequence mapped to the 22q12.1–q12.2 region (ZyOrange fluorochrome label) harboring the *KREMEN1* gene, which was used as an internal control region to detect large chromosome 22q deletions. FISH was performed on interphase nuclei using the Histology FISH accessory kit (Dako, Glostrup, Denmark) according to the manufacturer's protocol [23], as previously described [7]. For each slide, a minimum of 100 intact nuclei within the tumor area previously marked by the pathologist were scored using a BX41 fluorescence microscope (Olympus, Tokyo, Japan) at 100× magnification, and visible alteration in at least 10% of the cells was considered a positive result. Nuclei with no signal or signals in overlapping nuclei were considered non-informative and were not analyzed. A Color View III CCD camera soft imaging system (Olympus) was used to capture images, which were subsequently analyzed with CytoVision imaging software version 7.5 (Leica Biosystem Richmond Inc., Richmond, IL, USA). The presence of two green signals and two orange signals in a 1:1 ratio was considered the normal copy number pattern; any FISH signals differing from this pattern were classified as altered. The detection of one green signal and one orange signal indicated a monoallelic co-deletion of *SMARCB1* and the control region, which was classified as a monoallelic 22q large deletion, and the presence of additional copies of both green and orange signals indicated a copy number gain (CNG) of chromosome 22.

OS was defined as the time between the date of diagnosis and the date of death or the last follow-up, and DFS was defined as the time between the first disease relapse or metastasis and the last follow-up. Descriptive statistics were used to report patient and clinical characteristics. All the continuous data were expressed as the means and the standard deviations of the means; the categorical data were expressed as frequencies and percentages. Fisher's chi-square exact test was used to analyze dichotomous variables. Pearson's chi-square exact test was performed to investigate categorical variables. Kaplan–Meier survival analyses with the log-rank test were performed to assess the influence of the different parameters on OS and DFS. For all the tests, $p < 0.05$ was considered as statistically significant. All the statistical analyses were performed using SPSS v.19.0 (IBM Corp., Armonk, NY, USA).

3. Results

Table 1 summarizes the main clinicopathological features of 89 patients with conventional spinal chordomas.

Table 1. Clinicopathological features of 89 patients with conventional spinal chordomas.

Parameters	All Samples (n = 89)
Gender (N, %)	
Male	51 (57.3%)
Female	38 (42.7%)
Age (median, range in years)	61 (17–86)
Age (N, %)	
≤60 years	42 (47.2%)
>60 years	47 (52.8%)
Tumor size (N, %)	
<5 cm	36 (40.4%)
≥5 cm	39 (43.9%)
Not available	14 (15.7%)
Tumor localization	
Cervical–thoracic–lumbar region	43 (48.3%)
Sacrococcygeal region	46 (51.7%)
Surgical margin	
Adequate	45 (50.6%)
Inadequate	25 (28.1%)
Not available	19 (21.3%)
Ki-67 index (median, range)	3 (1–12)
Ki-67 index (N, %)	
≤3%	43 (48.3%)
>3%	37 (41.6%)
Not evaluable	9 (10.1%)
SMARCB1/INI1 immunohistochemical expression (N, %)	
Positive	52 (58.4%)
Positive/negative	37 (41.6%)
Previous surgery	
No	53 (59.6%)
Yes	21 (23.6%)
Not available	15 (16.9%)
Previous treatment	
No	59 (66.3%)
Yes	14 (15.7%)
Not available	16 (18%)
Type of surgery	
En bloc resection	54 (60.7%)
Other surgery	16 (18%)
No surgery	19 (21.3%)
Charlson Comorbidity Index (CCI)	
Mean (SD)	4.1 (0.260)

The dataset included 51 (57.3%) males and 38 (42.7%) females, with a median age at diagnosis of 61 years (range 17–86). Clinically, 43 (48.3%) tumors were located in the cervical–thoracic–lumbar region (mobile spine), while 46 (51.7%) were located in the sacrococcygeal region. The median tumor size at presentation was 5.9 cm (range 1.4–16 cm). The mean CCI of the population was 4.1. Twenty-one patients (23.6%) underwent previous surgical treatment, and 14 patients (15.7%) underwent previous systemic therapy and/or radiotherapy for the same tumor.

Among the 70 patients who underwent surgical resection, 45 patients (50.6%) had adequate surgical margins (wide and radical), while 25 (28.1%) had inadequate surgical margins (intralesional and marginal), according to the Enneking classification [20] (Table S1). Among the remaining 19 inoperable patients, 12 were treated with carbon ion therapy, 3 with proton therapy, and 1 with radiation and chemotherapy; for 3 patients only biopsy information was available without follow-up data. Of the cases with inadequate margins, nine cases were localized at the cervical region, seven cases were localized at the thoracic–lumbar region (six patients were previously treated with surgery at other centers), and nine cases were localized at the sacrococcygeal region (three patients were previously treated with surgery at other centers). When feasible, a classification according to the WBB system [21] was performed and all 10 tumors analyzed had very large extensions with both extra-osseous and intracanal components (Table 2), which did not allow resection with wide margins.

Table 2. The WBB classification of patients with surgical inadequate margins.

Case Number	Tumor Localization	WBB Classification	Revision Surgery
1	L3	layers A–E; zones 12–1	NO
2	sacrum	n.a.	NO
7	C4–C5	layers C–E, zones 8–5	NO
15	sacrum	n.a.	NO
19	sacrum	n.a.	NO
25	L5	n.a.	YES
29	C2	layers A–E; zones 11–7	NO
34	C3	layers A–E; zones 2–8	NO
35	C2	layers A–E; zones 9–4	NO
40	L3	n.a.	YES
42	sacrum	n.a.	YES
44	C2–C3	layers A–E; zones 6–2	NO
45	L4–L5	n.a.	YES
48	L2	n.a.	YES
52	sacrum	n.a.	YES
58	T2–T3	n.a.	YES
64	T9	layers A–E; zones 9–1	YES
66	C2	layers A–E; zones 7–4	NO
68	C2	layers A–E; zones 11–5	NO
71	C5–C6	n.a.	YES
72	coccyx	n.a.	NO
73	sacrum	n.a.	YES
78	sacrum	n.a.	NO
79	C1–C2	layers A–E; zones 6–3	NO
89	sacrum	n.a.	NO

n.a. = not applicable, because of localization on the sacrococcygeal region or because of the absence of pre-operative imaging.

The median Ki-67 labelling index was 3% (range 1–12%), excluding nine non-evaluable cases (absence of positive internal controls in normal bone marrow cells). The SMARCB1/INI1 immunohistochemical analyses revealed a partial loss of SMARCB1/INI1 (range 10–80%) in 37 (41.6%) patients, while 52 (58.4%) patients exhibited complete protein expression in all neoplastic cells (Table S1). In the 37 patients with focal SMARCB1/INI1 loss, 2 different staining patterns were identified: 27 cases had a mosaic expression pattern (with mixed negative and positive nuclei), while 10 cases had a clonal expression pattern (with separate fully negative and fully positive high-magnification fields) (Figure 1A,B). The four poorly differentiated chordomas exhibited complete loss of SMARCB1/INI1 in all the evaluated neoplastic cells.

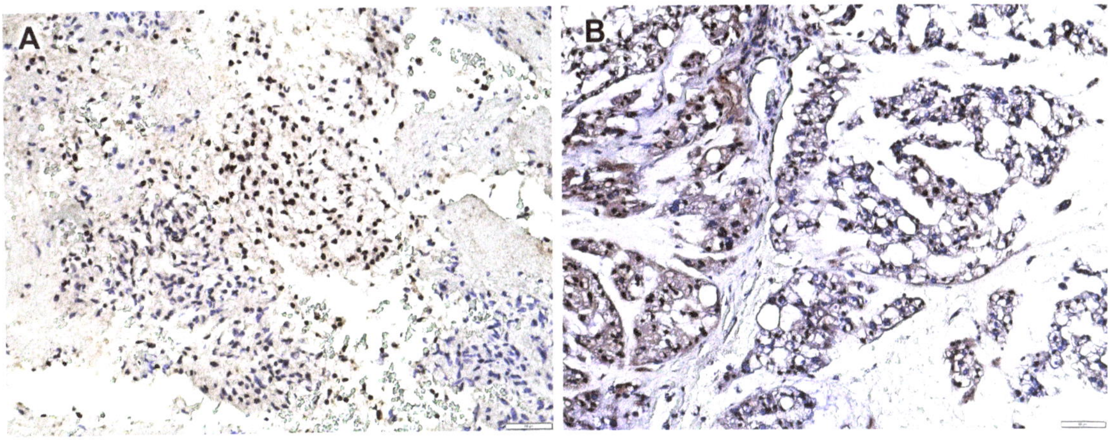

Figure 1. (**A**) Case n.25 showing clonal expression of SMARCB1/INI1; (**B**) case n.44 showing mosaic expression of SMARCB1/INI1.

Partial loss of the immunohistochemical expression of SMARCB1/INI1 was significantly associated with localization in the cervical–thoracic–lumbar region (p = 0.033) and inadequate surgical margins (p = 0.007). No significant associations were found with gender, age at diagnosis, tumor size, or Ki67 index (Table 3).

Table 3. Clinicopathological features according to SMARCB1/INI1 immunohistochemical expression.

	SMARCB1/INI1 + (n = 52)	SMARCB1/INI +/− (n = 37)	*p*-Value
Gender (N, %)			
Male	28 (53.8%)	23 (62.2%)	0.516
Female	24 (46.2%)	14 (37.8%)	
Age (median, range in years)	61.5 (28–86)	59 (17–79)	0.511
Age (N, %)			
≤60 years	22 (42.3%)	20 (54.1%)	0.291
>60 years	30 (57.7%)	17 (45.9%)	
Tumor size (N, %)			
<5 cm	20 (38.5%)	16 (43.2%)	
≥5 cm	20 (38.5%)	19 (51.4%)	0.818
Not available	12 (23%)	2 (5.4%)	
Tumor localization			
Cervical–thoracic–lumbar region	20 (38.5%)	23 (62.2%)	0.033
Sacrococcygeal region	32 (61.5%)	14 (37.8%)	
Surgical margin			
Adequate	30 (57.7%)	15 (40.5%)	
Inadequate	8 (15.3%)	17 (46%)	0.007
Not available	14 (27%)	5 (13.5%)	
Ki-67 index (median, range in percentage)	3 (1–12%)	3 (1–9%)	0.459
Ki-67 index (N, %)			
≤3%	26 (50%)	17 (46%)	
>3%	24 (46.2%)	13 (35%)	0.817
Not evaluable	2 (3.8%)	7 (19%)	

Statistically significant p values are shown in red color.

The FISH analysis performed on 37 conventional spinal chordoma patients with focal SMARCB1/INI1 loss revealed three possible molecular patterns (Figure 2).

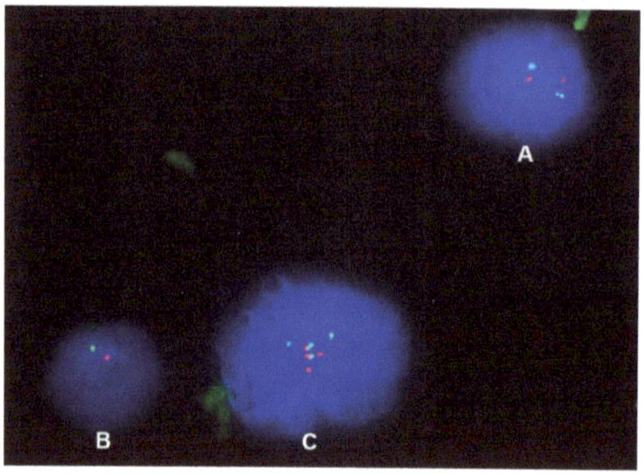

Figure 2. (**A**) Normal nucleus, with two signals for the control region (orange) and two signals for the *SMARCB1* gene (green); (**B**) nucleus with monoallelic deletion, with only one signal for the control region (orange) and only one signal for the *SMARCB1* gene (green); (**C**) nucleus with CNG, with three or more signals for both the control region (orange) and *SMARCB1* gene (green).

Monoallelic deletion of the *SMARCB1* gene associated with co-deletion of the control region was observed in 16 cases of conventional chordoma (range 26–94%) (Figure 3A,B); 5 of these also had nuclei with additional copies of both signals (Figure 3C,D). One case exhibited only nuclei with CNG and none with deletions. Due to poor tissue quality, 20 samples did not show hybridized signals and were considered inadequate for FISH scoring (Table S1). Considering the two different staining patterns of focal SMARCB1/INI1 expression, all 10 cases with mosaic patterns had a monoallelic 22q deletion (range 30–94%), 3 of these cases also had nuclei with CNG of both signals; 5 of 6 cases with clonal patterns had a monoallelic 22q deletion (range 26–81%); 2 of these cases also had nuclei with extra copies of *SMARCB1* and the control region, whereas 1 case had only nuclei with CNG of both signals.

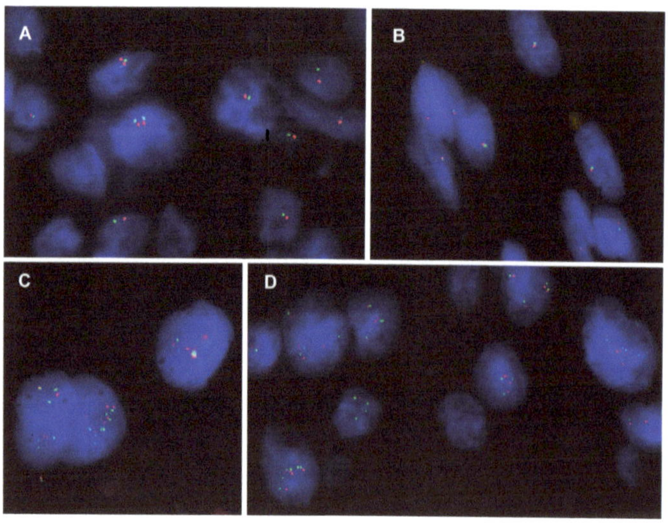

Figure 3. (**A**,**B**) Nuclei with monoallelic co-deletion of the *SMARCB1* gene and the control region from cases n.58 and n.21, respectively; (**C**,**D**) nuclei with CNG from cases n.37 and n.66, respectively.

In the four cases of poorly differentiated chordoma, FISH analyses revealed biallelic *SMARCB1* deletions in two cases, a monoallelic deletion in one case, and a pattern with a monoallelic *SMARCB1* deletion associated with an additional control region signal in one case. The average follow-up duration after treatment completion was 66 months (range 2–148). The 5-year OS and 5-year DFS rates were 90.8% (SE 3.6%) and 54.9% (SE 6%), respectively. Univariate analysis revealed worse overall survival for patients older than 60 years ($p = 0.046$). The risk of local recurrence or metastasis was greater for patients with a tumor in the cervical–thoracic–lumbar region ($p = 0.017$), for those with inadequate surgical margins ($p = 0.009$), and for patients who underwent a previous surgery for the same tumor ($p < 0.0005$) (Table 4; Figures 4 and 5). Moreover, the presence of comorbidities significantly affected both OS and DFS, as shown in Tables 4 and 5.

Table 4. Results from univariate Kaplan–Meier models for OS and DFS.

	5 Years—OS % (SE)	*p*-Value	5 Years—DFS % (SE)	*p*-Value
Entire sample	90.8% (3.6%)		54.9% (6%)	
Gender (N, %)				
Male	91% (5%)	0.731	51.1% (8.1%)	0.728
Female	89.7% (5.6%)		59.8% (8.8%)	
Age (N, %)				
≤60 years	96.8% (3.2%)	0.046	51.9% (8.3%)	0.907
>60 years	85.1% (6.2%)		58.3% (8.5%)	
Tumor size (N, %)				
<5 cm	90.5% (5.2%)	0.800	52.7% (9%)	0.486
≥5 cm	94.4% (5.4%)		49.7% (9.3%)	
Tumor localization				
Cervical–thoracic–lumbar region	87.7% (5.8%)	0.477	44.2% (8.5%)	0.017
Sacrococcygeal region	94.6% (3.7%)		64.8% (8.1%)	
Surgical margin				
Adequate	96.8% (3.2%)	0.065	61% (8%)	0.009
Inadequate	82.2% (9.3%)		23.2% (10.4%)	
Ki-67 index (N, %)				
≤3%	89.6% (5.7%)	0.648	60.5% (7.9%)	0.125
>3%	96.7% (3.3%)		47.3% (9.1%)	
SMARCB1/INI1 immunohistochemical expression (N, %)				
Positive	94.8% (3.6%)	0.210	58.6% (8.8%)	0.275
Positive/negative	85.5% (6.8%)		49.4% (9.1%)	
Previous surgery				
No	88.9% (4.8%)	0.98	66.3% (7.5%)	<0.0005
Yes	93.7% (7.4%)		25.3% (10.4%)	
Previous treatment				
No	90.6% (4.5%)	0.858	54.4% (7.4%)	0.56
Yes	90.0% (9.5%)		58.4% (14.5%)	
Type of surgery				
En bloc resection	88.7% (4.8%)	0.693	61.6% (7.2%)	0.899
Other	90.0% (9.5%)		44.7% (17.1%)	
Charlson Comorbidity Index (CCI)				
≤4	92.3%	0.076	63.0% (7.3%)	0.011
>4	83.8%		39.3% (10.2%)	

Statistically significant *p* values are shown in red color.

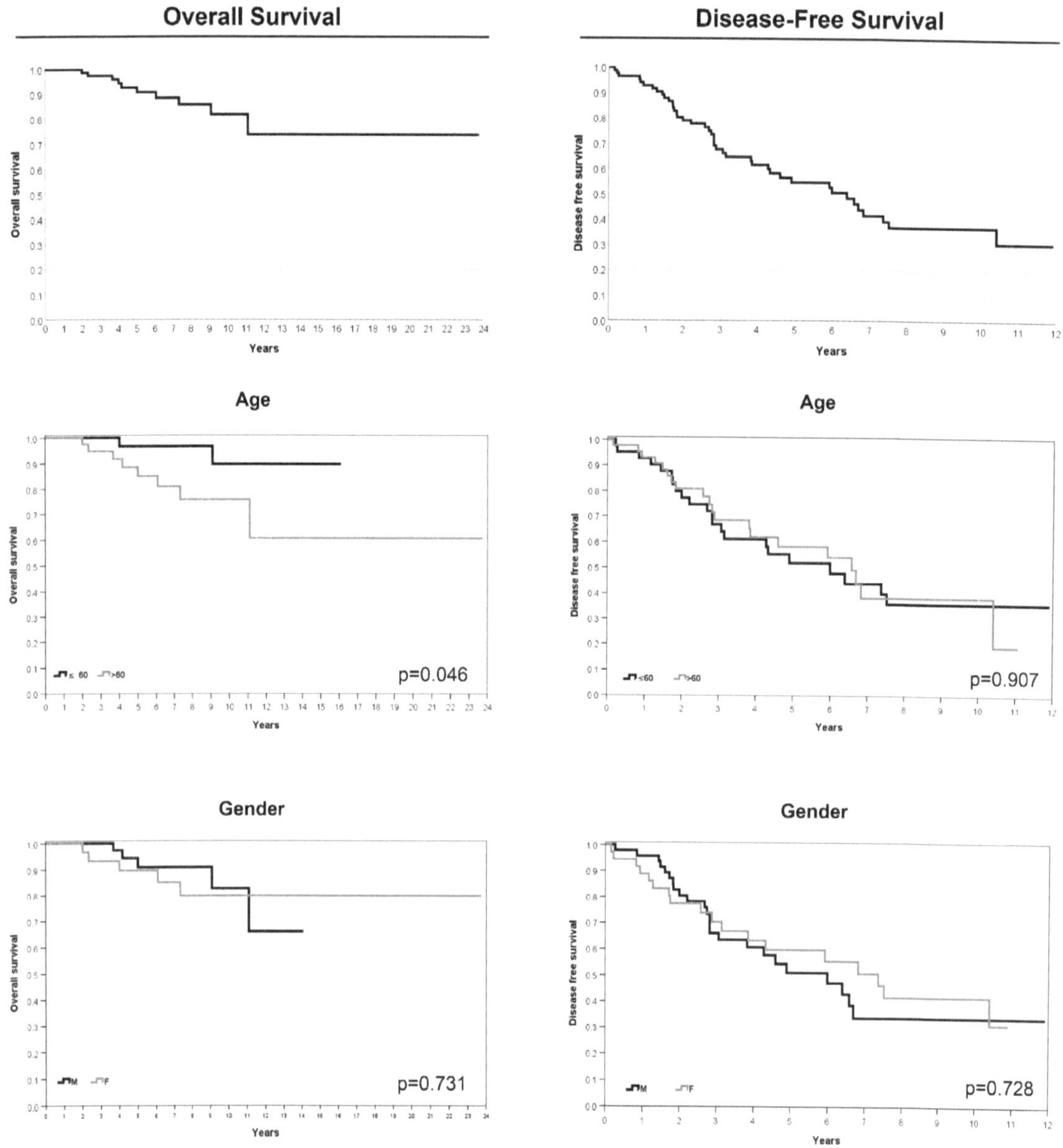

Figure 4. Kaplan–Meier survival analyses for age and gender features.

The results of the multivariate analysis demonstrated that the inadequate surgical margin and an age older than 60 years significantly impaired the OS (Table 6). The risk of local recurrence or metastases was increased by a higher Ki67 index, by an inadequate surgical margin, and by a high CCI: with the same surgical margin and Ki67 scores, the increase of 1 unit of the CCI increases the risk by 40.5% (Table 7). It should be noted that the CCI includes the age, and all patients older than 60 years have a CCI higher than 4.

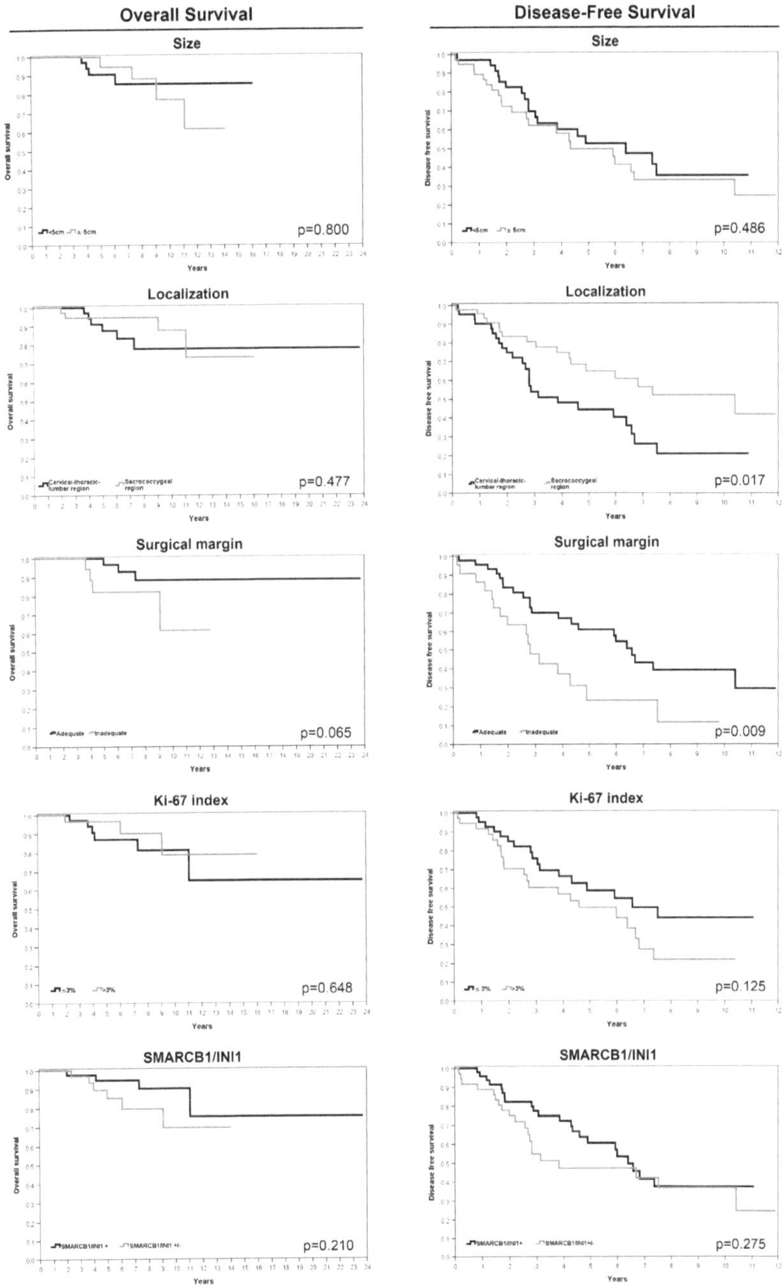

Figure 5. Kaplan–Meier survival analyses for size and tumor localization, surgical margins, Ki-67 index, and SMARCB1/INI1 immunohistochemical expression.

Table 5. Univariate analysis for CCI as continuous variable.

		p-Value	HR	95.0% CI Inferior	Superior
5 years—OS	**CCI**	0.043	1.694	1.018	2.820
		***p*-Value**	**HR**	**95.0% CI Inferior**	**Superior**
5 years—DFS	**CCI**	0.078	1.222	0.978	1.528

Statistically significant *p* values are shown in red color.

Table 6. Multivariate analysis for overall survival.

		p-Value	HR	95.0% CI Inferior	Superior
Phase 1	**CCI**	0.788	1.110	0.519	2.373
	margin (1 vs. 0) *	0.006	29.965	2.619	342.854
	age (>60 vs. ≤60)	0.050	19.600	1.001	383.640
Phase 2	**margin (1 vs. 0) ***	0.006	30.049	2.634	342.745
	age (>60 vs. ≤60)	0.012	24.592	2.019	299.586

* 0 = adequate margin; 1 = inadequate margin. Statistically significant *p* values are shown in red color.

Table 7. Multivariate analysis for disease-free survival.

		p-Value	HR	95.0% CI Inferior	Superior
Phase 1	**Ki67**	0.037	1.216	1.012	1.461
	margin (1 vs. 0) *	0.036	2.501	1.060	5.904
	localization	0.233	0.530	0.187	1.504
	previous surgery	0.321	1.489	0.678	3.270
	CCI	0.008	1.526	1.119	2.079
	type of surgery (other)	0.556	0.681	0.189	2.447
	type of Surgery(en bloc resection)	0.868	0.913	0.313	2.664
Phase 2	**Ki67**	0.033	1.216	1.016	1.455
	margin (1 vs. 0) *	0.026	2.598	1.119	6.032
	localization	0.210	0.548	0.214	1.403
	previous surgery	0.279	1.529	0.709	3.298
	CCI	0.004	1.513	1.141	2.007
Phase 3	**Ki67**	0.032	1.216	1.017	1.453
	margin (1 vs. 0) *	0.018	2.771	1.195	6.429
	localization	0.203	0.547	0.216	1.383
	CCI	0.004	1.517	1.143	2.013
Phase 4	**Ki67**	0.061	1.188	0.992	1.421
	margin (1 vs. 0) *	0.019	2.667	1.173	6.059
	CCI	0.004	1.502	1.142	1.976

* 0 = adequate margin; 1 = inadequate margin. Statistically significant *p* values are shown in red color.

4. Discussion

Conventional spinal chordoma is a rare, slow-growing, locally aggressive malignant neoplasm [1,2]. In recent years, an increasing number of tumors, including poorly differentiated chordomas, have been found to exhibit complete loss of SMARCB1/INI1 protein expression. In many patients, molecular analyses of the *SMARCB1* gene revealed a biallelic deletion [3,11]. Recently, conventional skull base chordomas have also been investigated by immunohistochemistry, and partial loss of SMARCB1/INI1 was identified [7]. In our study, the immunohistochemical pattern of SMARCB1/INI1 in conventional spinal chordomas was analyzed for the first time, and partial loss of SMARCB1/INI1 was observed in 41.6% of cases. In particular, two distinct expression patterns were detected, mosaic and, less frequently, clonal, confirming what has been previously reported on conventional skull base chordomas [7]. From a molecular perspective, several types of genetic alterations have been described among tumors with focal expression, but the most frequent is the monoallelic deletion of a portion of the long arm of chromosome 22 (involving *SMARCB1*) [7,10,16]. However, the genomic studies in the literature revealed that the loss of chromosome 22 or the monoallelic deletion of *SMARCB1* is rare in conventional spinal chordomas [17,18]. In our series, we genetically investigated only conventional chordomas with impaired SMARCB1/INI1 pattern expression, and in 43.2% of the feasible cases, a monoallelic co-deletion of the *SMARCB1* gene and the control region was observed. To evaluate the *SMARCB1* locus at chromosome 22q, we used FISH analysis with a CE-IVD probe. Due to cross-hybridization of chromosome 22 alpha satellites to other centromeric regions, a probe mapped to the 22q12.1-q12.2 region was used as an internal control, which has already been proven to be a reliable control for investigating large deletions [24]. Heterozygous partial deletion of the long arm of chromosome 22 was confirmed as the main molecular mechanism underlying the focal expression of the SMARCB1/INI1 protein. Specifically, the chordomas with mosaic SMARCB1/INI1 expression showed mainly monoallelic 22q deletion, whereas the cases with clonal SMARCB1/INI1 expression were associated with different types of genetic patterns. Nuclei with additional copies of the *SMARCB1* gene and 22q12 control region were also frequently detected in several subclones of cases with deletion, confirming a previously described event [7,16,19]. However, point mutations in *SMARCB1* were not investigated in our study, and epigenetic alterations or post-translational modifications might play an additional role in interpreting the large genetic variability associated with the phenotypic expression of SMARCB1/INI1. We observed that partial loss of SMARCB1/INI1 was significantly associated with the cervical–thoracic–lumbar region ($p = 0.033$) and inadequate surgical margins ($p = 0.007$), suggesting that partial loss of the protein might be associated with increased clinical aggressiveness. A possible reason for the correlation between partial SMARCB1/INI1 loss and inadequate margins could be the major extra-osseous and intracanal involvement of the tumors in the mobile spine, thus increasing the difficulty in obtaining adequate surgical margins. Indeed, 37.5% of patients with inadequate surgical margins were treated for local recurrence of the tumor. The statistical analysis, moreover, indicated the localization in the mobile spine and the presence of surgical inadequate margins as negative prognostic factors in terms of the disease-free survival ($p = 0.017$ and $p = 0.009$, respectively), unlike the cases located in the skull base, where no correlations were found between the partial loss of SMARCB1/INI1 and the clinicopathological parameters evaluated [7]. The multivariate analyses revealed the most crucial factors to be monitored for patient prognosis. The presence of inadequate surgical margins was confirmed as the prevalent risk factor both for OS and DFS; moreover, an age older than 60 years also significantly impaired the OS, whereas DFS was also associated with a high Ki67 index and by a high CCI.

Due to the difficulty in surgically eradicating tumors and the known resistance of chordoma to common chemotherapies [25,26], new molecular targets are being investigated to properly treat these tumors [15]. Increasing knowledge of SMARCB1/INI1 function has enabled the identification of specific targets, including the EZH2 gene. This target is a catalytic subunit of the polycomb repressive complex 2 (PRC2), which plays a role in the chromatin regulation, in cell fate determination, and in cellular differentiation and is often up-regulated in tumors with a loss of SMARCB1/INI1 [8,27,28]. An increase in EZH2 expression correlates with tumor aggressiveness [28], and specifically, this mechanism has been associated with the progression of chordomas [29]. Thus, clinical trials on inhibitors of the EZH2 enzyme are currently underway in tumors with complete loss of SMARCB1/INI1 expression, including poorly differentiated chordomas (ClinicalTrials.gov Identifiers: NCT02601950 and NCT05407441) [30–32]. These trials show the safety tolerability and effectiveness of the drug, with the possibility of use in other types of malignancies [2,3,28]; specifically, the potential use of EZH2 inhibitors could also be promising for patients with partial SMARCB1/INI1 loss, but it needs further exploration.

5. Conclusions

In conclusion, we retrospectively analyzed 89 cases of conventional spinal chordoma, and two distinct expression patterns (mosaic and clonal) of partial SMARCB1/INI1 loss were observed. The most frequent molecular alteration detected in conventional chordoma was the monoallelic deletion of the 22q locus (including *SMARCB1* gene). Partial loss of SMARCB1/INI1 was significantly associated with location in the mobile spine and inadequate surgical margins. Inadequate surgical margins, a high Ki67 index, a high CCI, and an age older than 60 years were also associated with a worse prognosis. Treatments with inhibitors of the EZH2 enzyme are currently ongoing in tumors with complete loss of SMARCB1/INI1 expression; therefore, tumors with partial loss of SMARCB1/INI1 could also have therapeutic implications.

Supplementary Materials: The following supporting information can be downloaded at: https://www.mdpi.com/article/10.3390/cancers16162808/s1.

Author Contributions: Conceptualization, A.R.; formal analysis, E.P.; investigation, M.M., S.C., G.G. and S.B.; resources, G.B., A.G. and R.G.; data curation, C.G., L.E.N., C.A., A.R, M.M., M.G. and C.F.; writing—original draft, M.M., S.C., C.G. and E.P.; writing—review and editing, A.R., M.G., S.B., G.M., G.G., S.C., M.M., C.G., C.F. and R.G.; funding acquisition, S.A. All authors have read and agreed to the published version of the manuscript.

Funding: This research was supported by funds for selected research topics from the Fondazione CARISBO Project (#19344).

Institutional Review Board Statement: The study was conducted according to the guidelines of the Declaration of Helsinki and approved by the Ethics Committee of Area Vasta Emilia Centro on 27/04/2023 (protocol # CE AVEC: 312/2023/Oss/IOR).

Informed Consent Statement: Informed consent was obtained from all subjects involved in the study.

Data Availability Statement: The original contributions presented in the study are included in the article. Further inquiries can be directed to the corresponding authors.

Acknowledgments: We are grateful to Muscolo skeletal tumor biobank – biobanca dei tumori muscoloscheletrici (BIOTUM)—member of the CRB-IOR—which provided us with the biological samples. The authors thank Cristina Ghinelli for her help in the graphic design and Monica Contoli for recovery of archival material.

Conflicts of Interest: The authors declare no conflicts of interest.

References

1. WHO Classification of Tumours Editorial Board (Ed.) *WHO Classification of Tumours 5th Edition: Soft Tissue and Bone Tumours*, 5th ed.; WHO Press: Geneva, Switzerland, 2020.
2. Wedekind, M.F.; Widemann, B.C.; Cote, G. Chordoma: Current status, problems, and future directions. *Curr. Probl. Cancer* **2021**, *45*, 100771. [CrossRef]
3. Ulici, V.; Hart, J. Chordoma. *Arch. Pathol. Lab. Med.* **2022**, *146*, 386–395. [CrossRef]
4. Vujovic, S.; Henderson, S.; Presneau, N.; Odell, E.; Jacques, T.S.; Tirabosco, R.; Boshoff, C.; Flanagan, A.M. Brachyury, a crucial regulator of notochordal development, is a novel biomarker for chordomas. *J. Pathol.* **2006**, *209*, 157–165. [CrossRef]

5. Shih, A.R.; Chebib, I.; Deshpande, V.; Dickson, B.C.; Iafrate, A.J.; Nielsen, G.P. Molecular characteristics of poorly differentiated chordoma. *Genes Chromosomes Cancer* **2019**, *58*, 804–808. [CrossRef]
6. Mobley, B.C.; McKenney, J.K.; Bangs, C.D.; Callahan, K.; Yeom, K.W.; Schneppenheim, R.; Hayden, M.G.; Cherry, A.M.; Gokden, M.; Edwards, M.S.B.; et al. Loss of SMARCB1/INI1 expression in poorly differentiated chordomas. *Acta Neuropathol.* **2010**, *120*, 745–753. [CrossRef]
7. Righi, A.; Cocchi, S.; Maioli, M.; Zoli, M.; Guaraldi, F.; Carretta, E.; Magagnoli, G.; Pasquini, E.; Melotti, S.; Vornetti, G.; et al. SMARCB1/INI1 loss in skull base conventional chordomas: A clinicopathological and molecular analysis. *Front. Oncol.* **2023**, *13*, 1160764. [CrossRef]
8. Kalimuthu, S.N.; Chetty, R. Gene of the month: SMARCB1. *J. Clin. Pathol.* **2016**, *69*, 484–489. [CrossRef]
9. Centore, R.C.; Sandoval, G.J.; Soares, L.M.M.; Kadoch, C.; Chan, H.M. Mammalian SWI/SNF Chromatin Remodeling Complexes: Emerging Mechanisms and Therapeutic Strategies. *Trends Genet.* **2020**, *36*, 936–950. [CrossRef]
10. Kohashi, K.; Oda, Y. Oncogenic roles of SMARCB1/INI1 and its deficient tumors. *Cancer Sci.* **2017**, *108*, 547–552. [CrossRef]
11. Pawel, B.R. SMARCB1-deficient Tumors of Childhood: A Practical Guide. *Pediatr. Dev. Pathol.* **2018**, *21*, 6–28. [CrossRef]
12. Parker, N.A.; Al-Obaidi, A.; Deutsch, J.M. SMARCB1/INI1-deficient tumors of adulthood. *F1000Research* **2020**, *9*, 662. [CrossRef]
13. Chitguppi, C.; Rabinowitz, M.R.; Johnson, J.; Bar-Ad, V.; Fastenberg, J.H.; Molligan, J.; Berman, E.; Nyquist, G.G.; Rosen, M.R.; Evans, J.E.; et al. Loss of SMARCB1 Expression Confers Poor Prognosis to Sinonasal Undifferentiated Carcinoma. *J. Neurol. Surg. Part B Skull Base* **2020**, *81*, 610–619. [CrossRef]
14. Wang, J.; Andrici, J.; Sioson, L.; Clarkson, A.; Sheen, A.; Farzin, M.; Toon, C.W.; Turchini, J.; Gill, A.J. Loss of INI1 expression in colorectal carcinoma is associated with high tumor grade, poor survival, BRAFV600E mutation, and mismatch repair deficiency. *Hum. Pathol.* **2016**, *55*, 83–90. [CrossRef]
15. Chen, S.; Ulloa, R.; Soffer, J.; Alcazar-Felix, R.J.; Snyderman, C.H.; Gardner, P.A.; Patel, V.A.; Polster, S.P. Chordoma: A Comprehensive Systematic Review of Clinical Trials. *Cancers* **2023**, *15*, 5800. [CrossRef]
16. Wen, X.; Cimera, R.; Aryeequaye, R.; Abhinta, M.; Athanasian, E.; Healey, J.; Fabbri, N.; Boland, P.; Zhang, Y.; Hameed, M. Recurrent loss of chromosome 22 and SMARCB1 deletion in extra-axial chordoma: A clinicopathological and molecular analysis. *Genes Chromosomes Cancer* **2021**, *60*, 796–807. [CrossRef]
17. Choy, E.; MacConaill, L.E.; Cote, G.M.; Le, L.P.; Shen, J.K.; Nielsen, G.P.; Iafrate, A.J.; Garraway, L.A.; Hornicek, F.J.; Duan, Z. Genotyping cancer-associated genes in chordoma identifies mutations in oncogenes and areas of chromosomal loss involving CDKN2A, PTEN, and SMARCB1. *PLoS ONE* **2014**, *9*, e101283. [CrossRef]
18. Wang, L.; Zehir, A.; Nafa, K.; Zhou, N.; Berger, M.F.; Casanova, J.; Sadowska, J.; Lu, C.; Allis, C.D.; Gounder, M.; et al. Genomic aberrations frequently alter chromatin regulatory genes in chordoma. *Genes Chromosomes Cancer* **2016**, *55*, 591–600. [CrossRef]
19. Curcio, C.; Cimera, R.; Aryeequaye, R.; Rao, M.; Fabbri, N.; Zhang, Y.; Hameed, M. Poorly differentiated chordoma with whole-genome doubling evolving from a SMARCB1- deficient conventional cordoma: A case report. *Genes Chromosomes Cancer* **2021**, *60*, 43–48. [CrossRef]
20. Enneking, W.F.; Spanier, S.; Goodman, M.A. A system for the surgical staging of musculoskeletal sarcoma. *Clin. Orthop. Relat. Res.* **1980**, *153*, 106–120. [CrossRef]
21. Boriani, S.; Weinstein, J.N.; Biagini, R. Primary bone tumors of the spine. Terminology and surgical staging. *Spine* **1997**, *22*, 1036–1044. [CrossRef]
22. Charlson, M.E.; Pompei, P.; Ales, K.L.; MacKenzie, C.R. A new method of classifying prognostic comorbidity in longitudinal studies: Development and validation. *J. Chronic Dis.* **1987**, *40*, 373–383. [CrossRef] [PubMed]
23. Cocchi, S.; Gamberi, G.; Magagnoli, G.; Maioli, M.; Righi, A.; Frisoni, T.; Gambarotti, M.; Benini, S. CIC rearranged sarcomas: A single institution experience of the potential pitfalls in interpreting CIC FISH results. *Pathol. Res. Pract.* **2022**, *231*, 153773. [CrossRef] [PubMed]
24. Huang, S.C.; Zhang, L.; Sung, Y.S.; Chen, C.L.; Kao, Y.C.; Agaram, N.P.; Antonescu, C.R. Secondary EWSR1 gene abnormalities in SMARCB1-deficient tumors with 22q11-12 regional deletions: Potential pitfalls in interpreting EWSR1 FISH results. *Genes Chromosomes Cancer* **2016**, *55*, 767–776. [CrossRef] [PubMed]
25. Yakkioui, Y.; van Overbeeke, J.J.; Santegoeds, R.; van Engeland, M.; Temel, Y. Chordoma: The entity. *Biochim. Biophys. Acta* **2014**, *1846*, 655–669. [CrossRef] [PubMed]
26. Stacchiotti, S.; Sommer, J.; Chordoma Global Consensus Group. Building a global consensus approach to chordoma: A position paper from the medical and patient community. *Lancet Oncol.* **2015**, *16*, e71–e83. [CrossRef]
27. Duan, R.; Du, W.; Guo, W. EZH2: A novel target for cancer treatment. *J. Hematol. Oncol.* **2020**, *13*, 104. [CrossRef]
28. Rosen, E.Y.; Shukla, N.N.; Bender, J.L.D. EZH2 inhibition: It's all about the context. *J. Natl. Cancer Inst.* **2023**, *115*, 1246–1248. [CrossRef]
29. Ma, X.; Qi, S.; Duan, Z.; Liao, H.; Yang, B.; Wang, W.; Tan, J.; Li, Q.; Xia, X. Long non-coding RNA LOC554202 modulates chordoma cell proliferation and invasion by recruiting EZH2 and regulating miR-31 expression. *Cell Prolif.* **2017**, *50*, e12388. [CrossRef] [PubMed]
30. Passeri, T.; Dahmani, A.; Masliah-Planchon, J.; Naguez, A.; Michou, M.; El Botty, R.; Vacher, S.; Bouarich, R.; Nicolas, A.; Polivka, M.; et al. Dramatic In Vivo Efficacy of the EZH2-Inhibitor Tazemetostat in PBRM1-Mutated Human Chordoma Xenograft. *Cancers* **2022**, *14*, 1486. [CrossRef]

31. Italiano, A.; Soria, J.C.; Toulmonde, M.; Michot, J.M.; Lucchesi, C.; Varga, A.; Coindre, J.M.; Blakemore, S.J.; Clawson, A.; Suttle, B.; et al. Tazemetostat, an EZH2 inhibitor, in relapsed or refractory B-cell non-Hodgkin lymphoma and advanced solid tumours: A first-in-human, open-label, phase 1 study. *Lancet Oncol.* **2018**, *19*, 649–659. [CrossRef]
32. Gounder, M.M.; Zhu, G.; Roshal, L.; Lis, E.; Daigle, S.R.; Blakemore, S.J.; Michaud, N.R.; Hameed, M.; Hollmann, T.J. Immunologic Correlates of the Abscopal Effect in a SMARCB1/INI1-negative Poorly Differentiated Chordoma after EZH2 Inhibition and Radiotherapy. *Clin. Cancer Res.* **2019**, *25*, 2064–2071. [CrossRef]

MDPI AG
Grosspeteranlage 5
4052 Basel
Switzerland
Tel.: +41 61 683 77 34

Cancers Editorial Office
E-mail: cancers@mdpi.com
www.mdpi.com/journal/cancers

www.ingramcontent.com/pod-product-compliance
Lightning Source LLC
Chambersburg PA
CBHW041411210926
51850CB00013B/3238
* 9 7 8 3 7 2 5 8 3 3 7 3 3 *